FA00

OXFORD MONOGRAPHS
ON METEOROLOGY

Editor

P. A. SHEPPARD

OXFORD MONOGRAPHS ON METEOROLOGY

Editor P. A. SHEPPARD

———

THE PHYSICS OF CLOUDS

By B. J. MASON

ATMOSPHERIC RADIATION

I
THEORETICAL BASIS

BY

R. M. GOODY

*Abbott Lawrence Rotch Professor of
Dynamic Meteorology, and
Director of the Blue Hill Observatory
Harvard University*

OXFORD
AT THE CLARENDON PRESS
1964

Oxford University Press, Amen House, London E.C.4

GLASGOW NEW YORK TORONTO MELBOURNE WELLINGTON
BOMBAY CALCUTTA MADRAS KARACHI LAHORE DACCA
CAPE TOWN SALISBURY NAIROBI IBADAN ACCRA
KUALA LUMPUR HONG KONG

PRINTED IN GREAT BRITAIN

PREFACE

THE first intention for this book was a single volume embracing all aspects of 'Atmospheric Radiation', to be written jointly by Dr. G. D. Robinson and myself. Dr. Robinson's interests lie mainly in the field of non-spectral instrumentation and radiation climatology, while mine are in spectroscopy and the theory of radiative transfer, so the division of labour provided no problem. As I progressed with the fundamental theory, however, it became apparent that it could not be condensed beyond a certain extent if the book were to give a connected account on the level of sophistication required by, say, a Ph.D. candidate. Our plans were therefore altered, and Part I, 'Theoretical Basis', now appears as a first volume.

I have attempted to make a deductive presentation based upon the fundamental laws of physics. In my view, no other course was possible. In many branches of atmospheric physics the logical connexion between the observed phenomenon and the fundamental laws cannot yet be traced, and postulates based on physical insight have to be introduced at many levels. If these postulates are accepted the student may gain a false sense of understanding the 'physics of the problem', but later, when he probes deeper, his insecurity will trouble and deter him. Many current publications on atmospheric radiation would read very differently had the writer been more aware of the relationship of his work to fundamental concepts.

Such weaknesses have become particularly clear in recent work on planetary atmospheres. Most of the needed fundamental ideas have been investigated in the earth's atmosphere, and yet astronomers and physicists working in this area must often start anew because the special postulates (e.g. thermodynamic equilibrium) of the 'meteorological literature' make it difficult to extend their ideas. It is my hope, on the contrary, that the work described in this book can immediately be applied to Mars and Venus, and will have some relevance to studies of the outer planets.

One of my prejudices should be mentioned explicitly. This subject is weighed down by a huge, unpublished literature, partly classified. The United States is perhaps the principal source through its 'project reports', but other countries also contribute to this grey literature. I believe that it should not be recognized in the same sense as publication in a book or commercially available journal, so that authors may be encouraged to submit their papers to scrutiny by their colleagues, or lose recognition. I have therefore attempted to avoid reference to unpublished material but, in a few cases where important data are not elsewhere available, I have been prepared to abandon my principles.

As far as I am aware I have made reference to all of the essential published work in the English language, and to many important works in French, German, and Russian. No claim is made to completeness however, and, in particular, I have made no attempt to bring the references up to date in proof. They should, however, be satisfactory up to the summer of 1962, when the manuscript was handed over to the publishers.

I am indebted to many of my colleagues for reviews of early manuscripts, and particularly to Dr. W. S. Benedict for his comments on Chapters 3 and 4. Many errors have been detected by students, while using the manuscript as course material. Miss Lucy Straub produced an excellent manuscript, and Messrs. H. E. Goody and R. S. Lindzen have assisted with the proof-reading.

I wish to acknowledge permission from D. van Nostrand Co. Inc. and Dr. G. Herzberg to reproduce a number of figures from Dr. Herzberg's books. For part of the time spent on this book I was working on a research grant from the United States National Science Foundation. Finally, it gives me great pleasure to thank the Clarendon Press for their courtesy, and to acknowledge the high quality of their work.

R. M. G.

Cambridge, Mass.
January 1964

CONTENTS

PLATES

1

INTRODUCTION

1.1. The nature of the problem

EARTH, like the other inner planets, receives virtually all of its energy from space in the form of solar electromagnetic radiation. Its total heat content does not vary greatly, indicating a close overall balance between the absorbed solar radiation and the diffuse stream of low-temperature thermal radiation emitted by the planet. The transformation of the incident solar radiation into scattered and thermal radiation, and the consequent thermodynamic effects on the earth's gaseous envelope, are the subject of this book.

The scope must, however, be narrowed for, in its broadest interpretation, our title could embrace the photochemistry of the atmosphere and many of the topics usually discussed in books dealing with the upper atmosphere. By concentrating attention upon the thermodynamic aspects of the problem this difficulty of selection usually resolves itself. For example, the absorption of energy accompanying ionization will be treated where it compares with other energy sources; on the other hand, the electrical effects of this ionization are not considered for their own sake. Similarly, the properties of the airglow are relevant to the extent that the emitted energy may possibly influence the heat balance at the levels concerned, but not in the detail which has made the subject an important and interesting field of study in its own right.

The irradiation† at mean solar distance—*the solar constant*—is approximately 2·0 cal cm⁻² min⁻¹, giving a mean flux of solar energy, perpendicular to the earth's surface, of about 0·5 cal cm⁻² min⁻¹ (the factor 4 is the ratio of surface area to cross-section for a sphere). Of this, approximately 35 per cent is reflected from the atmosphere (including clouds), 50 per cent reaches the earth's surface directly or after scattering, and 45 per cent is absorbed at the surface. The ratio of outward to inward flux of solar radiation is known as the *albedo*, and we speak of the albedo of the earth as a whole, and of different surfaces such as cloud, snow, sea, or grassland. The albedo is a function of the frequency of the radiation (the *monochromatic albedo*) but the word is

† A brief review of the status of solar physics, as it concerns the sun as a source of radiation, is given in Appendix 14.

B

also used to denote an average, weighted according to the energy distribution of the sun. In this latter sense the albedo of the earth as a whole is believed to be about 0·4, so that an average of 0·6×0·5 or 0·3 cal cm^{-2} min^{-1} is available for heating, directly and indirectly, the earth and the atmosphere.

The redistribution of this energy by dynamical and radiative processes and its ultimate return to space as low-temperature *terrestrial radiation* is the most important problem treated in this book, and is, indeed, one of the central problems of meteorology. This problem is related mainly to conditions in the troposphere and lower stratosphere, which contain most of the atmospheric mass. There are, however, other problems of equal interest connected with the small amounts of ultra-violet radiation which can be absorbed high in the atmosphere, where the density is low and the resulting thermal effects can be large. This brings us to problems unfamiliar in the lower atmosphere, such as the emission of non-thermal radiation and the response of the atmosphere to small fluctuations in the output of solar energy.

Assuming the earth to radiate as a black body in the infra-red spectrum, we can compute the general level of terrestrial temperatures. The energy absorbed by the earth is

$$\text{absorbed energy} = f\pi r^2(1-a), \tag{1.1}$$

where f = solar constant (2·0 cal cm^{-2} min^{-1}),

   r = earth's radius,

   a = albedo for solar radiation (0·4).

The energy emitted by the earth is

$$\text{emitted energy} = 4\pi r^2 \sigma \theta_e^4, \tag{1.2}$$

where σ = Stefan–Boltzmann constant,

   θ_e = equivalent emission temperature of the earth.

If the planet is in a steady state, we may equate (1.1) and (1.2) obtaining a value of 246·5° K for the equivalent emission temperature. The computed temperature is lower than the average for the earth's surface but is close to the average temperature of the atmosphere itself, indicating that a considerable fraction of the terrestrial radiation comes from the atmosphere and not from the surface; a deduction which is confirmed by a cursory examination of the atmospheric absorption spectrum.

At a temperature of 246·5° K radiation of wavelengths less than about 4 μ† is of negligible energy, and since the solar flux carries little energy

† The micron (μ) and other spectrographic units are defined and discussed in Appendix 3.

at greater wavelengths, it is possible and convenient to treat the solar and terrestrial fluxes separately. Let us first briefly consider absorption of the terrestrial component. The principal gaseous constituents of the atmosphere (O_2, N_2, A) are almost transparent to wavelengths longer than 4μ, but minor polyatomic constituents such as H_2O, CO_2, O_3, N_2O, CO, CH_4 have intense and complex absorption spectra in this region, and are present in sufficient quantities to absorb a considerable proportion of the terrestrial radiation. Dust, haze, and particularly clouds also absorb strongly throughout the infra-red spectrum and modify the field of terrestrial radiation.

Clouds, ground, and atmosphere do not differ greatly in temperature and it follows from Kirchhoff's laws that each element of the atmosphere absorbs and emits similar amounts of radiation. Terrestrial radiation is therefore passed from layer to layer in the atmosphere, creating a transfer problem of considerable complexity. The problem differs in the upper atmosphere because emission does not obey Kirchhoff's laws if the pressure is sufficiently low. Approximate graphical and numerical solutions have been made of significant atmospheric problems, and idealized models can sometimes be treated analytically.

Absorption by minor polyatomic gases is complex, each band consisting of many lines whose individual shapes can be important when computing the radiative transfer. Much effort has been expended in the laboratory in attempts to obtain the data required for atmospheric computations. If the detailed data are used in entirety even a relatively simple atmospheric problem would be beyond the scope of an electronic computer. Methods are therefore required which treat correctly only the essential statistics of the problem. The development of such methods has led to a fairly complete understanding of the problems connected with absorption along a constant-pressure path. Absorption along a real atmospheric path, where pressure, temperature, and composition all vary, presents new problems not all of which have been solved.

An attempt has been made in Fig. 1.1 to give a general picture of the importance of different absorptions in the lower atmosphere in mid-latitudes. An indication of the energy absorbed by the stratosphere and troposphere may be obtained by multiplying (a) by (c) or (a) by $\{(b)-(c)\}$ respectively.

Most of the solar absorption in the stratosphere is by the ultra-violet Hartley bands and the visible Huggins bands of ozone. At higher levels, in the ionosphere, the small amount of solar energy below about $0\cdot2\,\mu$ is absorbed mainly by molecular oxygen. In the troposphere depletion of sunlight is principally by a group of near infra-red bands of water vapour. For terrestrial radiation, water vapour is the dominant component of the troposphere, although carbon dioxide also plays a part. In the stratosphere water vapour, carbon dioxide, and ozone are of comparable importance while the strong carbon dioxide band at $15\,\mu$ is the main radiator throughout the mesosphere.

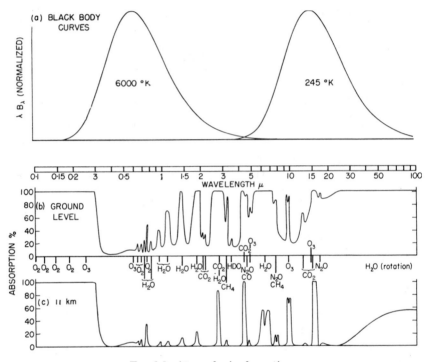

Fig. 1.1. Atmospheric absorptions.

(a) Black-body curves for 6000° K and 245° K. (b) Atmospheric gaseous absorption spectrum for a solar beam reaching ground level. (c) The same for a beam reaching the temperate tropopause. The axes are chosen so that areas in (a) are proportional to radiant energy. Integrated over the earth's surface and over all solid angles the solar and terrestrial fluxes are equal; consequently, the two black-body curves are drawn with equal areas beneath them. An absorption continuum has been drawn beneath bands in (b). This is partly hypothetical because it is difficult to distinguish from the scattering continuum, particularly in the visible and near infra-red spectrum. Conditions are typical of mid-latitudes and for a solar elevation of 40° or diffuse terrestrial radiation.

In addition to the absorptions of Fig. 1.1, both streams of radiation are scattered and absorbed by dust, haze, molecules, and clouds. The theory of molecular scattering and scattering by water droplets is well understood. Dust and haze cannot, however, be treated with the same precision, and their amounts are variable and difficult to relate to other physical phenomena.

The motions, temperatures, chemical composition, and cloud amount of the atmosphere are closely interrelated by complex mechanisms. Ideally all properties of the atmosphere should be predictable starting from a physical model, imposing only the boundary condition of a given flux of solar radiation. Very little advance in this direction has yet been made, although it is possible that electronic computers may yield some results in the near future. Some simple model situations are treated in this volume, but most investigations have attempted to isolate the radiative aspects by concentrating attention on the radiative heating as if it were a distinct and significant atmospheric parameter.

Two different classes of study have resulted from these two approaches, one synthetic but limited in scope, the other descriptive but aiming at completeness. The former, typified by the early work of Gold and Emden, approaches from the standpoint of local radiative equilibrium in a cloudless atmosphere of known composition, and inquires how far the structure of the real atmosphere can be explained in this way. The other, of which Simpson's work forms the best-known example, accepts the observed atmospheric thermal structure, and attempts to work out the resulting field of radiation. The methods are complementary; each throws some light on the nature of the hydrodynamic processes, the former by the ways in which the deduced structure differs from that observed, the latter by the computed departures from local radiative equilibrium.

Most of this book is devoted to the methods of analysis of the radiation problem, comparatively little to the synthesis and still less to purely empirical data. Since the analysis involves many different physical ideas and mathematical methods our presentation is necessarily fragmentary and non-systematic, and moreover, for reasons which are not strictly logical, it has been necessary to divide the material into two parts. It may therefore help the reader to indicate briefly how the different chapters fit into the scheme of ideas which has been outlined.

The remainder of Chapter 1 provides necessary background information to our problem in the form of a discussion of the thermal structure and chemical composition of the atmosphere. Chapter 2 outlines the formal mathematical theory required to handle problems of radiative transfer. Chapter 3 describes the physical processes of absorption, and Chapter 4 discusses methods which isolate from this mass of detail its essential statistical features. Chapter 5 then gives quantitative data for application to atmospheric problems. These three chapters having provided the required information on gaseous absorption, Chapter 6 is concerned with the mathematical problems of computing radiative heating in an atmosphere of arbitrary structure. Dust and haze are not considered at this point, and cloud only in an idealized form. Chapter 7 outlines the theory of scattering by small particles. Chapter 8 considers the theory of planetary atmospheres in radiative equilibrium, and Chapter 9 outlines the few results available on the dynamics of radiating fluids.

Part 2 will be concerned with instruments, observations, and the application of theoretical methods to the earth's radiation field. The observational material divides into measurements of the solar constant, measurements of the total radiation flux at different levels in the atmosphere, and measurements involving spectral resolution. The observed properties of cloud and aerosol and the problem of radiative transfer in a scattering medium will be treated in separate chapters. In the final chapter of Part 2 we will make an overall assessment of the radiation balance of the earth and all levels of the atmosphere, based on all available observational and theoretical information.

1.2. The thermal structure of the atmosphere

The average thermal structure of the atmosphere does not differ greatly from year to year. It is therefore valuable to think in terms of climatological mean conditions, with superposed short-period fluctuations. Fig. 1.2 gives a recent estimate of the mean conditions below 30 km in the four seasons along longitude 80° W. in the northern hemisphere. For the purposes of our discussion these cross-sections can often be taken to represent the conditions at all longitudes and for both hemispheres, although the distribution of land and sea does influence the climatological mean to a measurable extent.

The following features of the diagram are noteworthy. In the lowest 2 or 3 km the structure is complicated, with inversions at many latitudes. Above 3 km the atmosphere has remarkably regular features up to the tropopause. The equally spaced isotherms indicate a constant lapse rate of about 6·5 °K km^{-1}, independent of both season and latitude. At the tropopause a remarkable and sudden change to near-isothermal or inversion conditions takes place. The tropopause is usually complex somewhere between latitude 30° and 60°, where the high tropical tropopause (c. 17 km) overlaps the low arctic tropopause (down to 7 km). On daily maps a similar multiple tropopause usually appears

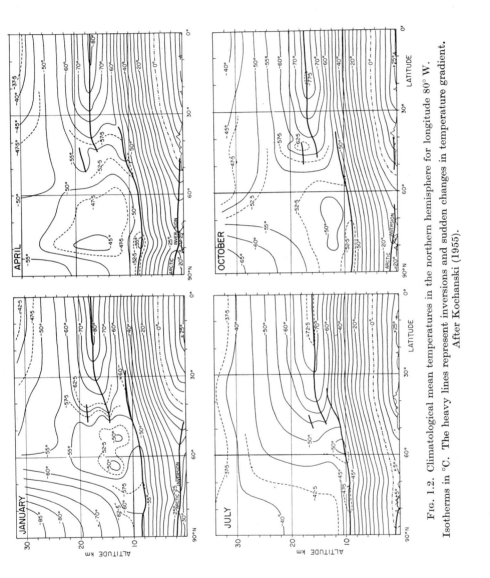

FIG. 1.2. Climatological mean temperatures in the northern hemisphere for longitude 80° W. Isotherms in °C. The heavy lines represent inversions and sudden changes in temperature gradient. After Kochanski (1955).

in association with the polar front, but the position varies so much from day to day that it cannot be seen in climatological, averaged data. Conditions in the stratosphere differ markedly from those in the troposphere. The base of the stratosphere is coldest in the tropics where the base of the troposphere is warmest, and vice versa in polar regions. The seasonal temperature wave changes phase discontinuously at the

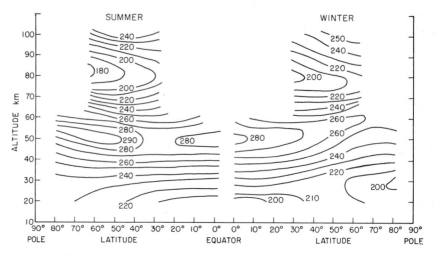

FIG. 1.3. Mean temperature (°K) between 20 km and 100 km.
After Murgatroyd (1957).

tropopause. These and other properties of the stratosphere are discussed in greater detail in books devoted to the subject. In Appendix 5 are given details of temperature, density, and pressure in a number of *standard atmospheres*, one of which can usually be chosen to approximate conditions at any required latitude and season.

Above 30 km the data are more sparse. Fig. 1.3 shows one attempt to depict climatological mean values, based both on directly observed temperatures and inference from the observed wind field. Tabulated data are given in Appendix 5. The main feature of Fig. 1.3 is the temperature maximum at all latitudes in the region of 50 km. The temperature minimum at 80 km may be deeper (possibly as low as 130° K) than is indicated in this diagram. Above this minimum the rise of temperature into the thermosphere is well established.

Above 100 km the physical state of the atmosphere is debatable. Rocket and satellite data yield densities from which temperatures can be determined if the molecular weight of the atmosphere is known, but at these levels dissociation and diffusive separation lead to great changes

in the composition with height. It is probable that the temperature
increases with height up to about 250 km, above which height the
thermal conductivity of air is so large that a near-isothermal state is
probable. Whether this constant temperature is as low as 500° K or
as high as 2000° K is uncertain; the diurnal variation is large. Fig. 1.4

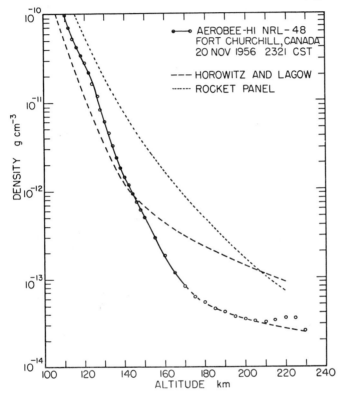

FIG. 1.4. Density as a function of height up to 230 km.
The points indicate data observed on 20 November 1956, 2321 cst, at Fort
Churchill, Canada. The dotted and broken lines are two other published curves
based on less complete data. After Newell (1960).

shows some density data below 230 km, while Appendix 5 gives a model
atmosphere extending to 1000 km altitude.

Short-period variations of temperature about the mean values can
be large in the troposphere and stratosphere where measurements are
available. They are probably large at other levels also. The detailed
temperature records of the lower atmosphere are essential data for the
weather forecaster and are far too extensive to discuss here. The reader
will, however, have experienced the magnitude and speed of the variations, and many textbooks are available if details are needed.

The lowest one or two kilometres of the atmosphere present many problems of their own. In particular, the lowest 100 m have been given much attention and many measurements are available. Conditions in the boundary layer are controlled largely by the absorption of solar radiation at the earth's surface and diurnal changes can be large.

1.3. The chemical composition of the atmosphere

A summary of the available data on the chemical composition of dry air is given in Table 1.1. Relative isotopic concentrations for atoms of significance for the radiation problem are given in Table 1.2. From these figures the totals in Table 1.1 can be broken down into isotopic components, if required.

TABLE 1.1

The composition of dry air

Molecule	Fraction by volume in the troposphere	Comments
N_2	$7 \cdot 8084 \times 10^{-1}$	Photochemical dissociation high in ionosphere. Mixed at lower levels.
O_2	$2 \cdot 0946 \times 10^{-1}$	Photochemical dissociation above 95 km. Mixed at lower levels.
A	$9 \cdot 34 \times 10^{-3}$	Mixed up to 110 km; diffusive separation above.
CO_2	$3 \cdot 3 \times 10^{-4}$	Slightly variable. Mixed up to 100 km; dissociated above.
Ne	$1 \cdot 818 \times 10^{-5}$	Mixed up to 110 km; diffusive separation above.
He	$5 \cdot 24 \times 10^{-6}$	Mixed up to 110 km; diffusive separation above.
CH_4	$1 \cdot 6 \times 10^{-6}$	Mixed in troposphere; oxidized in stratosphere; dissociation in mesosphere.
Kr	$1 \cdot 14 \times 10^{-6}$	Mixed up to 110 km; diffusive separation above.
H_2	5×10^{-7}	Mixed in troposphere and stratosphere, dissociated above.
N_2O	$3 \cdot 5 \times 10^{-7}$	Slightly variable at surface. Continuous dissociation in stratosphere and mesosphere.
CO	7×10^{-8}	Variable combustion product.
O_3	$\sim 10^{-8}$	Highly variable; photochemical origin.
NO_2 NO	0 to 2×10^{-8}	Industrial origin in troposphere. Photochemical origin in mesosphere and ionosphere.

Many other trace gases are present in the atmosphere, some of which (e.g. Xe, SO_2, NH_3, Rn, etc.) have been extensively studied, but none influences the radiation fluxes to a significant extent. In addition to these gases the atmosphere contains solid matter in suspension, whose concentration and composition are very variable. Water vapour is discussed in the text; the above figures apply only to dry air.

A few additional remarks about the constituents of principal importance to radiation studies are necessary. Molecular oxygen is decomposed photochemically above 90 km, but turbulent mixing causes departures from strict radiation equilibrium. Some measurements of

TABLE 1.2

Isotopic abundances in nature

Isotope	Percentage relative abundance
H^1	99·9851
D^2	0·0149
C^{12}	98·892
C^{13}	1·108
O^{16}	99·758
O^{17}	0·0373
O^{18}	0·2039
N^{14}	99·631
N^{15}	0·369

Almost all terrestrial hydrogen is combined in the form of water. Since HHO and HDO have different vapour pressures, the relative concentration $D^2:H^1$ can vary from phase to phase by as much as 10 per cent. Small differences in the concentration of oxygen isotopes also occur.

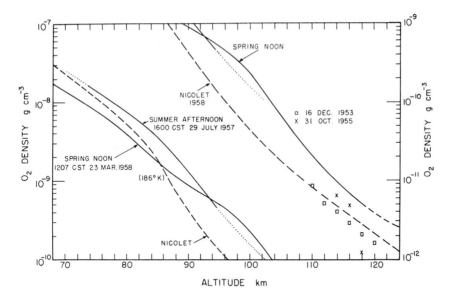

Fig. 1.5. Density of molecular oxygen as a function of altitude measured at Fort Churchill, Canada.

The curve marked 'Nicolet' is based upon theoretical considerations. The onset of photochemical decomposition is indicated by a steepening of the slope at 86 km for the midsummer flight and 96 km for the early spring flight. Note the different ordinates for right and left groups of curves. After Friedman (1960).

molecular oxygen derived from solar absorption spectra are shown in Fig. 1.5.

The concentration of water vapour is mainly governed by condensation processes, and the gas is not mixed evenly with the main constituents below 15 km. The vapour pressure can vary over a very wide range indeed. In the troposphere the average relative humidity is close

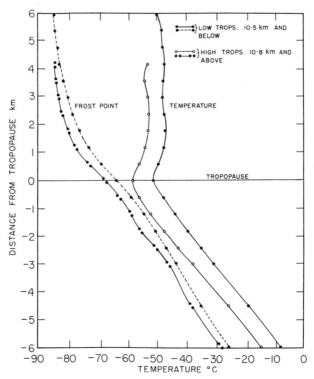

FIG. 1.6. Mean temperature and frost-point over England.
After Murgatroyd, Goldsmith, and Hollings (1955).

to 50 per cent. Thus for a ground temperature of 278° K the vapour pressure is close to 8 mb. Above 15 km relative humidity has fallen to about 1 per cent and the vapour pressure is about 2×10^{-4} mb, giving a range of vapour pressure of about 4×10^{4} accompanying a range of total pressure of 8. Measured frost-point temperatures below 15 km are shown in Fig. 1.6. These can be interpreted in terms of vapour pressures with the aid of the vapour-pressure data in Appendix 6. The parallel frost-point and temperature curves in the troposphere denote approximately constant relative humidity. The highest available routine humidity measurements from aircraft indicate a frost-point temperature at 15 km close to 189° K at all times and over a wide range of latitudes. These observations suggest that above this level water vapour maintains a constant mixing ratio until photochemical decomposition starts at around 60 km. Some recent measurements between 15 and 30 km seem to contradict this conclusion and indicate instead a slow increase of mixing ratio with height; the matter remains in some doubt at the present time.

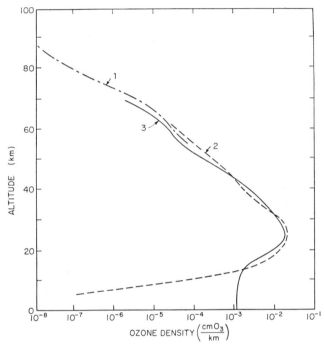

FIG. 1.7. Observed and computed ozone distributions.
1. Bates and Nicolet (theoretical). 2. Paetzold (theoretical). 3. Observed mean for middle latitudes. After Paetzold and Regener (1957).

The vertical distribution of ozone differs from that of other atmospheric gases, having a maximum concentration near 25 km and a maximum mixing ratio with air near 30 km (see Fig. 1.7). Ozone is formed photochemically in the atmosphere; rapidly above 30 km so that equilibrium with oxygen obtains when the sun is shining; slowly below this level so that in the lower stratosphere and troposphere its concentration is mainly dependent upon advection and mixing processes, and is consequently highly variable. This variability is reflected in the seasonal and meridional changes shown in Fig. 1.8. Note that the maximum ozone amount occurs in the polar night where none is formed by photochemical processes. Other ozone variations can be related to the passage of weather systems.

Carbon dioxide influences the radiation field at all levels below 100 km. Being chemically unreactive and having its main sources and sinks in biological processes at the earth's surface, it is to be expected that it will be mixed with the main atmospheric constituents. The few

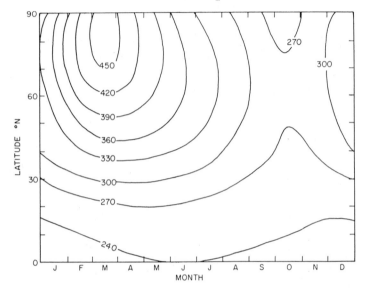

FIG. 1.8. Seasonal and meridional variations in the total ozone in a vertical atmospheric column.

The unit is 10^{-3} cm ozone at s.t.p. After Godson (1960).

stratospheric measurements which have been made confirm this supposition, giving a fraction by volume of $2 \cdot 6 \times 10^{-4}$, slightly lower than ground-level values. At the ground the concentration is variable, but on the average it appears to increase slowly with time in accordance with the increase of industrial effluent during the twentieth century (see Fig. 1.9). Above the dissociation level of molecular oxygen, carbon dioxide bands in the ultra-violet spectrum absorb solar radiation and the gas decomposes.

Methane and nitrous oxide probably resemble carbon dioxide in distribution, since both appear to be of biospheric origin. Somewhere in the mesosphere both will be destroyed by photochemical processes. Spectrographic observations confirm that both gases are mixed in the troposphere, while other methods of analysis indicate variability in the concentration near the ground.

Tropospheric carbon monoxide is a combustion product and is more variable in concentration than methane or nitrous oxide. Variations in the total atmospheric content of 0·04 to 0·07 cm s.t.p. and more have been reported. The vertical distribution has not been determined, but it is probably mixed with the main atmospheric constituents, except in the ionosphere where its concentration will be determined by the photochemical destruction of carbon dioxide.

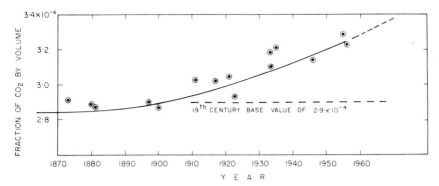

FIG. 1.9. The amount of CO_2 in the free air of the North Atlantic region. The full curve is obtained from measurements on fossil fuel. After Callendar (1958).

The final constituents of importance to radiation studies in a cloud-free atmosphere are dust and haze. The variability both in amount and composition of these components is such that few generalizations can be made, particularly about the distribution in the vertical. Visibility measurements reflect the aerosol concentration at ground level. The visual range can vary between a few cm and about 200 km depending upon the atmospheric conditions. Generally dust and haze concentrations decrease rapidly with height in the troposphere, with a scale height of the order of 1 km. Visible tops are formed at inversions. The sizes of the particles, which are usually hygroscopic, depend upon the relative humidity.

In the stratosphere the concentration of aerosol particles smaller than $0·1\,\mu$ continues to fall off with height, but larger particles ($0·1\,\mu$ to $1·0\,\mu$) increase to a maximum concentration near 20 km. At still higher levels (c. 85 km) the rare occurrence of noctilucent clouds indicates large localized increases in concentration.

The most important effects of aerosols concern the way in which they modify the field of solar radiation, but the possibility that aerosol may also influence the transfer of terrestrial radiation must also be borne in mind, particularly in the lower troposphere. Clouds have, of course, a very strong influence on both terrestrial and solar radiation, and although they are properly described as aerosols they are conveniently treated as an independent phenomenon.

BIBLIOGRAPHY

1.1. The nature of the problem

A number of monographs and review articles on atmospheric radiation are in existence, which the reader may wish to consult. In the English language the classical monograph is by Elsasser:

ELSASSER, W. M., 1942, *Heat transfer by infrared radiation in the atmosphere*, Harvard Meteorological Studies No. 6. Harvard Univ. Press.

Also in English are:

Compendium of meteorology, 1951, ed. T. F. Malone. Boston : American Meteorological Society, containing articles by S. FRITZ, F. MÖLLER, A. ÅNGSTROM, H. NEUBERGER, Z. SEKARA, and W. E. K. MIDDLETON ; and

GOODY, R. M., and ROBINSON, G. D., 1951. 'Radiation in the troposphere and lower stratosphere', *Quart. J. R. Met. Soc.* **77**, p. 151.

In the German literature there have been many excellent reviews, the most recent of which appear in

Handbuch der Physik, 1957, ed. S. Flügge. Band XLVIII: *Geophysik*. II. Berlin : Springer ; and

Handbuch der Geophysik, 1956, eds. F. Linke and F. Möller. Band VIII : *Physik der Atmosphäre*. I. Berlin : Borntraeger.

The former contains articles by F. MÖLLER, W. E. K. MIDDLETON, Z. SEKARA, and J. BRICARD. The latter is not yet completed and contains articles by R. MEYER, F. VOLZ, and F. LAUSCHER.

The only complete textbooks on atmospheric radiation are in Russian, but it is expected that an English translation will be available before the present volume is published :

KONDRATIEV, K. Y., 1954, *Radiation exchange in the atmosphere*. Leningrad : Hydrometeorological Publishers ; and

—— 1956, *The radiative energy of the sun*. Leningrad : Hydrometeorological Publishers.

1.2. The thermal structure of the atmosphere

Climatological mean temperatures below 31 km in the northern hemisphere have been assembled by

KOCHANSKI, A., 1955, 'Cross sections of the mean zonal flow and temperature along 80° W ', *J. Met.* **12**, p. 95.

Details of different *standard atmospheres* are to be found in an article by

WARES, G. W., 1960, *Handbook of geophysics*. New York : Macmillan, p. 1.1.

The most complete review of temperatures up to 100 km available at the time of writing is by

MURGATROYD, R. J., 1957, 'Winds and temperatures between 20 km and 100 km—a review', *Quart. J. R. Met. Soc.* **83**, p. 417.

More details about the tropopause and lower stratosphere are given by

GOODY, R. M., 1954, *The physics of the stratosphere*. Cambridge Univ. Press.

The most recent reviews of ionospheric data are by

NEWELL, H. F., 1960, 'The upper atmosphere studied by rockets and satellites', chapter 3 of *Physics of the upper atmosphere*, ed. J. A. Ratcliffe. New York: Academic Press; and

NICOLET, M., 1960, 'The properties and constitution of the upper atmosphere', ibid., chapter 2.

Some details of short-period temperature changes are to be found in any textbook on meteorology. The earth's boundary layer is usually treated as an independent problem, see

PRIESTLEY, C. H. B., 1959, *Turbulent transfer in the lower atmosphere*. Chicago Univ. Press;

LETTAU, H. H., and DAVIDSON, B. (eds.), 1957, *Exploring the atmosphere's first mile*, vols. i and ii. London: Pergamon; and

GEIGER, R., 1957, *The climate near the ground*. Harvard Univ. Press.

1.3. The chemical composition of the atmosphere

Review articles on the composition of the atmosphere have been written by

GLUECKAUF, E., 1951, 'The composition of atmospheric air', in *Compendium of meteorology*, ed. T. F. Malone. Boston: American Meteorological Society, and NICOLET, M. (1960) (§ 1.2).

Some rocket results on upper-atmospheric composition are given by

FRIEDMAN, H., 1960, 'The sun's ionizing radiations', chapter 4 of *Physics of the upper atmosphere*, ed. J. A. Ratcliffe. New York: Academic Press.

The photochemistry of minor atmospheric constituents is discussed by

BATES, D. R., and WITHERSPOON, A. E., 1952, 'The photochemistry of some minor constituents of the earth's atmosphere (CO_2, CO, CH_4, H_2O)', *M.N.R.A.S.* **112**, p. 101;

and the photochemistry of water vapour by

BATES, D. R., and NICOLET, M., 1950, 'The photochemistry of atmospheric water vapour', *J. Geophys. Res.* **55**, p. 301.

Measurements of the density of water vapour in the upper troposphere and lower stratosphere are presented by

BANNON, J. K., FRITH, R., and SHELLARD, H. C., 1952, *Humidity of the upper troposphere and lower stratosphere over southern England*, Geophysical Memoir No. 88. Meteorological Office, London: H.M.S.O.;

MURGATROYD, R. J., GOLDSMITH, P., and HOLLINGS, W. E. H., 1955, 'Some recent measurements of humidity from aircraft up to heights of about 50,000 ft. over southern England', *Quart. J. R. Met. Soc.* **81**, p. 533; and

HELLIWELL, N. C., 1958, 'Research at high altitudes by the Meteorological Research Flight', *Weather* **13**, p. 287.

For a discussion of observations indicating that humidity may increase with height above 15 km see

GUTNICK, M., 1961, 'How dry is the sky?', *J. Geophys. Res.* **66**, p. 2867.

Atmospheric ozone is the subject of many books and reviews, e.g.

PAETZOLD, H. K., and REGENER, E., 1957, 'Ozon in der Erdatmosphäre', *Handbuch der Physik*, ed. S. Flügge. Band XLVIII: *Geophysik*. II. Berlin: Springer, p. 370;

FABRY, CH., 1950, *L'ozone atmosphérique.* Paris: Centre National de la recherche scientifique; and

GÖTZ, F. P. W., 1951, 'Ozone in the atmosphere', in *Compendium of Meteorology*, ed. T. F. Malone. Boston: American Meteorological Society.

Recent measurements at high latitudes are given by

GODSON, W. L., 1960, 'Total ozone and the middle stratosphere over arctic and sub-arctic areas in winter and summer', *Quart. J. R. Met. Soc.* **86**, p. 301.

Data on the twentieth-century increase of carbon-dioxide concentration are given by

CALLENDAR, G. S., 1958, 'On the amount of carbon dioxide in the atmosphere', *Tellus*, **10**, p. 243.

Measurements of carbon monoxide by spectrographic methods are described by

SHAW, J. H., 1958, 'The abundance of atmospheric carbon monoxide above Columbus, Ohio', *Astrophys. J.* **128**, p. 428.

Data on methane and nitrous oxide are described by GOODY (1954) (§ 1.2) and by BATES and WITHERSPOON (1950).

Isotopic ratios are given by

EVANS, R. D., 1955, *The atomic nucleus.* New York: McGraw-Hill.

Data on atmospheric aerosol are reviewed by

MASON, B. J., 1957, *The physics of clouds.* Oxford Univ. Press; and

JUNGE, C. E., and MANSON, J. E., 1961, 'Stratospheric aerosol studies', *J. Geophys. Res.* **66**, p. 2163.

2

THEORY OF RADIATIVE TRANSFER

2.1. Definitions

2.1.1. *Intensity, flux, energy density*

IN common with astrophysical usage the word *intensity* will denote *specific intensity of radiation*, i.e. the flux of energy in a given direction per second per unit frequency (or wavelength) range per unit solid angle per unit area perpendicular to the given direction. In Fig. 2.1 the

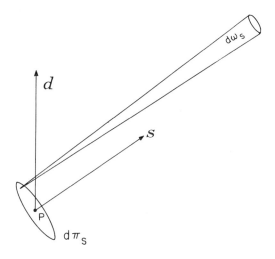

FIG. 2.1. Specific intensity of radiation.

point P is surrounded by a small element of area $d\pi_s$, perpendicular to the direction of the vector $\mathbf{s}$. From each point on $d\pi_s$ a cone of solid angle $d\omega_s$ is drawn about the $\mathbf{s}$ vector. The bundle of rays, originating on $d\pi_s$, and contained within $d\omega_s$, transports in time dt and in the frequency range ν to $\nu+d\nu$, the energy

$$E_\nu = I_\nu(P, \mathbf{s})\, d\pi_s\, d\omega_s\, d\nu dt, \tag{2.1}$$

where $I_\nu(P, \mathbf{s})$ is the specific intensity at the point P in the $\mathbf{s}$-direction. If I_ν is not a function of direction the intensity field is said to be *isotropic*; if I_ν is not a function of position the field is said to be *homogeneous*. If it is more convenient to use wavelength than frequency we have the alternative definition of intensity

$$I_\lambda \, d\lambda = I_\nu \, d\nu,$$

or
$$I_\lambda = \frac{\nu^2}{\mathbf{c}} I_\nu, \tag{2.2}$$

where **c** is the velocity of light.

The component of *flux* in the **d**-direction, $F_{\nu,d}(P)$, is defined as the total energy flowing across unit area perpendicular to **d**, per unit frequency interval. An infinitesimal area $d\pi_d$ has a projected area in the **s**-direction

$$d\pi_s = d\pi_d \cos(\mathbf{d}, \mathbf{s}),$$

where $(\mathbf{d}, \mathbf{s})$ denotes the angle between the two vectors. The energy flux across $d\pi_d$, integrated over all **s**-directions is, from the definitions of $I_\nu(P, \mathbf{s})$ and $F_{\nu,d}(P)$,

$$F_{\nu,d}(P) \, d\nu d\pi_d = \int I_\nu(P, \mathbf{s}) \cos(\mathbf{d}, \mathbf{s}) \, d\nu d\pi_d \, d\omega_s,$$

or
$$F_{\nu,d}(P) = \int I_\nu(P, \mathbf{s}) \cos(\mathbf{d}, \mathbf{s}) \, d\omega_s, \tag{2.3}$$

where the integral extends over all solid angles.

If **x**, **y**, **z** are three orthogonal unit vectors, we have the trigonometric identity

$$\cos(\mathbf{d}, \mathbf{s}) \equiv \cos(\mathbf{d}, \mathbf{x})\cos(\mathbf{s}, \mathbf{x}) + \cos(\mathbf{d}, \mathbf{y})\cos(\mathbf{s}, \mathbf{y}) + \cos(\mathbf{d}, \mathbf{z})\cos(\mathbf{s}, \mathbf{z}).$$

Substituting in (2.3),

$$F_{\nu,d}(P) = F_{\nu,x}(P)\cos(\mathbf{d}, \mathbf{x}) + F_{\nu,y}(P)\cos(\mathbf{d}, \mathbf{y}) + F_{\nu,z}(P)\cos(\mathbf{d}, \mathbf{z}). \tag{2.4}$$

Equation (2.4) is the transformation law for the components of the vector
$$\mathbf{F}_\nu(P) = \mathbf{x}F_{\nu,x}(P) + \mathbf{y}F_{\nu,y}(P) + \mathbf{z}F_{\nu,z}(P). \tag{2.5}$$

In the problems discussed in this book the horizontal flux components are zero or divergenceless from symmetry considerations. In either case it is only necessary to evaluate the vertical component of the flux, and the direction of the flux vector does not present a problem.

The sun's disk subtends, on the average, an angle of 32′ at the earth's surface; for most practical purposes it can be regarded as a parallel beam of radiation. The above definition of intensity is unnecessarily general for this case. Let us suppose that the sun's direction is that of the vector $\mathbf{s}_\odot$, and let its disk subtend a solid angle $d\omega_\odot$ at the earth. For unmodified solar radiation $I_\nu(P, \mathbf{s})$ is only non-zero if **s** is very close to $\mathbf{s}_\odot$. In most circumstances therefore we may replace the angle $(\mathbf{d}, \mathbf{s})$ by $(\mathbf{d}, \mathbf{s}_\odot)$. From (2.3),

$$F_{\nu,d}(P) \simeq \int_{\omega_s} I_\nu(P, \mathbf{s})\cos(\mathbf{d}, \mathbf{s}_\odot)\,d\omega_s$$

$$= \cos(\mathbf{d}, \mathbf{s}_\odot)\bar{I}_\nu(P)\,d\omega_\odot, \qquad (2.6)$$

where $\bar{I}_\nu$ is the mean value of the intensity, averaged over the sun's disk.

If we write
$$F_{\nu,d}(P) = f_\nu \cos(\mathbf{d}, \mathbf{s}_\odot),$$

then
$$f_\nu = \bar{I}_\nu \, d\omega_\odot, \qquad (2.7)$$

is the solar *irradiation*, a function of solar distance only.

We can now evaluate the *energy density* (u_ν) of a radiation field. Consider a cylinder, parallel to the $\mathbf{s}$-direction, of length ds and cross-section $d\pi_s$ (Fig. 2.2). A quantum of radiation travelling in the $\mathbf{s}$-direc-

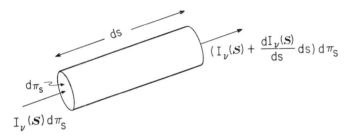

FIG. 2.2. Radiative heating.

tion will spend a time $dt = ds/\mathbf{c}$ in the cylinder, where $\mathbf{c}$ is the velocity of light. The total amount of energy in the cylinder made up of quanta travelling in the $\mathbf{s}$-direction within the solid angle $d\omega_s$ is that which crosses $d\pi_s$ in time dt:

$$I_\nu(P, \mathbf{s})\,d\pi_s\,d\nu d\omega_s\,dt = \frac{I_\nu(P, \mathbf{s})}{\mathbf{c}}\,d\pi_s\,d\nu d\omega_s\,ds.$$

The volume of the cylinder is $ds d\pi_s$ and these quanta contribute to the energy density an amount

$$\frac{I_\nu(P, \mathbf{s})}{\mathbf{c}}\,d\nu d\omega_s.$$

Integrating over all directions we find for the energy density in the frequency range ν to $\nu+d\nu$:

$$u_\nu(P)\,d\nu = \frac{d\nu}{\mathbf{c}}\int I_\nu(P, \mathbf{s})\,d\omega_s.$$

Hence
$$u_\nu(P) = \frac{4\pi}{\mathbf{c}} \bar{I}_\nu(P),$$
(2.8)

where
$$\bar{I}_\nu = \frac{1}{4\pi} \int I_\nu(\mathbf{s}) \, d\omega_s,$$
(2.9)

is the *average intensity* of the radiation field.

The divergence of the energy flux equals the rate at which energy is *added* to the field per unit volume, i.e. the rate at which energy is *lost* by the matter. Let h_ν be the rate per unit volume at which heat is gained by matter from radiation in unit frequency range

$$h_\nu = -\boldsymbol{\nabla} \cdot \mathbf{F}_\nu.$$
(2.10)

Expanding the right-hand side of (2.10) and using the definition of flux (2.3) we find

$$
\begin{aligned}
\boldsymbol{\nabla} \cdot \mathbf{F}_\nu &= \frac{\partial F_{\nu,x}}{\partial x} + \frac{\partial F_{\nu,y}}{\partial y} + \frac{\partial F_{\nu,z}}{\partial z} \\
&= \int d\omega_s \left\{ \frac{\partial I_\nu}{\partial x} \cos(\mathbf{s}, \mathbf{x}) + \frac{\partial I_\nu}{\partial y} \cos(\mathbf{s}, \mathbf{y}) + \frac{\partial I_\nu}{\partial z} \cos(\mathbf{s}, \mathbf{z}) \right\} \\
&= \int d\omega_s \frac{dI_\nu}{ds}.
\end{aligned}
$$
(2.11)

The meaning of (2.11) is made clear by Fig. 2.2. The energy *lost* by matter in the small cylinder is $\{dI_\nu(\mathbf{s})/ds\} \, dsd\pi_s \, d\omega_s$ per second per unit frequency range. Since $dsd\pi_s$ is the volume of the cylinder the heat loss per unit volume from radiation travelling in the $\mathbf{s}$ direction is $\{dI_\nu(\mathbf{s})/ds\} \, d\omega_s$, and (2.11) follows by integrating over all ω_s.

In later chapters we will frequently make use of the concept of *radiative equilibrium*, whereby there is no net energy exchange between matter and the radiation field

$$h = \int_0^\infty h_\nu \, d\nu = 0.$$
(2.12)

A special case of radiative equilibrium is *monochromatic radiative equilibrium* where

$$h_\nu = 0.$$
(2.13)

In defining radiative equilibrium and in other problems concerning atmospheric thermodynamics the relevant quantities are those which are integrated over the entire spectrum. This operation will be indicated by omitting the suffix ν from frequency-dependent quantities. Thus

$$I = \int_0^\infty I_\nu \, d\nu,$$

$$F = \int_0^\infty F_\nu \, d\nu,$$

$$u = \int_0^\infty u_\nu \, d\nu, \text{ etc.}$$

There are circumstances in which limited frequency ranges have particular physical significance, e.g. the frequency range embracing an isolated absorption band. Integrals over such ranges will be designated by a suffix other than ν. Thus

$$I_i = \int_i I_\nu \, d\nu,$$

$$I = \sum_i I_i, \text{ etc.}$$

2.1.2. *Extinction and emission*

We now require formal definitions of the interactions between matter and a radiation field. For most atmospheric problems a name can be assigned to a process without difficulty, and careful definitions may seem pedantic. However, if we wish to examine the whole range of atmospheric phenomena, without having to modify definitions, it pays to examine the situation closely.

The totality of interactions between radiation and matter will be classed as either *extinction* or *emission*. The two processes are distinguished by the sign of the change of radiant intensity as a result of the interaction. If the intensity *decreases* then we have *extinction*; if the intensity *increases* we have *emission*. No interaction at all can be pictured as the simultaneous extinction and emission of identical quanta, or alternatively as a case of vanishingly small interaction coefficients. This prosaic distinction between extinction and emission is the only one which applies to all phenomena described by the two terms.

The fundamental law of extinction is that of *Lambert*.† It states that the extinction process is linear independently in the intensity of radiation and in the amount of matter, provided that the physical state (i.e. temperature, pressure, composition) is held constant. The possible processes non-linear in the light intensity have been fully explored in recent years. The scattering of light by light is a tractable theoretical

† The name of *Bouguet* is more commonly used in European literature.

problem and the non-linear scattering from bound electrons can now be observed in the laboratory with the aid of coherent light amplifiers. The photon densities required to exhibit non-linear effects are greatly in excess of those in planetary atmospheres, and deviations from Lambert's law on this account are completely negligible. On the other hand, the optical properties of individual molecules are strongly influenced by the proximity of other molecules; the problem of pressure broadening of spectral lines, which will be discussed in Chapter 3, is one example. However, if we postulate that the matter is always in the same physical state, then the inter-molecular forces are fixed. Under this condition only a linear dependence of extinction on amount of matter is possible.

If da is the amount of matter per cm² along the direction in which the intensity is defined, then, by Lambert's law, the change of intensity in traversing this infinitesimal path is

$$dI_\nu \text{ (extinction)} = -e_\nu I_\nu \, da. \qquad (2.14)$$

The constant of proportionality e_ν is the *extinction coefficient*.

The argument that the extinction process is linear in the amount of matter applies with equal force to the emission process. As a formal statement, we may write

$$dI_\nu \text{ (emission)} = +e_\nu J_\nu \, da, \qquad (2.15)$$

defining the *source function* J_ν whose form will be the subject of later discussion.

The amount of matter per cm² can be specified in many ways, and each defines a different extinction coefficient, with different physical dimensions. If we use the path length (s) as a measure of a, we obtain the *volume extinction coefficient* e_v; the mass per cm² ($a = \rho s$) where ρ is the density, leads to the *mass extinction coefficient* e_m; the number of molecules per cm² ($a = ns$), where n is the number density of molecules, gives the *molecular extinction coefficient* e_n; finally, the virtual path length if the gas were to be reduced to s.t.p. ($a = sn/n_s$), where n_s is Loschmidt's number, yields the *extinction coefficient per cm s.t.p.* e_s. The interrelations between these coefficients are given in Appendix 3. We may note from (2.14) that the product ($e \, da$) is dimensionless. It is therefore convenient to introduce an increment of *optical path*

$$d\tau_\nu = -e_\nu \, da. \qquad (2.16)$$

The negative sign in (2.16) is related to the historical development of the subject in astrophysical literature; at a later stage we will take the

optical path itself to be a positive definite quantity, and accept whichever sign is appropriate in (2.16).

Our earlier statement that all interactions can be classed as extinction
or emission can now be summed up in the statement that any change
in intensity resulting from the interaction of matter and radiation must
be the sum of (2.14) and (2.15),

$$dI_\nu(P, \mathbf{s}) = dI_\nu(\text{extinction}) + dI_\nu(\text{emission})$$
$$= d\tau_\nu\{I_\nu(P, \mathbf{s}) - J_\nu(P, \mathbf{s})\},$$

and hence
$$\frac{dI_\nu(P, \mathbf{s})}{d\tau_\nu} = I_\nu(P, \mathbf{s}) - J_\nu(P, \mathbf{s}). \tag{2.17}$$

Equation (2.17) is known as the *equation of transfer*, and was first
given in this form by Schwarzschild. While it sets the pattern of the
formalism used in transfer problems, its physical content is very slight.
The physics is mainly contained in the definitions of the extinction
coefficient and the source function. One point must be kept in mind,
which is difficult to convey in a convenient notation, namely that the
l.h.s. of (2.17) is a differential along the **s**-direction.

Combining (2.10), (2.11), (2.14), and (2.17) we can now express the
heating rate in terms of the source function and extinction coefficient

$$h_\nu = -\int d\omega_s \frac{dI_\nu}{ds} = 4\pi e_{\mathrm{v},\nu}(\bar{I}_\nu - \bar{J}_\nu), \tag{2.18}$$

where
$$(\bar{I}_\nu, \bar{J}_\nu) = \int \frac{d\omega_s}{4\pi}(I_\nu, J_\nu). \tag{2.19}$$

Equation (2.18) expresses the heating rate as the difference between
the absorption from the mean radiation field and the mean emission.
The condition for monochromatic radiative equilibrium (2.13) is now

$$\bar{I}_\nu = \bar{J}_\nu. \tag{2.20}$$

We will find it convenient for some purposes to use only quantities
with the dimensions of intensity. The *heating function* H_i has such
dimensions, but is defined for a finite frequency interval only,

$$H_i = \frac{\int_i h_\nu \, d\nu}{2\pi \int_i e_{\mathrm{v},\nu} \, d\nu}. \tag{2.21}$$

Further subdivision of interactions between matter and the radiation
field depends upon the physical process involved, and particularly upon
changes of internal energy of the matter. If radiation interacts with
matter whose only mode of internal energy is translational, then the
interaction coefficients are very small in all circumstances discussed in

this book. For example, if a $0 \cdot 1 \, \mu$ quantum interacts with a free electron (Compton scattering), only 5×10^{-5} of its energy is transformed into kinetic energy, and this is the most efficient conversion of any relevance to the atmosphere. If there is no interaction with any form of internal energy the interaction is a *simple scattering process*. There is a close approach to *simple scattering* when the matter has only narrow (quantized) states of internal energy and the incident quanta have a frequency far from that of a possible transition.

All matter with which we will be concerned has electronic, vibrational, and rotational internal energy, and some small interaction at least will take place between these energy states and incident radiation. The process may still be classified as *scattering*, however. An interacting quantum will cause a transition to a higher, excited state. The excited state has a limited lifetime, and if the absorbed quantum is re-emitted with negligible conversion to translational energy, the process is one of *scattering*. If the transition to the lower state takes place in one step, then the emitted quanta will have a frequency identical to the absorbed quanta, and we refer to a process of *coherent scattering*. The molecule may, however, revert to the ground state by a cascade process through intermediate levels. The emitted quanta now differ in frequency from those incident, but the total energies of the extinction and emission processes are the same; this is *incoherent scattering*.

It often happens, however, that before the matter can re-emit, molecular collisions occur, during which non-radiating transitions (*de-activation*) can take place. The energy then ends up in other forms of internal energy. In the case of complete *thermodynamic equilibrium* the energy is shared equally among all the accessible degrees of freedom. Where energy is transferred to kinetic energy, the process is called *absorption*; the reverse process will be called *thermal emission*. Thermal emission and scattering are not mutually exclusive, and frequently occur simultaneously. We will see that circumstances occur when there is essentially no distinction, other than semantic, between them. Since all processes are linear, the extinction coefficient can be expressed as the sum of an *absorption coefficient* (k_ν) and a *scattering coefficient* (s_ν):

$$e_\nu = k_\nu + s_\nu. \tag{2.22}$$

One physical observation can usefully be introduced at this stage; namely, that the atmosphere is effectively *isotropic*. Exceptions exist, such as raindrops falling under gravity and the ionospheric propagation of radio waves in the earth's magnetic field, but these are of a relatively obvious character. Assuming isotropy and accepting the random nature of molecular agitation, there is no sufficient reason for absorption and thermal emission to be anything but isotropic.

Simple scattering, on the other hand, involves a very direct connexion between incident and emitted radiation. Since light has vector properties (see § 2.1.3) isotropy is no longer expected, and is not observed for scattering either by molecules or small particles. The more complex scattering process, which involves transitions between quantized states, may or may not preserve some memory of the vector properties of the incident photon. Thus, although it must be with circumspection, it is possible to think of absorption and scattering processes as differing in the symmetry of the source function, although this is neither as fundamental nor as useful a distinction as that adopted in this section.

If we consider carefully the meaning of the word 'heating' it is synonymous with an increase in the translational energy of matter. Our definitions of scattering and radiative equilibrium therefore correspond, and there is, for example, no formal distinction between absorption in radiative equilibrium and isotropic scattering.

2.1.3. *Simple scattering*

In the previous paragraph the vector properties of electromagnetic radiation were first mentioned. We will now discuss how to write the equation of transfer in a suitable matrix form.

The most general radiation field encountered in nature can be specified by four quantities, known as *Stokes parameters*, each with the physical dimensions of an intensity. We will denote these by $I_\nu^{(i)}$ $(i = 1, 2, 3, 4)$.

The most general polarization from a single source is elliptical. The polarization ellipse can be defined in terms of the intensities of two components polarized at right angles to each other, and the direction in space of the major or minor axis of the ellipse. Let l and $\mathbf{r}$ be two unit vectors forming an orthogonal set with $\mathbf{s}$, the direction of propagation. Let $I_\nu^{(l)}$ and $I_\nu^{(r)}$ be the intensities of the two polarized components of the beam. The total intensity is

$$I_\nu = I_\nu^{(l)} + I_\nu^{(r)}, \tag{2.23}$$

and both component intensities are determined if we also know

$$Q_\nu = I_\nu^{(l)} - I_\nu^{(r)}. \tag{2.24}$$

Let $\tan\beta$ equal the ratio of the axes of the polarization ellipse, and let χ be the angle between l and the major axis. We define

$$U_\nu = Q_\nu \tan 2\chi, \tag{2.25}$$

and

$$V_\nu = Q_\nu \tan 2\beta \sec 2\chi. \tag{2.26}$$

We can now make a variety of choices for the Stokes parameters: $(I_\nu, Q_\nu, U_\nu, V_\nu)$ is one possibility; $(I_\nu^{(l)}, I_\nu^{(r)}, U_\nu, V_\nu)$ is another. Four parameters are more than are necessary to specify an elliptically polarized beam, for which the following relation exists

$$I_\nu^2 = Q_\nu^2 + U_\nu^2 + V_\nu^2. \qquad (2.27)$$

A more general field may contain many independent, polarized components. If, as in nature, these possess no systematic phase relations, then the Stokes parameters are additive, but (2.27) no longer holds. Then all four parameters are required to define the field. *Natural* (or unpolarized) *light* has $I_\nu^{(l)} = I_\nu^{(r)}$, and therefore $Q_\nu = U_\nu = V_\nu = 0$.

If we now consider the equation of transfer (2.17), we need to make no modifications except to substitute $I_\nu^{(i)}$ and $J_\nu^{(i)}$ in place of I_ν and J_ν. In particular we do not need to modify our definition of the extinction coefficient, which is the same for all Stokes parameters if the medium is isotropic. The problem is to write the source function in a form which can be related to the *phase matrix* P_{ij}, which in turn can be calculated from electromagnetic theory.

When an incident bundle of radiation, characterized by the fluxes $I^{(i)}(\mathbf{s}) \, d\omega_s$ $(i = 1, 2, 3, 4)$, is scattered by an infinitesimally small quantity of matter we observe a new radiation field characterized by the vector $dI^{(j)}(\mathbf{d})$ (emission) $(j = 1, 2, 3, 4)$. According to Lambert's law, these quantities must be linearly related. Stating this relationship in its most general form

$$dI^{(j)}(\mathbf{d}) \text{ (emission)} = W_{ij}(\mathbf{s}, \mathbf{d}) I^{(i)}(\mathbf{s}) \, d\omega_s, \qquad (2.28)$$

where, following the sum rule for repeated indices, the r.h.s. is summed over all i. $W_{ij}(\mathbf{s}, \mathbf{d})$ is an intensity transformation matrix, whose form will be discussed at length in Chapter 7, and which according to Lambert's law must be proportional to the amount of matter. As a matter of definition, which is obviously pertinent, we may assume W_{ij} to be proportional to $s/4\pi$ where s is the scattering coefficient. Thus we can write

$$W_{ij}(\mathbf{s}, \mathbf{d}) = \frac{sda}{4\pi} P_{ij}(\mathbf{s}, \mathbf{d}), \qquad (2.29)$$

where the constant of proportionality P_{ij} is the phase matrix.

Substituting (2.29) in (2.28), and integrating over all angles of incidence, we find

$$dI^{(j)}(\mathbf{d}) \text{ (emission)} = sda \int \frac{d\omega_s}{4\pi} P_{ij}(\mathbf{s}, \mathbf{d}) I^{(i)}(\mathbf{s}), \qquad (2.30)$$

and hence from (2.15) the source function is

$$J^{(j)}(\mathbf{d}) = \frac{s}{e} \int P_{ij}(\mathbf{s},\mathbf{d}) I^{(i)}(\mathbf{s}) \frac{d\omega_s}{4\pi}. \qquad (2.31)$$

2.2. Thermal emission

2.2.1. *Thermodynamic equilibrium*

Thermodynamic equilibrium describes the state of matter and radiation inside a constant temperature enclosure. The equilibrium is complete for, according to the second law of thermodynamics, no thermal changes can be produced by any mechanism without the intervention of external forces. The implications are far-reaching and were first described by Kirchhoff in 1882. Radiation inside the enclosure is known as *black body* or *enclosed* radiation. The latter is the better name, but the former is more commonly used.

Kirchhoff's deductions were as follows.

(*a*) In each separate, homogeneous medium within the enclosure the radiation is *homogeneous, unpolarized,* and *isotropic.*

(*b*) The source function is equal to the intensity,

$$I_\nu = J_\nu.$$

(*c*) $\mathbf{c}'^2 I_\nu$ is the same in all media in the enclosure, where $\mathbf{c}'$ is the velocity of light in the medium concerned.

(*d*) As a direct consequence of (*c*), $\mathbf{c}'^2 I_\nu$ must be a universal function of temperature (θ) only, and will be written $\mathbf{c}^2 B_\nu$ where $\mathbf{c}$ is the velocity of light *in vacuo.*

A further consequence of the existence of strict thermodynamic equilibrium is that B_ν must have the form

$$B_\nu = \frac{\nu^3}{\mathbf{c}^2} F\left(\frac{\theta}{\nu}\right), \qquad (2.32)$$

where F is an unknown function. The proof of this statement is to be found in standard textbooks on thermodynamics. Differentiating (2.32) with respect to ν, we can show that the turning points of the function B_ν are determined by the ratio θ/ν, or the product $\theta\lambda$, only. This is *Wien's displacement law.*

Integrating (2.32) over all frequencies and making the substitution $x = \theta/\nu$, there results

$$\int_0^\infty B_\nu \, d\nu = \int_0^\infty \frac{\theta^4}{\mathbf{c}^2 x^5} F(x) \, dx = \pi \boldsymbol{\sigma} \theta^4. \qquad (2.33)$$

This is *Stefan–Boltzmann radiation law*, and σ is the *Stefan–Boltzmann constant*.

Investigation of the function F was the historical reason for the development of quantum theory. Planck's theory leads to the expression

$$B_\nu(\theta) = \frac{2h\nu^3}{c^2(e^{h\nu/k\theta}-1)},\tag{2.34}$$

where

$$h = \text{Planck's constant},$$
$$k = \text{Boltzmann's constant}.$$

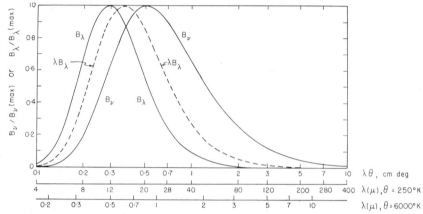

FIG. 2.3. The Planck function.
The curve with ordinate λB_λ gives areas proportional to energies.

Using equation (2.2) we may write alternatively

$$B_\lambda = \frac{\nu^2}{c} B_\nu = \frac{C_1\lambda^{-5}}{(e^{C_2/\lambda\theta}-1)},\tag{2.35}$$

where

$$C_1 = 2\pi hc^2,\tag{2.36}$$

and

$$C_2 = hc/k,\tag{2.37}$$

are known as the *first* and *second radiation constants* respectively.

The form of the Planck function (2.34) is shown in Fig. 2.3. Tabulated values are given in Appendix 7. B_ν has a single maximum at

$$\frac{\theta c}{\nu} = 0\cdot50990\pm0\cdot00004 \text{ cm deg},$$

while the maximum of B_λ is at

$$\theta\lambda = 0\cdot28975\pm0\cdot00003 \text{ cm deg}.$$

The curve with ordinate λB_λ in Fig. 2.3 illustrates an important practical point in atmospheric computations. A black body at 6000° K has only 0·4 per cent of its energy at wavelengths longer than $5\,\mu$;

a 250° K black body, on the other hand, has only 0·4 per cent of its energy at wavelengths shorter than $5\,\mu$. For most practical purposes therefore the solar and terrestrial radiation streams can be treated independently.

The Planck function behaves very differently in its two wings. As $\lambda \to \infty$ or $\nu \to 0$,

$$B_\nu \to 2k\theta\nu^2/c^2, \qquad (2.38)$$

$$B_\lambda \to 2k\theta c/\lambda^4. \qquad (2.39)$$

(2.38) and (2.39) are known as the *Rayleigh–Jeans distribution*. As $\lambda \to 0$ or $\nu \to \infty$,

$$B_\nu \to \frac{2h\nu^3}{c^2}\,e^{-h\nu/k\theta}, \qquad (2.40)$$

$$B_\lambda \to \frac{2hc^2}{\lambda^5}\,e^{-hc/\lambda k\theta}, \qquad (2.41)$$

which is the *Wien distribution*.

2.2.2. *Breakdown of thermodynamic equilibrium*

In strict thermodynamic equilibrium the source function depends upon the temperature, frequency, and velocity of light only. This greatly reduces the problems involved in a solution of the equation of transfer, and it is therefore important to make an objective assessment of the applicability of this idealization under atmospheric conditions. Before Milne's analysis of the problem in 1930 it was assumed that the Planck function was always the correct source function for thermal emission, for reasons which appear to have been based on the following argument. Consider a small element of matter inside a constant-temperature enclosure, where it absorbs and emits according to Planck's and Kirchhoff's laws. If we extract this element of matter from the enclosure without changing its physical state, then the only change is in the radiation incident upon it. Since we may suppose that the emission is a property of the matter alone, it follows that the source function should not be altered by this change.

The argument is, however, fallacious, as Einstein demonstrated, because of the existence of *stimulated emission*, i.e. emission by a molecule which is stimulated by the impact of a photon and depends upon the intensity of the radiation environment. Stated in this way, it now becomes a little surprising that computations based upon Planck's function are ever satisfactory outside a constant temperature enclosure. Thermodynamic reasoning has been exhausted, and for further discussion we must now turn to an examination of the mechanism, based upon the methods of statistical mechanics.

Molecules can possess internal energy in four forms: translational, electronic, vibrational, and rotational. The last three are quantized and can take part in the exchange of energy between matter and the radiation field. The wave function for an isolated molecule is usually factorized into translational, electronic, vibrational, and rotational wave functions, each with its associated energy. This implies that the four modes are independent and that energy transfer can only take place between them during *collisions*, which will be regarded as occupying a negligible fraction of the total time. The average time required to transfer energy from one mode to another by collisions is the *relaxation time η*, which is inversely proportional to the number of collisions per second.

Before discussing the breakdown of thermodynamic equilibrium we must introduce one further property of matter in a constant-temperature enclosure: *Boltzmann's law*, which relates the number densities of molecules, $n(E)$, to the energy, E:

$$\frac{n(E_1)}{n(E_2)} = \frac{g_1}{g_2} e^{(E_2-E_1)/k\theta},\qquad(2.42)$$

where g_1 and g_2 are the statistical weights of the two states. (2.42) applied to the translational energy leads to *Maxwell's distribution of velocities*, and the following relation for the translational energy

$$\bar{E}_t = \tfrac{3}{2}nk\theta,\qquad(2.43)$$

where n is the number density of molecules, and the bar denotes a state of thermodynamic equilibrium. In principle the translational energy is an observable quantity and if we substitute this for $\bar{E}_t$ in (2.43) we have a formal *definition of temperature*, whether or not thermodynamic equilibrium prevails. To make the arbitrary nature of this definition apparent we may use the term *kinetic temperature*. A sufficient (and the only likely) condition for (2.43) to be true is that Boltzmann's law applies to all translational energies. There are other ways in which we can define a temperature. For any pair of levels we may write from (2.42),

$$\theta_{12} = \frac{E_2-E_1}{k\ln\{n(E_2)g_1/n(E_1)g_2\}}.\qquad(2.44)$$

This procedure is only useful if a group of levels has θ_{12} in common; a circumstance which is, however, not unusual for the reason that translational, rotational, vibrational, and electronic levels adjust themselves to equilibrium at very different rates. Thus it is possible to have one or more groups of levels (e.g. translational and rotational) in thermo-dynamic equilibrium (i.e. obeying Boltzmann's law) while others (e.g. vibrational and electronic) are not. It is usual to indicate equilibrium among a restricted group of levels by the use of such terms as *rotational temperature, vibrational temperature*, etc.

The kinetic temperature has a special significance because the transla-tional modes adjust more readily to equilibrium than any others, since every collision (by definition) makes some adjustment. We can, in fact, relate the existence of a kinetic temperature to the existence of an atmosphere. Let us consider the upper levels of the ionosphere, where diffusion is the most important mixing process. At these levels disturb-ance of equilibrium among translational states is caused by the arrival of molecules or atoms from levels with differing kinetic temperatures. A relevant non-dimensional parameter is clearly

$$A = \frac{l}{\theta}\frac{d\theta}{dz}, \tag{2.45}$$

where l is the molecular mean-free-path, and z is the altitude. If $A \ll 1$ we have, in effect, an unlimited, isothermal region which is the required condition for Boltzmann's law. If $A \gg 1$, on the other hand, a Maxwell distribution is clearly impossible. A gradient of $1°$ K km^{-1} at a tem-perature of $1000°$ K gives $A \simeq l \times 10^{-8}$. If l approaches one scale height ($H \simeq 50$ km at these levels) then molecules can leave the atmosphere without performing collisions on the way. This is the definition of the lower limit of the *exosphere*, which may conveniently be taken as the upper limit of the atmosphere. At this level $A \simeq 0.05$ and equilibrium may be expected between the translational modes. The kinetic tem-perature is therefore a valuable concept at all levels considered in this book.

In subsequent sections we will consider a model of a gas whose translational and rotational states are in thermodynamic equilibrium, and we will examine the conditions under which thermodynamic equi-librium prevails between the vibrational states. We will show later that rotational states are always in equilibrium when the problem of the vibrational states is important; when it is unimportant the rota-tional problem is also unimportant, although for different reasons. With the notable exception of the atomic oxygen coronal lines which we will discuss in a later chapter, thermal radiation from electronic levels is not important in the atmosphere. The model chosen is there-fore relevant to atmospheric problems.

2.2.3. *The interaction of matter and radiation*†

In our model molecular vibrations will be assumed to be simple-harmonic. In nature, such vibrations have strong anharmonicities which will give rise to interactions between the normal modes; nevertheless, the simple theory of the harmonic oscillator should give a good qualitative picture of the relevant physical features of the problem. A harmonic oscillator can have several independent modes with energy levels

$$E_v = (v + \tfrac{1}{2})h\nu_0, \tag{2.46}$$

where v is the (integral) vibrational quantum number and ν_0 the fundamental frequency of the mode. Absorption of a quantum of energy $h\nu_0$ takes place by means of a transition $(v-1) \to v$, and emission by a transition $v \to (v-1)$. During the transition the rotational energy can also change; this energy is defined by a rotational quantum number which goes from J' in the upper vibrational state to J in the lower. The frequency emitted or absorbed (ν) differs slightly from ν_0 because of the change of rotational energy during the transition. A schematic representation of some possible transitions between vibrational and rotational levels is shown in Fig. 2.4. Since the rotational levels are in thermodynamic equilibrium (by assumption) we require no further information about them. In order to define the rate of absorption we introduce a rate coefficient $B(v-1, v)b(J, J')$. Two processes are involved in emission: *spontaneous emission* (whose rate is determined by a coefficient $A(v, v-1)a(J', J)$) is a property of the molecules and is not influenced by the ambient radiation field; *induced emission* (whose rate is determined by a coefficient $B(v, v-1)b(J', J)$) results from interaction between molecules and the radiation field. These three rate coefficients are known as the *Einstein coefficients* and are defined by the following statements:

spontaneous emissions per c.c. per second $(v, J') \to (v-1, J)$

$$= n(v)f(J')a(J', J)A(v, v-1); \tag{2.47}$$

induced emissions per c.c. per second $(v, J') \to (v-1, J)$

$$= n(v)f(J')b(J', J)B(v, v-1)u_\nu; \tag{2.48}$$

absorptions per c.c. per second $(v-1, J) \to (v, J')$

$$= n(v-1)f(J)b(J, J')B(v-1, v)u_\nu; \tag{2.49}$$

† In the interests of continuity of ideas we will make use of terms which are not fully discussed until Chapter 3; most of these terms should, however, be familiar.

where $n(v)$ = number density of molecules with vibrational quantum
          number, v,
  $f(J)$ = fraction with rotational quantum number, J,
    u_ν = number density of ν-quanta.

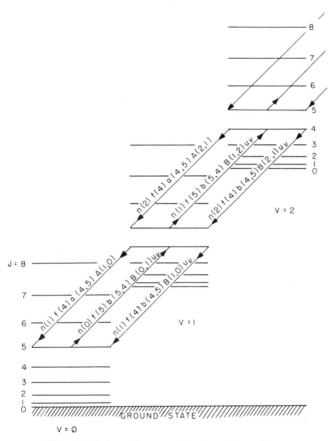

Fig. 2.4. Schematic energy levels and transitions.

In these equations we are assuming sharp levels, with finite energy flux
in an infinitesimal range of frequency; a transformation to more realistic
circumstances will be made later. Factorizing the coefficients into J-
and v-dependent components is permissible if vibrational and rotational
wave functions are separable. There is no loss of generality in assuming

$$\sum_J a(J', J) = \sum_{J'} b(J', J) = \sum_{J'} b(J, J') = 1. \qquad (2.50)$$

In general, (2.47), (2.48), and (2.49) do not exactly balance and there
is a net transfer of energy between the matter and radiation field. If

we consider a state of complete thermodynamic equilibrium, however, a balance must exist or matter in a constant-temperature enclosure could heat up or cool down. Using a bar over a quantity to denote thermodynamic equilibrium we have

$$\bar{n}(v)f(J')\{a(J',J)A(v,v-1)+b(J',J)B(v,v-1)\bar{u}_v\}$$
$$= \bar{n}(v-1)f(J)b(J,J')B(v-1,v)\bar{u}_v. \quad (2.51)$$

In (2.51) $f(J')$ does not require a bar because we have postulated that the rotational levels are always in thermodynamic equilibrium. The Einstein coefficients are a property of matter and also do not need to be distinguished by a bar under equilibrium conditions. From Boltzmann's law (2.42) we can write

$$\frac{\bar{n}(v)f(J')}{\bar{n}(v-1)f(J)} = \frac{g(J')}{g(J)}\exp\left(-\frac{h\nu}{k\theta}\right), \quad (2.52)$$

where we have made use of Planck's relation

$$h\nu = E(v,J')-E(v-1,J). \quad (2.53)$$

It is known from quantum-mechanical considerations that all vibrational levels have equal statistical weight and therefore g is a function only of the rotational quantum number.

Substituting (2.52) in (2.51), we find

$$\bar{u}_v = \frac{a(J',J)A(v,v-1)}{\{g(J)/g(J')\}b(J,J')B(v-1,v)e^{h\nu/k\theta}-b(J',J)B(v,v-1)}. \quad (2.54)$$

But, from (2.34) and (2.8) we have

$$\bar{u}_v = \frac{\delta_v}{e^{h\nu/k\theta}-1}, \quad (2.55)$$

where δ_v is a function of frequency only. (2.54) and (2.55) are consistent only if

$$b(J',J)B(v,v-1) = \frac{g(J)}{g(J')}b(J,J')B(v-1,v), \quad (2.56)$$

and

$$a(J',J)A(v,v-1) = \delta_v b(J',J)B(v,v-1). \quad (2.57)$$

With the aid of (2.56) and (2.57) we can now eliminate two of the Einstein coefficients from (2.47), (2.48), and (2.49), and form a balance equation which expresses the rate at which quanta enter or leave a bundle $I_v(s)\,d\omega_s$. The energy density caused by this bundle is $I_v(s)\,d\omega_s/c$. Only a fraction $d\omega_s/4\pi$ of the spontaneous emission will go into the solid angle $d\omega_s$, but, for reasons which are beyond the scope of this book to discuss, the induced emission is identical in all respects to the incident radiation; it is indistinguishable in practice from a negative absorption. Using this information we may equate the gain of quanta per cm³ by the beam ($= \{dI_\nu(\mathbf{s})/ds\}d\omega_s$, see discussion of (2.11)) to $\{(2.47)+(2.48)-(2.49)\}$. There results

$$\frac{dI_\nu(\mathbf{s})}{ds}\,d\omega_s = -\sum_{v=1}^{\infty} n(v-1)f(J)b(J, J')B(v-1, v)\frac{I_\nu(\mathbf{s})}{c}\,d\omega_s +$$

$$+ \sum_{v=1}^{\infty} n(v)f(J')b(J, J')B(v-1, v)\frac{g(J)}{g(J')}\frac{I_\nu(\mathbf{s})}{c}\,d\omega_s +$$

$$+ \sum_{v=1}^{\infty} n(v)f(J')b(J, J')B(v-1, v)\frac{g(J)}{g(J')}\delta_\nu\frac{d\omega_s}{4\pi}, \qquad (2.58)$$

where sums have been taken over all vibrational levels, since these are uniformly spaced (see § 3.3) and each pair of levels can absorb and emit identical quanta. We now introduce the total vibrational energy (omitting zero-point energy), expressed in numbers of quanta

$$E_v = \sum_{v=1}^{\infty} vn(v), \qquad (2.59)$$

and make use of a property of the simple-harmonic oscillator which is derived in elementary textbooks on quantum mechanics:

$$B(v, v-1) = vB(0, 1). \qquad (2.60)$$

If we now recall that $f(J)$ is determined by Boltzmann's law, (2.58) is found to simplify to

$$\frac{dI_\nu(\mathbf{s})}{ds} = -nk_{n,\nu}I_\nu(\mathbf{s}) + n\bar{k}_{n,\nu}B_\nu E_v/\bar{E}_v, \qquad (2.61)$$

where

$$k_{n,\nu} = \frac{B(0, 1)}{c}b(J, J')f(J)\left(1 + \frac{E_v/\bar{E}_v\{\exp(-h\nu_0/k\theta) - \exp(-h\nu/k\theta)\}}{1 - \exp(-h\nu_0/k\theta)}\right),$$
$$(2.62)$$

$$\bar{k}_{n,\nu} = \frac{B(0,1)}{c}b(J, J')f(J)\frac{1 - \exp(-h\nu/k\theta)}{1 - \exp(-h\nu_0/k\theta)}, \qquad (2.63)$$

and

$$\bar{E}_v = n\left\{\exp\left(\frac{h\nu_0}{k\theta}\right) - 1\right\}^{-1}, \qquad (2.64)$$

is the vibrational energy in a state of thermodynamic equilibrium. Comparing the equations (2.17) and (2.61) and using the definitions of source function and absorption coefficient, it follows that $k_{n,\nu}$ is the molecular absorption coefficient under the conditions of observation, and that $(\bar{k}_{n,\nu}/k_{n,\nu})B_\nu(E_v/\bar{E}_v)$ is the source function.

Before proceeding we must resolve two difficulties of interpretation. Firstly, from this point on we will treat I_ν, k_ν, $\check{k}_\nu$, etc. as continuous functions of ν, as was implied in their definitions in §§ 2.1.1 and 2.1.2. The derivation of (2.61) was, however, based upon sharp energy levels. In practice vibration-rotation lines are very narrow and there is little difficulty in accepting this transformation, although it is difficult to perform with mathematical rigour; we will not therefore complicate the derivation to the extent necessary to accommodate continuous energy states but will simply assume that it can be done. Having returned to continuous definitions we can write

$$\delta_\nu = \frac{8\pi\nu^2}{c^3}, \tag{2.65}$$

which could not be done before because our definition of intensity was inconsistent with the concept of sharp levels.

Secondly, we must note the formal requirement for two absorption coefficients. The first, k_ν, is measured under the conditions actually occurring in the atmosphere, including the same ambient radiation field. These conditions cannot easily be simulated in the laboratory where vibrational and rotational levels are always close to thermodynamic equilibrium; then $E = \bar{E}$ and $k_{n,\nu} = \check{k}_{n,\nu}$. Thus, (2.63) is the quantity normally measured in the laboratory, but it differs from the required absorption coefficient. Fortunately, we can neglect this difference. In the first place, it is partly artificial, arising from the supposedly identical quanta from ground state ($v = 0$ to 1) and upper state (e.g. $v = 1$ to 2) transitions. In practice, the lines arising from these transitions are not superposed exactly and can be distinguished, with the result that the distinction between k_ν and $\check{k}_\nu$ ceases to exist. Further, in the far wings of the $15\,\mu$ CO_2 band (probably the worst case), $k_\nu/\check{k}_\nu$ varies from 0·985 to 1·015 only, as $E_\nu/\bar{E}_\nu$ varies over the wide range of 0 to 2; the question is therefore unimportant, even if real.

It now remains to determine the ratio $E_\nu/\bar{E}_\nu$. It is possible to make a theoretical analysis of the rate at which energy is transferred from vibration to translation when E_v differs from $\bar{E}_v$ (there is by definition no net transfer if the translational and vibrational modes are both in thermodynamic equilibrium) and analysis based upon an assumption of weak collisional interaction gives

$$\frac{dE_v}{dt} = -\frac{1}{\eta}(E_v - \bar{E}_v). \tag{2.66}$$

We will not discuss this theory because the approximation of weak interactions is poor and we are able as an alternative to appeal to experiment for justification for (2.66). The experiments concerned are those on the frequency dispersion of supersonic sound waves in gases, which can yield values for the relaxation time η. Unfortunately, only those modes which contribute to the specific heat can be investigated, which normally means only the lowest-frequency mode of the molecule. However, in principle, η as defined by (2.66) can be measured in the laboratory.

Now, if the matter and radiation field are in a steady state, the rate of increase of internal energy for one mode must equal the total radiative heating rate associated with the fundamental band for this mode, which we will designate by the subscript i. From (2.11), (2.61), and (2.66) we find the condition for a steady state

$$\frac{E_\nu}{\bar{E}_\nu} = \frac{\int d\omega \int_i d\nu\, nk_{n,\nu}\, I_\nu + \bar{E}_\nu/\eta_i}{\int d\omega \int_i d\nu\, nk_{n,\nu}\, B_\nu + \bar{E}_\nu/\eta_i}. \tag{2.67}$$

If we define a time constant

$$\phi_i = \frac{\bar{E}_\nu}{\int d\omega \int_i d\nu\, nk_{n,\nu}\, B_\nu}, \tag{2.68}$$

the source function becomes

$$J_\nu = B_\nu \left(\frac{1}{\eta_i} + \frac{1}{\phi_i}\right)^{-1} \left(\frac{1}{\eta_i} + \frac{1}{\phi_i}\frac{\int d\omega \int_i d\nu\, nk_{n,\nu}\, I_\nu}{\int d\omega \int_i d\nu\, nk_{n,\nu}\, B_\nu}\right). \tag{2.69}$$

If we again make use of the assumption that the rotational levels are in thermodynamic equilibrium, it can be shown that

$$\phi_i = \frac{1}{A(1,0)}, \tag{2.70}$$

is the *natural lifetime* of the first excited state.

The source function can be written in an alternative, approximate form which is important for application to atmospheric problems. The heating rate (2.18) for a single band is

$$h_i = \int d\omega \int_i d\nu\, nk_{n,\nu}\, I_\nu - \int d\omega \int_i d\nu\, nk_{n,\nu}\, J_\nu. \tag{2.71}$$

Eliminating the first term on the r.h.s. of (2.71) using (2.69), there results

$$J_\nu = B_\nu \left(\frac{\phi_i}{\phi_i + \eta_i} + \frac{\eta_i}{\phi_i + \eta_i} \frac{h_i + \int d\omega \int dv\, nk_{n,\nu} J_\nu}{\int d\omega \int dv\, nk_{n,\nu} B_\nu} \right). \qquad (2.72)$$

It follows from (2.72) that J_ν, like B_ν, is isotropic, and also has the same frequency dependence. Let us now assume that the frequency range i is so narrow that B_ν can be taken to have its mean value B_i at all frequencies. The integrations in (2.72) are now trivial and, from the definitions of the heating function H_i (2.21) we find

$$J_i = B_i + \frac{\eta_i H_i}{2\phi_i}. \qquad (2.73)$$

2.2.4. *Discussion of the source function*

We have derived a source function consistent with the simple model adopted, and can now discuss its physical significance; for the discussion we will also use a result obtainable from (2.61) and the subsequent discussion

$$J_\nu = B_\nu \frac{E_\nu}{\overline{E}_\nu}. \qquad (2.74)$$

We can now consider the following circumstances.

(i) *A constant-temperature enclosure.* If we set $I_\nu = B_\nu$ in (2.69), we find $J_\nu = B_\nu$. This confirms that our discussion is consistent with Kirchhoff's laws. However, from (2.71) we now have

$$h_i = 0.$$

There is, of course, little interest in a system in which no thermal or mechanical changes can take place.

(ii) *The energy levels are populated by collisions.* If $\eta_i \ll \phi_i$ the frequency of transitions caused by collisions greatly exceeds that by radiative processes. Under this circumstance, (2.69) reduces to $J_\nu = B_\nu$. This is the most important of Kirchhoff's laws, and the others also follow. However, the restriction to zero heating rate as found in (i), does not apply. Although it has not been brought out in the analysis, it also follows under this condition that the vibrational levels are populated according to Boltzmann's law. One aspect is illustrated by (2.74) which shows that the total vibrational energy is the same as for thermodynamic equilibrium.

(iii) *The energy levels are populated by radiative processes.* If $\eta_i \gg \phi_i$, from (2.69)

$$J_\nu \to B_\nu \frac{\int d\omega \int_i d\nu\, nk_{n,\nu} I_\nu}{\int d\omega \int_i d\nu\, nk_{n,\nu} B_\nu}.$$

Substituting in (2.71) we find

$$h_i \to 0.$$

According to the discussion in § 2.1.2 this defines an isotropic scattering process. It is not, however, a case of coherent scattering. Energy balance applies only over the band as a whole and not for each frequency individually. The assumption that the rotational energy levels are in thermodynamic equilibrium implies the continual redistribution of rotational quanta while the absorption and re-emission are taking place.

The important features of this analysis are contained in the above three propositions, although they may be stated in many different guises. Their significance in terms of atmospheric processes depends on the magnitude of the parameter η_i/ϕ_i. ϕ_i presents no difficulties. From (2.68) and (2.64), and assuming a vibrational band narrow compared with the Planck function, as in the derivation of (2.73), we find

$$\phi_i^{-1} = 8\pi \frac{\nu^2}{c^2} S_{n\,i}, \tag{2.75}$$

where

$$S_{n,i} = \int_i k_{n,\nu}\, d\nu, \tag{2.76}$$

is the molecular band intensity.† The behaviour of this quantity as a function of the state variables is known, and will be discussed in Chapter 3. For our present purposes we only need know that it can be measured under conditions similar to those occurring in the atmosphere. Some results for active fundamentals of H_2O, CO_2, and O_3 are shown in Table 2.1. For all practical purposes these may be assumed not to vary throughout the atmosphere. Vibrational relaxation times for collisions between atmospheric absorbing gases and air molecules are not so well determined. Values of 1.5×10^{-5} s and 2×10^{-6} s for $\eta(H_2O, \nu_2)$ and $\eta(CO_2, \nu_2)$ have been assigned at 220° K and 1 atmosphere

† Because (2.47), (2.48), and (2.49) are linear in the Einstein coefficients it can be simply demonstrated that (2.76), (2.75), and (2.70) also hold for each individual rotation line in the spectrum. The integral in (2.76) now runs over *one line only* and the resulting quantity S_l is known as the *line intensity*. ϕ_l is the lifetime of the upper state of the transition. Clearly

$$S_i = \sum S_l, \tag{2.76 a}$$

where the sum is over all lines in the band.

pressure of air. Uncertainties as to the absolute value of η are offset by its sensitivity to the total pressure ($\eta \sim p^{-1}$) and the rapid change with height of this parameter. From Table 2.1 we find that $\eta_i/\phi_i = 3.4 \times 10^{-5}$ for the ν_2 bands of both H_2O and CO_2 for a pressure of 10^6 dyne cm^{-2}, and therefore $\eta_i/\phi_i = 1$ at 34 dyne cm^{-2}. In middle latitudes this pressure occurs at an altitude of about 74 km. Below this altitude $\eta_i/\phi_i \ll 1$ and we may expect a close approach to thermodynamic equilibrium.

TABLE 2.1

Natural lifetimes of vibrational states

	Vibrational mode	ν_0/c (cm^{-1})	$S_{n,i}$ (cm^2 s^{-1})	ϕ_i (s)
H_2O	ν_1	3657	5.1×10^{-9}	0.58
	ν_2	1595	1.26×10^{-7}	0.12
	ν_3	3756	9.9×10^{-8}	0.029
CO_2	ν_2	667	2.37×10^{-7}	0.37
	ν_3	2349	3.03×10^{-6}	0.0024
O_3	ν_3	1042	4.08×10^{-7}	0.09

To conclude this section we will briefly consider electronic transitions and rotational transitions. Permitted electronic transitions have radiative lifetimes of the order of 10^{-8} s. On the other hand, since the energy jumps in transitions are normally large, the relaxation times will be considerably longer than vibrational relaxation times ($\eta \gg 10^{-6}$ s at s.t.p.). For permitted transitions, therefore, $\eta/\phi \gg 10^2$ even at the ground. Only resonant scattering can be expected. On the other hand, we will discuss in Part 2 the forbidden transitions between the $^3P_1$ and $^3P_2$ ground states of atomic oxygen. ϕ is about 10^4 s. η is unknown, but at 300° K $h\nu \simeq 0.8k\theta$, suggesting a relaxation time comparable to the time between collisions. The time between collisions is only as great as 10^4 s for altitudes of the order of 1000 km. This transition should therefore obey Kirchhoff's law at all levels below the exosphere; a point of importance to the heat balance of the ionosphere. The electronic transitions which we shall treat seem to fall into extreme categories in this way without difficulty and the problem of a complete equation of transfer is not important.

Now consider a pressure, assumed to be well below that required for vibrational relaxation, at which departures from a Boltzmann distribution in the rotational levels first appear. The source function will then be (2.69), and, according to (iii) above, we may expect negligible thermal interaction between matter and radiation field. At a slightly lower pressure the rotational levels will no longer be populated by collisions but by interaction with the radiation field. If a steady-state is set up between matter and radiation without reference to collisions, then no transfer of quanta into translational energy is needed to support it, i.e. there is no net transfer of energy between the radiation field and the matter. This is a general argument, which leads, for vibrational levels, to the result expressed in proposition (iii). The only difference is now that the result applies to each rotational transition separately and the scattering is coherent. The change is a very slight one, for both cases lead to the result that heating is small when η for the vibrational levels is much greater than ϕ.

We must now justify the assertion that rotational relaxation takes place at a lower pressure than vibrational relaxation. The magnitude of the ratio η/ϕ is critical and it must be shown to be smaller for rotational transitions. It is a matter of observation that the natural lifetimes of rotational levels tend to be longer than those of vibrational levels. The lifetime of a combined transition is closest to the shorter of the two involved. It follows that, for vibration-rotation bands, ϕ is approximately the same for a rotational level as for a vibrational level. On the other hand, rotational relaxation times differ greatly from vibrational relaxation times. Anderson's theory of line broadening (see § 3.6) shows that rotational relaxation times are related to line widths, which are observable quantities. On this basis it appears that $\eta = 10^{-10}$ s at s.t.p. is a typical value for a rotational level. This suggests a ratio η/ϕ for rotational states of the order of 10^{-9} at s.t.p. as compared to 10^{-4} to 10^{-5} for vibrational states, confirming the hypothesis upon which our argument was based.

The pure rotation band of water vapour has to be considered on a different basis. Natural lifetimes can be computed from the matrix elements for a rigid rotator (see § 3.4); many are close to 0·1 s, while for small J-values they may be as large as 10 s. The relaxation time quoted (about 10^{-10} s at s.t.p.) is a reasonable average for the whole band, and η/ϕ therefore lies in the range 10^{-9} to 10^{-11} at s.t.p. Thermodynamic equilibrium will therefore obtain for all rotational levels at pressures above 10^{-3} dyne cm^{-2} and for none at pressures below 10^{-5} dyne cm^{-2}. For atmospheric pressures between these two values the water-vapour rotation band would pose a formidable problem were it not for photochemical decomposition which starts at much higher pressures (§ 1.3). A pressure of 10^{-3} dyne cm^{-2} corresponds to an altitude of about 150 km, and at this level it is probable that no water-vapour molecules exist. Kirchhoff's law can therefore be used at all levels where water vapour is likely to be of consequence.

2.3. The integral equations

2.3.1. *Introduction*

For the rest of this chapter we will be concerned with formal mathematical manipulation of the equation of transfer needed for subsequent chapters. No new physics is introduced and the reader may if he wishes treat the remaining paragraphs of this chapter as appendixes, to be consulted when the equations are required. There are, of course, physical grounds for choosing the particular models treated. In §§ 2.3.3 and 2.3.4 the model of a plane-parallel atmosphere is introduced, first for an isotropic thermal source function with uniform isotropic boundary conditions, and second for a scattering source function with a nearly parallel incident flux as an upper boundary condition. These two models are used to treat terrestrial and solar radiation respectively and owe their usefulness to the effective independence of the two radiation fields. Discussion of the detailed correspondence between the models and the actual atmosphere will, however, be left for Chapter 6.

The purpose of our treatment is to separate physical ideas from formal operations. This makes it easier to decide at the end of an involved calculation whether a critical result has a physical origin or expresses a mathematical approximation. Furthermore, it makes it possible to reformulate physical ideas without repeating the computations.

Taken by themselves the mathematical problems of radiative transfer have an elegance which has attracted the attention of mathematicians, and has resulted in the publication of treatises based on the minimum physical hypothesis. While these treatises are not sufficient in themselves for the largely practical purposes of this book, it should, nevertheless, not be forgotten that mathematical elegance can often be equated to simplicity and simplicity (at least for computational purposes) to accuracy. We can see no purpose therefore in failing to take advantage of the available formalism, provided it is appreciated where new physical ideas have been introduced, and what their nature may be.

2.3.2. *The general solution*

The increment of optical path has been defined in (2.16). The *optical path* along a ray trajectory from point 1 to point 2 is

$$\tau(1,2) = \left| \int_{1}^{2} e \, da \right|. \tag{2.77}$$

Note that we have defined τ to be *positive definite*.

Consider the path of integration shown in Fig. 2.5. The equation of transfer at P' is

$$\frac{dI_\nu(P',\mathbf{s})}{d\tau_\nu} = I_\nu(P',\mathbf{s}) - J_\nu(P',\mathbf{s}). \tag{2.78}$$

(N.B. The sign of the l.h.s. depends upon the sign of $d\tau_\nu/ds$, and therefore upon the way that Fig. 2.5 is drawn.)

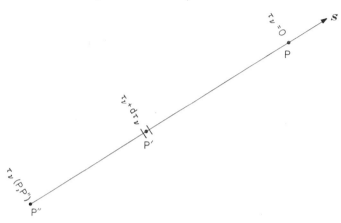

FIG. 2.5. Path of integration.

Multiplying both sides of (2.78) by $e^{-\tau_\nu}$ we find

$$\frac{d\{I_\nu(P',\mathbf{s})e^{-\tau_\nu}\}}{d\tau_\nu} = -e^{-\tau_\nu}J_\nu(P',\mathbf{s}).$$

Integrating from P to P'',

$$I_\nu(P'',\mathbf{s})e^{-\tau_\nu(P,P'')} - I_\nu(P,\mathbf{s}) = -\int_0^{\tau_\nu(P,P'')} J_\nu(P',\mathbf{s})e^{-\tau_\nu}\,d\tau_\nu. \tag{2.79}$$

Equation (2.79) can be evaluated if the source function between P and P'' is known, and also the incident intensity at P''. Atmospheric problems are frequently posed in such a form, and this solution is widely used. It states that the intensity at P is made up of the intensity imposed at P'', attenuated according to Lambert's law, together with contributions from each intervening, radiating element, attenuated by the appropriate optical path.

We can simplify (2.79) by observing that the equation of transfer must also be obeyed inside the boundaries, which we may take to be infinitely thick, optically. Consequently, if P'' tends to infinity, so will $\tau_\nu(P, P'')$ and (2.79) becomes

$$I_\nu(P,\mathbf{s}) = \int_0^\infty J_\nu(P',\mathbf{s})e^{-\tau_\nu}\,d\tau_\nu. \tag{2.80}$$

We can recover (2.79) if we write

$$J_\nu(P', \mathbf{s}) = I_\nu(P'', \mathbf{s}) \quad \text{for} \quad \tau_\nu \geqslant \tau_\nu(P, P''). \tag{2.81}$$

Thus a boundary condition on the intensity can be introduced into the integral equation by assuming a constant source function from the boundary to infinity. This procedure is widely used in atmospheric computations.

For some special problems in Chapter 9 we will require a volume integral for $\bar{I}_\nu(P)$. First we define $\mathbf{s}$ to be the *radius vector* from the point P; it is therefore the reverse of the vector in Fig. 2.5, and (2.80) becomes

$$I_\nu(P, -\mathbf{s}) = \int_0^\infty J_\nu(P', -\mathbf{s}) e^{-\tau_\nu} \, d\tau_\nu,$$

where

$$d\tau_\nu = +e_{\mathbf{v},\nu}(P') \, ds.$$

From the definition of $\bar{I}_\nu$ (2.9) we have

$$\bar{I}_\nu(P) = \frac{1}{4\pi} \int_0^\infty e_{\mathbf{v},\nu}(P') \, ds \int d\omega_s \, J_\nu(P', -\mathbf{s}) e^{-\tau_\nu}.$$

A volume element at the point P' (distance s from P) is (Fig. 2.6)

$$dV(P') = \pi s^2 \, ds \, d\omega_s, \tag{2.82}$$

and therefore

$$\bar{I}_\nu(P) = \int \frac{e_{\mathbf{v},\nu}(P') J_\nu(P', -\mathbf{s}) e^{-\tau_\nu(P,P')}}{4\pi s^2} \, dV(P'), \tag{2.83}$$

where the integral extends over all space.

2.3.3. *Thermal radiation in a stratified atmosphere*

We will now consider an isotropic source function in an atmosphere for which absorption coefficient and temperature are functions of the vertical coordinate (z) alone (a *stratified atmosphere*, see Fig. 2.7).

An important parameter for this problem is the optical path for a ray passing vertically out to space from a point under consideration (the *optical depth*). The optical depth at z is $\tau_\nu(z, \infty)$. If $\xi \, (= \cos \zeta)$ is the cosine of the zenith angle of the vector $\mathbf{s}$, we have (see Fig. 2.7)

$$\tau(P, P') = \left| \frac{\tau(z', \infty) - \tau(z, \infty)}{\xi} \right| = \frac{|\tau(z', \infty) - \tau(z, \infty)|}{|\xi|}. \tag{2.84}$$

In order to simplify the notation we may write

$$\tau_\nu(z', \infty) = t,$$

$$\tau_\nu(z, \infty) = \tau. \tag{2.85}$$

J_ν is isotropic both within and without the boundaries and we may write (2.80) in the form

$$I_\nu(\tau, \xi) = \int_{|t-\tau|=0}^{|t-\tau|=\infty} J_\nu(t) \exp\left(-\frac{|t-\tau|}{|\xi|}\right) \frac{d\,|t-\tau|}{|\xi|}, \qquad (2.86)$$

with the convention that when z' is inside a boundary the source function is equal to the surface emission. Consider an upward travelling

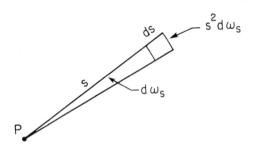

Fig. 2.6. Element of solid angle.

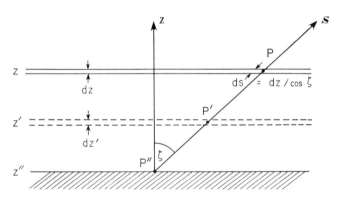

Fig. 2.7. A stratified atmosphere.

stream of radiation when the lower boundary $(z = z'')$ is a black body with temperature θ^*. Then (see Fig. 2.7)

$$J_\nu(t) = B_\nu(\theta^*) \quad \text{for} \quad |t-\tau| \geqslant \tau(z'', z). \qquad (2.87)$$

For a downward travelling stream, on the other hand, the appropriate boundary condition for a planetary atmosphere is that no thermal radiation is incident upon the atmosphere from outer space. Then we have

$$J_\nu(t) = 0 \quad \text{for} \quad |t-\tau| \geqslant \tau_\nu(z, \infty). \qquad (2.88)$$

Since boundary conditions differ for upward and downward streams we will use indices $(+)$ and $(-)$ to distinguish them. Both I_ν^+ and I_ν^- are positive quantities under all circumstances.

The boundary condition is not the only difference between I^+ and I^-. Since we are using $|t-\tau|$ as a variable the distinction between $t > \tau$ and $t < \tau$ is lost. There are, therefore, two relationships between J_ν and $|t-\tau|$, one of which is to be used to evaluate $I^+ (t > \tau)$ and one to

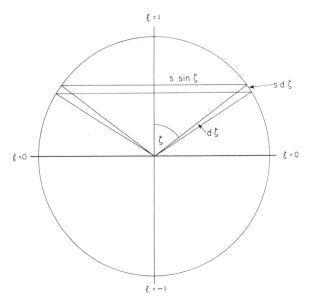

FIG. 2.8. Evaluation of the flux.

evaluate $I^- (t < \tau)$. To make this point clear $(+)$ and $(-)$ indices will also be used when necessary on $J_\nu(t)$.

The isotropy of $J_\nu(t)$ enables us to perform angular integrations in terms of exponential integrals of order n (see Appendix 8),

$$E_n(x) = \int_1^\infty \frac{e^{-wx}}{w^n}\, dw. \tag{2.89}$$

From Fig. 2.8 an element of solid angle is

$$d\omega = \frac{2\pi s \sin \zeta . s d\zeta}{s^2} = -2\pi\, d\xi, \tag{2.90}$$

and the angular integration runs from $\xi = +1$ to $\xi = 0$ in the positive hemisphere and $\xi = 0$ to $\xi = -1$ in the negative hemisphere.

The mean intensity (2.9) can now be written

$$\bar{I}_\nu(\tau) = \frac{1}{4\pi} \int I_\nu(\tau, \xi)\, d\omega$$

$$= \tfrac{1}{2} \int_0^1 I_\nu^+(\tau, \xi)\, d\xi + \tfrac{1}{2} \int_{-1}^0 I_\nu^-(\tau, \xi)\, d\xi. \qquad (2.91)$$

Substituting from (2.86)

$$\bar{I}_\nu(\tau) = \tfrac{1}{2} \int_0^\infty J_\nu^+(t) E_1(|t-\tau|)\, d|t-\tau| + \tfrac{1}{2} \int_0^\infty J_\nu^-(t) E_1(|t-\tau|)\, d|t-\tau|, \qquad (2.92)$$

or, using a property of E_1 and E_2 which is described in Appendix 8,

$$= \tfrac{1}{2} \int_0^1 J_\nu^+(t)\, dE_2(|t-\tau|) + \tfrac{1}{2} \int_0^1 J_\nu^-(t)\, dE_2(|t-\tau|). \qquad (2.93)$$

Similarly we find for the flux (2.3)

$$F_\nu(\tau) = \int I_\nu(\tau, \xi)\xi\, d\omega$$

$$= 2\pi \int_0^1 I_\nu^+(\tau, \xi)\xi\, d\xi + 2\pi \int_{-1}^0 I_\nu^-(\tau, \xi)\xi\, d\xi$$

$$= F^+(\tau) + F^-(\tau). \qquad (2.94)$$

Substituting from (2.86),

$$F_\nu(\tau) = 2\pi \int_0^1 d|\xi| \int_0^\infty J_\nu^+(t) e^{-|t-\tau|/|\xi|}\, d|t-\tau| -$$

$$- 2\pi \int_0^1 d|\xi| \int_0^\infty J_\nu^-(t) e^{-|t-\tau|/|\xi|}\, d|t-\tau|$$

$$= 2\pi \int_0^\infty J_\nu^+(t) E_2(|t-\tau|)\, d|t-\tau| -$$

$$- 2\pi \int_0^\infty J_\nu^-(t) E_2(|t-\tau|)\, d|t-\tau|. \qquad (2.95)$$

From a property of E_2 and E_3 given in Appendix 8

$$F_\nu(\tau) = 2\pi \int_0^{1/2} J_\nu^+(t)\, dE_3(|t-\tau|) - 2\pi \int_0^{1/2} J_\nu^-(t)\, dE_3(|t-\tau|). \qquad (2.96)$$

We may note that since the flux is a vector quantity, F^+ is positive definite while F^- is negative definite, i.e. the sign follows that of ξ.

Finally, from (2.18), (2.21), and (2.93) we can express the heating function in the form

$$H_i(\tau) = \left(\int_i e_{v,\nu} \, dv \right)^{-1} \int_i e_{v,\nu} \left\{ -2J_\nu(\tau) + \int_0^1 J_\nu^+(t) \, dE_2(|t-\tau|) + \right.$$

$$\left. + \int_0^1 J_\nu^-(t) \, dE_2(|t-\tau|) \right\} dv. \quad (2.97)$$

2.3.4. *Solar radiation in a stratified atmosphere*

Equation (2.80) is the general solution to all radiative transfer problems, given the appropriate source function and boundary conditions. If we treat the solar and thermal radiation fields as separable, the source function involves only scattering processes, and in practice attention is usually concentrated on simple scattering by molecules and aerosol particles. The appropriate source function is therefore given by (2.31). The lower boundary condition will express a reflectivity condition involving, in general, a reflection matrix.

This is over complex for real geophysical situations, however, and a more commonly used boundary condition assumes that the surface has uniform brightness and that the albedo is

$$a_\nu = \frac{F_\nu^+(z'')}{F_\nu^-(z'')}. \quad (2.98)$$

This is equivalent to requiring that

$$J_\nu(t) = \frac{a_\nu}{\pi} F_\nu^-(z'') \quad \text{for} \quad z \leqslant z''. \quad (2.99)$$

The upper boundary condition expresses the fact that an unpolarized and nearly parallel beam of solar radiation strikes the atmosphere at an angle defined by $(\xi_\odot, \varphi_\odot)$, where φ is an azimuth angle. Some instruments are able to discriminate in favour of the incident beam by recording radiation only in the neighbourhood of the sun. Consequently it is of some value to assume strictly parallel incident radiation and to state the equation of transfer in two parts, one valid for angles close to $(\xi_\odot, \varphi_\odot)$ and one for all other angles.

The equation of transfer (2.17) can be given in terms of the optical depth by means of the transformation

$$d\tau_\nu = \frac{d\tau_\nu(z', \infty)}{\xi}, \quad (2.100)$$

which follows from (2.84) if we note that $\tau(z', \infty) > \tau(z, \infty)$ for $\xi > 1$

and vice versa for $\xi < 1$. If we now change $\mathbf{s}$, $\mathbf{d}$, $d\omega_s$, $d\omega_d$ in (2.31) to (ξ, φ), (ξ', φ'), $d\omega$, and $d\omega'$ respectively, we find

$$\xi \frac{dI_\nu^{(i)}(\xi, \varphi; z')}{d\tau_\nu(z', \infty)} = I_\nu^{(i)}(\xi, \varphi; z') - \frac{s_\nu}{e_\nu} \int P_{ij}(\xi, \varphi; \xi', \varphi') I_\nu^{(j)}(\xi', \varphi'; z') \frac{d\omega'}{4\pi}.$$

$$(2.101)$$

Now consider a small solid angle $d\omega_\odot$ surrounding the direction $(\xi_\odot, \varphi_\odot)$. Integrate both sides of (2.101) over $d\omega_\odot$ and let $d\omega_\odot$ become very small. The second term on the r.h.s. will tend to zero, as will contributions from the other two terms provided that no source of 'parallel' radiation is involved. Such a source carries a finite irradiation $f_\nu^{(i)}(z')$ in an infinitesimally small solid angle, so that from (2.7)

$$\lim_{d\omega_\odot \to 0} \int_{d\omega_\odot} I_\nu^{(i)}(z') \, d\omega = f_\nu^{(i)}(z') \neq 0.$$

The differential equation governing the 'parallel' irradiation is therefore

$$\xi \frac{df_\nu^{(i)}(z')}{d\tau_\nu(z', \infty)} = f_\nu^{(i)}(z'),$$

$$(2.102)$$

or, integrating from $z' = z$ to $z' = \infty$ and remembering that ξ is negative for the solar beam

$$f_\nu^{(i)}(z) = f_\nu^{(i)}(\infty) e^{-|\tau_\nu(z, \infty)|/|\xi|},$$

$$(2.103)$$

where $f_\nu^{(i)}(\infty)$ is the irradiation outside the earth's atmosphere.

It is appropriate to point out here that the laboratory experimenter attempts to reproduce the conditions of (2.103) by placing an absorption tube in a collimated beam formed from a small source, which is ultimately focused on the slit of a spectrophotometer. A quantity proportional to the irradiation is recorded and by alternately filling and emptying the absorption tube the optical path and hence the absorption coefficient can be determined.

Returning now to (2.101) we will redefine $I_\nu^{(i)}$ so that it is continuous near $(\xi_\odot, \varphi_\odot)$ and therefore does not include the direct, 'parallel' beam. Equation (2.101) is still satisfactory except for that part of the integral over ω' near to $(\xi_\odot, \varphi_\odot)$. This contributes a term

$$\frac{s_\nu}{e_\nu} \int_{d\omega_\odot} P_{ij}(\xi, \varphi; \xi', \varphi') I^{(j)}(\xi', \varphi'; z') \frac{d\omega'}{4\pi} = \frac{s_\nu}{e_\nu} P_{ij}(\xi, \varphi; \xi_\odot, \varphi_\odot) \frac{f^{(j)}(z')}{4\pi}.$$

According to our redefinition of the intensity to exclude direct radiation,

this term would be missed; it must therefore be added explicitly to the
equation of transfer, which now becomes

$$\xi\frac{dI_\nu^{(i)}(\xi,\varphi;z')}{d\tau_\nu(z',\infty)} = I_\nu^{(i)}(\xi,\varphi;z') - \frac{s_\nu}{e_\nu}\int\limits_{\omega'} P_{ij}(\xi,\varphi;\xi',\varphi')I^{(j)}(\xi',\varphi';z')\frac{d\omega'}{4\pi} -$$

$$-\frac{s_\nu}{e_\nu}P_{ij}(\xi,\varphi;\xi_\odot,\varphi_\odot)\frac{f^{(j)}(\infty)}{4\pi}e^{-|\tau_\nu(z',\infty)|/|\xi|}. \quad (2.104)$$

The upper boundary condition on the scattered intensity is now the
same as for thermal radiation, i.e. that there is no scattered radiation
from outer space,

$$I_\nu^-(z') \to 0 \quad \text{as} \quad z' \to \infty \quad \text{and} \quad \tau_\nu(z',\infty) \to 0. \quad (2.105)$$

2.4. Approximate and numerical methods

Methods used to solve the equation of transfer depend upon the
problem posed, and a wide variety is to be found in texts devoted to
the subject. Here we will briefly review a miscellany, which happen to
have been used effectively for important atmospheric problems. New
problems may require new methods and the reader who has a specific
problem in mind is advised to consult other sources (see Bibliography).

2.4.1. *Eddington's approximation*

This approximation is only suitable for a stratified medium, and it
owes its usefulness to the heating rate being a function of the angular
average intensity and the angular average source function (2.18). It is
consequently unlikely that the details of the angular distribution of I
and J need be known with accuracy, and we may hope that a satisfactory approximation can be obtained by assuming a simple angular
form with only one parameter to determine the upward intensity, and
one to determine the downward intensity. The simplest assumption is
(Fig. 2.9)

$$\left.\begin{aligned} I_\nu^+(\xi,\tau) &= I_\nu^+(\tau) \\ I_\nu^-(\xi,\tau) &= I_\nu^-(\tau) \end{aligned}\right\}. \quad (2.106)$$

If τ_ν is the optical depth the equation of transfer is, from (2.100) and
(2.17),

$$\xi\frac{dI_\nu(\xi)}{d\tau_\nu} = I_\nu(\xi) - J_\nu. \quad (2.107)$$

Multiplying both sides by $d\omega$, integrating over all solid angles, and
using (2.106) with the definition of flux (2.3), we have, for an isotropic
source function,

$$\frac{dF_\nu}{d\tau_\nu} = 2\pi(I_\nu^+ + I_\nu^-) - 4\pi J_\nu. \quad (2.108)$$

Multiplying (2.107) by $\xi \, d\omega$ and integrating over all solid angles,

$$\frac{2\pi}{3}\frac{d}{d\tau_\nu}(I_\nu^+ + I_\nu^-) = F_\nu. \tag{2.109}$$

Eliminating $(I_\nu^+ + I_\nu^-)$ between (2.108) and (2.109),

$$\frac{d^2 F_\nu}{d\tau_\nu^2} - 3F_\nu = -4\pi\frac{dJ_\nu}{d\tau_\nu}. \tag{2.110}$$

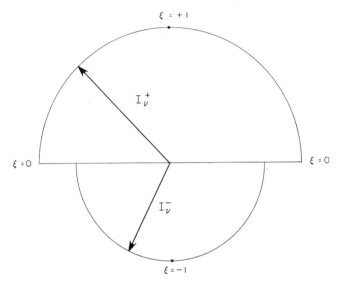

FIG. 2.9. Eddington's idealization of the radiation field in a stratified medium.

There is a special case of (2.110) for a medium in monochromatic radiative equilibrium. From (2.10) and (2.13) applied to a stratified medium

$$\mathbf{\nabla.F}_\nu = \frac{dF_\nu}{dz} \propto \left|\frac{dF_\nu}{d\tau_\nu}\right| = 0.$$

Hence, for monochromatic radiative equilibrium,

$$\frac{dJ_\nu}{d\tau_\nu} = +\frac{3}{4\pi}F_\nu \text{ (const.).} \tag{2.111}$$

Equation (2.110), being of second order, requires two boundary conditions. The difficulty in stating these conditions lies in ensuring that they are consistent with the Eddington approximation. According to the approximation, we have from the definition of flux

$$F_\nu = \pi(I_\nu^+ - I_\nu^-). \tag{2.112}$$

Eliminating either I_ν^+ or I_ν^- between (2.112) and (2.108) we find:

at $\tau_\nu = \tau_\nu^*$ (lower boundary)

$$I_\nu^+(\tau_\nu^*) = \frac{F_\nu(\tau_\nu^*)}{2\pi} + J_\nu(\tau_\nu^*) + \frac{1}{4\pi}\left(\frac{dF_\nu}{d\tau_\nu}\right)_{\tau_\nu=\tau_\nu^*}, \qquad (2.113)$$

at $\tau_\nu = 0$ (upper boundary)

$$I_\nu^-(\tau_\nu) = -\frac{F_\nu(0)}{2\pi} + J_\nu(0) + \frac{1}{4\pi}\left(\frac{dF_\nu}{d\tau_\nu}\right)_{\tau_\nu=0}. \qquad (2.114)$$

We will examine two special cases.

(i) *Monochromatic radiative equilibrium* with black bodies emitting $B_\nu^*(0)$ and $B_\nu^*(\tau^*)$ at the upper and lower boundaries. The third terms on the r.h.s. of (2.113) and (2.114) vanish, hence

$$F_\nu/2\pi = J_\nu(0) - B_\nu^*(0) = B_\nu^*(\tau_\nu^*) - J_\nu(\tau_\nu^*). \qquad (2.115)$$

Note that if the medium is in thermodynamic equilibrium ($J_\nu = B_\nu$), equation (2.115) requires a discontinuity in B_ν (i.e. a *temperature discontinuity*) at both boundaries.

(ii) *Continuity of temperature*, maintained by non-radiative effects, with an equilibrium source function ($J_\nu = B_\nu$), and boundary reflectivity α_ν.

At $\tau_\nu = \tau_\nu^*$, $I_\nu^+(\tau_\nu^*)$ is the sum of a reflected component $\alpha_\nu I_\nu^-(\tau_\nu^*)$ and an emitted component $(1-\alpha_\nu)B_\nu^*(\tau_\nu^*)$;

$$I_\nu^+(\tau_\nu^*) = \alpha_\nu I_\nu^-(\tau_\nu^*) + (1-\alpha_\nu)B_\nu^*(\tau_\nu^*).$$

Using (2.112), we find

$$I_\nu^+(\tau_\nu^*) = B_\nu^*(\tau_\nu^*) - \frac{\alpha_\nu}{1-\alpha_\nu}\frac{F_\nu(\tau_\nu^*)}{\pi}. \qquad (2.116)$$

Similarly, at $\tau_\nu = 0$

$$I_\nu^-(0) = B_\nu^*(0) + \frac{\alpha_\nu}{1-\alpha_\nu}\frac{F_\nu(0)}{\pi}. \qquad (2.117)$$

By hypothesis $B_\nu(0) = B_\nu^*(0)$, $B_\nu(\tau^*) = B_\nu^*(\tau^*)$ and $B_\nu = J_\nu$. Using (2.113) and (2.114) we find

at $\tau_\nu = \tau_\nu^*$,
$$\left(\frac{dF_\nu}{d\tau_\nu}\right)_{\tau_\nu=\tau_\nu^*} = -2\left(\frac{1+\alpha_\nu}{1-\alpha_\nu}\right)F_\nu(\tau_\nu^*), \qquad (2.118)$$

at $\tau_\nu = 0$,
$$\left(\frac{dF_\nu}{d\tau_\nu}\right)_{\tau_\nu=0} = +2\left(\frac{1+\alpha_\nu}{1-\alpha_\nu}\right)F_\nu(0). \qquad (2.119)$$

Equations (2.118) and (2.119) are singular if $\alpha_\nu = 1$ (mirror boundaries) and in this case the appropriate boundary conditions are

$$F_\nu(\tau_\nu^*) = F_\nu(0) = 0. \qquad (2.120)$$

Historically, an approximation of the same general type as the Eddington approximation was proposed earlier by both Schwarzschild and Schuster. Equations (2.106) were replaced by

$$\left. \begin{array}{l} I_\nu^+(\xi, \tau_\nu) = I_\nu^+(\tau_\nu)\delta(1, \xi) \\ I_\nu^-(\xi, \tau_\nu) = I_\nu^-(\tau_\nu)\delta(-1, \xi) \end{array} \right\} , \qquad (2.121)$$

where $\delta(1, \xi)$ and $\delta(-1, \xi)$ are δ-functions. With these equations, (2.108) is unaltered, but (2.109) loses the factor 3 in the denominator on the l.h.s. Thus the resulting differential equation differs only by factors of order unity from (2.110), and some of the difference may be taken up by modifying the definition of optical depth to

$$\tau_\nu' = \sqrt{3}\tau_\nu. \qquad (2.122)$$

2.4.2. Approximate differential equations

The Eddington approximation is only valid for a stratified medium; under certain limiting conditions, however, equations of a similar nature hold without restrictions on the geometry of the medium. We must suppose that there is a characteristic scale l associated with the medium (e.g. the wavelength of an harmonic wave), which has the following properties:

(i) fluctuations in any space-averaged quantity tend to zero as the linear dimensions of the volume of integration become larger than l;

(ii) to order of magnitude we can replace d^n/ds^n by $(1/l)d^{n-1}/ds^{n-1}$ (in boundary-layer problems of fluid flow, this defines what is known as a *smooth* function).

We will restrict our attention to thermodynamic equilibrium ($J_\nu = B_\nu$) and consider (2.18) which expresses the heating rate as the balance between the angle-averaged absorption and emission. If $e_{v,\nu} l \ll 1$ the medium is transparent over many scale lengths. Contributions to the first term on the r.h.s. of (2.18) originate largely from distances of the order of $e_{v,\nu}^{-1}$ (= the mean-free-path of the radiation). The angle average can be pictured as an average over a sphere of this radius, which corresponds to an average over many scale lengths. According to assumption (i) the first term will therefore not reflect fluctuations in the physical state of the medium, which are, however, intrinsic in the second term. Consequently, if δ indicates a small space variation and B_ν and I_ν are of the same order

$$\delta h_\nu \simeq -4\pi e_{v,\nu} \delta B_\nu. \qquad (2.123)$$

Here we have assumed that $\delta e_{v,\nu} = 0$, i.e. the fluid has constant composition. This is not a necessary assumption, but we will retain it in order to simplify the discussion. Equation (2.123) is a formal statement of Newton's law of cooling, which we will call the *transparent approximation*.

The other extreme assumption (the *opaque approximation*) is that $e_{v,v} l \gg 1$, and major contributions to the local intensity can only come from within a distance which is small compared with l. According to condition (ii) above, it is now meaningful to make a Taylor expansion of B_v provided only that B_v and τ_v are continuous functions of the space variable. The opaque approximation and the requirements for a Taylor expansion can be satisfied in the body of a fluid, but not near to boundaries where discontinuities and large temperature gradients may exist. We must therefore assume that we are dealing with a phenomenon which is unaffected by events at the (distant) boundaries, and this must be proved *a posteriori* in any application of the approximation.

The Taylor expansion of $B_v(\tau_v)$ is

$$B_v(\tau_v) = B_v(0)+\tau_v\left(\frac{dB_v}{d\tau_v}\right)_0 +\frac{\tau_v^2}{2!}\left(\frac{d^2B_v}{d\tau_v^2}\right)_0 +\frac{\tau_v^3}{3!}\left(\frac{d^3B_v}{d\tau_v^3}\right)_0 +\cdots . \quad (2.124)$$

Substituting (2.124) in (2.80) and integrating over all space

$$I_v(0,\mathbf{s}) = B_v(0)-\left(\frac{dB_v}{d\tau_v}\right)_0 +\left(\frac{d^2B_v}{d\tau_v^2}\right)_0 -\left(\frac{d^3B_v}{d\tau_v^3}\right)_0 +\cdots . \quad (2.125)$$

Multiplying by $d\omega$ and integrating over all solid angles

$$4\pi\bar{I}_v(0) = \int B_v(0)\,d\omega - \int \left(\frac{dB_v}{d\tau_v}\right)_0 d\omega+$$
$$+ \int \left(\frac{d^2B_v}{d\tau_v^2}\right)_0 d\omega - \int \left(\frac{d^3B_v}{d\tau_v^3}\right)_0 d\omega+\cdots . \quad (2.126)$$

The first differential operator in (2.126) may be written

$$\frac{d}{d\tau_v} = \xi_x\frac{\partial}{\partial\tau_{v,x}}+\xi_y\frac{\partial}{\partial\tau_{v,y}}+\xi_z\frac{\partial}{\partial\tau_{v,z}}, \quad (2.127)$$

where ξ_x, ξ_y, and ξ_z are the three direction cosines and $d\tau_{v,x} = -e_{v,v}\,dx$, $d\tau_{v,y} = -e_{v,v}\,dy$, and $d\tau_{v,z} = -e_{v,v}\,dz$. B_v is isotropic, and this isotropy is not affected by the three partial differential operators on the r.h.s. of (2.127). Thus the angular dependence of these three terms is that of ξ_x, ξ_y, and ξ_z respectively. The direction cosines are positive in one hemisphere and negative in the other; hence

$$\int \xi_x\,d\omega = \int \xi_y\,d\omega = \int \xi_z\,d\omega = 0, \quad (2.128)$$

and the second term on the r.h.s. of (2.126) is zero. The third term, however, involves terms such as $\int \xi_x\xi_y\,d\omega$, which are zero and terms such as $\int \xi_x^2 \, d\omega$ which equal $4\pi/3$. Applying the operator (2.127) twice, we find

$$\int \frac{d^2 B_\nu}{d\tau_\nu^2} \, d\omega = \frac{4\pi}{3}\left(\frac{\partial^2 B_\nu}{\partial \tau_{\nu,x}^2} + \frac{\partial^2 B_\nu}{\partial \tau_{\nu,y}^2} + \frac{\partial^2 B_\nu}{\partial \tau_{\nu,z}^2}\right)$$

$$= \frac{4\pi}{3e_{\text{v},\nu}} \nabla^2 B_\nu. \tag{2.129}$$

By similar arguments it can be shown that every second term on the r.h.s. of (2.126) is zero. According to the property (ii) of the scale length l, we may write

$$\frac{d^4 B_\nu}{d\tau_\nu^4} \simeq \frac{1}{e_{\text{v},\nu}^2 l^2} \frac{d^2 B_\nu}{d\tau_\nu^2} \ll \frac{d^2 B_\nu}{d\tau_\nu^2},$$

and neglect all terms in (2.126) after the third.

From (2.18) and (2.126) we now find

$$h_\nu = \frac{4\pi}{3} \frac{1}{e_{\text{v},\nu}} \nabla^2 B_\nu. \tag{2.130}$$

2.4.3. Approximate forms for the absorption coefficients

Stellar spectra exhibit strong continua and it was a logical step in the development of astrophysics to examine the hypothesis of *grey radiation* for which the extinction coefficient is independent of frequency. The idea can be readily extended to *semi-grey radiation*, for which two coefficients define the properties of two characteristically different spectral regions (e.g. the terrestrial radiation and the solar radiation).

The earth's atmosphere differs greatly from a grey absorber, however, and relations derived on the assumption of grey radiation have, in general, little quantitative meaning. Nevertheless, the extreme simplicity of the equations permits a qualitative examination of radiative effects in circumstances where a detailed analysis is not possible.

In two circumstances a complicated spectrum can be treated rigorously by the grey approximation, and an equivalent grey absorption coefficient $\bar{k}$ defined; these are when computing radiative heating if the opaque or the transparent approximations are valid. In (2.123) and (2.130) we can express the frequency-integrated heating in terms of the frequency-integrated source function if the mean absorption coefficients

$$\bar{k}_P = \frac{\int\limits_0^\infty k_\nu \, \delta B_\nu \, d\nu}{\int\limits_0^\infty \delta B_\nu \, d\nu}, \tag{2.131}$$

$$\text{or} \qquad (\bar{k}_R)^{-1} = \frac{\int\limits_0^\infty (k_\nu)^{-1} \nabla^2 B_\nu \, d\nu}{\int\limits_0^\infty \nabla^2 B_\nu \, d\nu}, \qquad (2.132)$$

respectively, are employed.

Below 120 km in the earth's atmosphere temperature varies by only ± 20 per cent around its mean value, and

$$\delta B_\nu \simeq \frac{\overline{dB_\nu}}{d\theta} \, \delta\theta,$$

where the over bar denotes a mean. Then

$$\nabla^2 B_\nu \simeq \frac{\overline{dB_\nu}}{d\theta} \nabla^2\theta,$$

$$\text{and} \qquad \bar{k}_P = \frac{\int\limits_0^\infty k_\nu (\overline{dB_\nu/d\theta}) \, d\nu}{\int\limits_0^\infty (\overline{dB_\nu/d\theta}) \, d\nu}, \qquad (2.133)$$

$$(\bar{k}_R)^{-1} = \frac{\int\limits_0^\infty (k_\nu)^{-1} (\overline{dB_\nu/d\theta}) \, d\nu}{\int\limits_0^\infty (\overline{dB_\nu/d\theta}) \, d\nu}. \qquad (2.134)$$

Equation (2.134) is commonly used in astrophysics under the name of the *Rosseland mean*. (2.133) is not in common use in astrophysics, although a similar mean with B_ν in place of $dB_\nu/d\theta$ is used, and is known as the *Planck mean*. Since the difference is small we will use the same name for $\bar{k}_P$.

The *Chandrasekhar mean*

$$\bar{k}_C = \frac{\int\limits_0^\infty k_\nu F_\nu \, d\nu}{\int\limits_0^\infty F_\nu \, d\nu}, \qquad (2.135)$$

is also used in astrophysical problems. It is the optimum choice for an atmosphere in radiative equilibrium, departing only slightly from grey absorption. Since the weighting function (F_ν) is unknown *a priori*, it is less convenient than the Planck and Rosseland means.

2.4.4. *The method of discrete ordinates*

In its most general form this method involves replacing integrals by sums over a finite number of discrete ordinates. Any of the integral equations discussed in earlier sections may be treated in this way, and since they are linear there results a system of linear simultaneous equations, whose solution involves no difficulties in principle, although it can be complicated in practice. The crux of the problem is to devise the most *efficient* method of numerical integration. Since all the functions involved behave in a sufficiently regular manner, almost any one of the standard methods of numerical quadrature will give good results if the ordinates are sufficiently closely spaced. However, each additional ordinate involves a large increase in computational labour, and the problem is to achieve a given result with the least effort.

A typical integral in a stratified medium may be replaced by a sum over $2m$ $(j = \pm 1, \pm 2, ..., \pm m)$ ordinates in the following way:

$$\int_{-1}^{+1} f(\xi)\,d\xi \simeq \sum_j a_j f(\xi_j), \qquad (2.136)$$

where the a_j are weighting coefficients which can be computed if the ξ_j are known, and

$$\sum_j a_j = 2. \qquad (2.137)$$

If $f(\xi)$ is a polynomial of degree less than m, any method of numerical quadrature gives exact results. Gauss's method, for which the ξ_i are zeros of the Legendre polynomial $P_m(\xi)$, gives exact results for polynomials of degree $2m$ and has been used extensively by Chandrasekhar (see Bibliography). Points of division and weights for Gaussian quadrature of an integral like (2.136) are given in Table 2.2.

TABLE 2.2

Weights and divisions for Gaussian quadrature

1st approximation	$\xi_1 = 0\cdot5773503$	$a_1 = 1$
2nd approximation	$\xi_1 = 0\cdot3399810$	$a_1 = 0\cdot6521452$
	$\xi_2 = 0\cdot8611363$	$a_2 = 0\cdot3478548$
3rd approximation	$\xi_1 = 0\cdot2386192$	$a_1 = 0\cdot4679139$
	$\xi_2 = 0\cdot6612094$	$a_2 = 0\cdot3607616$
	$\xi_3 = 0\cdot9324695$	$a_3 = 0\cdot1713245$
4th approximation	$\xi_1 = 0\cdot1834346$	$a_1 = 0\cdot3626838$
	$\xi_2 = 0\cdot5255324$	$a_2 = 0\cdot3137066$
	$\xi_3 = 0\cdot7966665$	$a_3 = 0\cdot2223810$
	$\xi_4 = 0\cdot9602899$	$a_4 = 0\cdot1012285$

To illustrate the use of the method of discrete ordinates we will briefly discuss the problem of radiative equilibrium in a stratified, grey-absorbing medium. The equation of transfer is (2.107) (without ν-suffixes) and this has to be combined with the condition for radiative equilibrium (2.20). Using the approximation (2.136) we obtain $2m$ linear equations with constant coefficients

$$\xi_i \frac{dI(\xi_i, \tau)}{d\tau} \simeq I(\xi_i, \tau) - \tfrac{1}{2} \sum_j a_j I(\tau, \xi_i). \qquad (2.138)$$

The solution to these equations has been shown to be

$$I(\xi_i) = \frac{3F}{4\pi} \left(\sum_{\alpha=1}^{m-1} \frac{L_\alpha e^{-c_\alpha\tau}}{1 + \xi_i c_\alpha} + \sum_{\alpha=1}^{m-1} \frac{L_{-\alpha} e^{+c_\alpha\tau}}{1 - \xi_i c_\alpha} + \tau + \xi_i + Q \right), \qquad (2.139)$$

where Q, L_α, and $L_{-\alpha}$ are $2m-1$ constants of integration to be determined from the boundary conditions. The source function for this solution is

$$J(\tau) = \frac{3}{4} \frac{F}{\pi} \left(\sum_{\alpha=1}^{m-1} L_\alpha e^{-c_\alpha\tau} + \sum_{\alpha=1}^{m-1} L_{-\alpha} e^{+c_\alpha\tau} + \tau + Q \right)$$

$$= \frac{3}{4} \frac{F}{\pi} \{\tau + q(\tau)\}, \qquad (2.140)$$

where $q(\tau)$ is *Hopf's function*.

The application of boundary conditions is straightforward, although complicated if many ordinates are involved. If the medium is bounded below by a black body emitting I^* at $\tau = \tau^*$, and by space above, the following relation holds independently of the number of ordinates:

$$F = \frac{4\pi}{3} \frac{I^*}{\tau^* + 2Q}. \qquad (2.141)$$

The accuracy of the method of discrete ordinates is high. A comparison for a semi-infinite atmosphere ($\tau^* = \infty$) with an exact solution shows that with four ordinates $q(\tau)$ varies from $0 \cdot 5774$ at $\tau = 0$ to $0 \cdot 7069$ at $\tau = \infty$, while the exact solution gives $0 \cdot 5773$ at $\tau = 0$ and $0 \cdot 7104$ at $\tau = \infty$. The Eddington approximation for the same problem gives $q(\tau) = 0 \cdot 6667$ for all τ.

To complete this brief review of the method of discrete ordinates, we may mention the so-called *infinite approximation*, which under certain circumstances can be obtained from (2.139) by proceeding to the limit $m \to \infty$. While the solution is of analytical interest the computation of a numerical result follows a process of successive approximations equivalent to increasing the number of discrete ordinates. In practice therefore the method of discrete ordinates is not superseded by the formal solution. For a semi-infinite atmosphere the result is

$$J(\tau) = \frac{3}{4} \frac{F}{\pi} \left(\tau + 0 \cdot 7104 - \frac{1}{2\sqrt{3}} \int_0^1 \frac{e^{-\tau/\xi'} d\xi'}{H(\xi')Z(\xi')} \right). \qquad (2.142)$$

Tabulations of the functions $H(\xi')$ and $Z(\xi')$ are available.

The infinite approximation for a finite atmosphere of thickness τ^* has not been obtained on a rigorous basis. The following expression has been proposed by analogy with (2.142)

$$J(\tau)$$

$$= \frac{3}{4}\frac{F}{\pi}\left\{\tau + Q(\tau^*) - L(\tau^*)\left(\int_0^1 \frac{\exp(-\tau/\xi')}{H(\xi')Z(\xi')}\,d\xi' - \int_0^1 \frac{\exp\{-(\tau^*-\tau)/\xi'\}}{H(\xi')Z(\xi')}\,d\xi'\right)\right\},$$
$$(2.143)$$

where $Q(\tau^*)$ and $L(\tau^*)$ are very slowly varying functions of τ^*, whose values are determined by the condition for constant flux; tables of these functions have been published (see Bibliography).

2.4.5. *The principles of invariance*

A different approach to scattering problems has been proposed by Ambarzumian and developed by Chandrasekhar. It seeks certain simple statements about the radiation field, not derived from the equation of transfer, but of equal physical validity. We will describe the principles of this method because it has provided a complete solution to the problem of molecular scattering in a stratified atmosphere of finite optical depth.

Let a parallel beam, with irradiation f, fall on a plane-parallel scattering atmosphere of optical depth τ^* in a direction specified by $(\xi_\odot, \varphi_\odot)$ (see Fig. 2.10).

The problem is to relate $I^+(0; \xi, \varphi)$ and $I^-(\tau^*; \xi, \varphi)$ to $f(\xi_\odot, \varphi_\odot)$ without a complete solution for $0 < \tau < \tau^*$. We define the *reflection function* $S(\tau^*; \xi, \varphi; \xi_\odot, \varphi_\odot)$ by the relation

$$I^+(0; \xi, \varphi) = \frac{f}{4\pi\xi}S(\tau^*; \xi, \varphi; \xi_\odot, \varphi_\odot), \qquad (2.144)$$

and the *transmission function*† $T(\tau^*; \xi, \varphi; \xi_\odot, \varphi_\odot)$ by the relation

$$I^-(\tau^*; \xi, \varphi) = \frac{f}{4\pi\xi}T(\tau^*; \xi, \varphi; \xi_\odot, \varphi_\odot). \qquad (2.145)$$

Both will later be related to the scattering matrix. If the intensities in (2.144) and (2.145) are replaced by Stokes parameters then S and T must be replaced by matrices S_{ij} and T_{ij}.

The principles of invariance are a series of common-sense relations between the scattering and transmission functions and the intensities

---

† In Chapter 3 we will use this name and symbol for a different quantity. Both definitions are hallowed by usage in different fields and we will trust to the reader's perception to distinguish them.

at $\tau = 0$, $\tau = \tau^*$, and some intermediate, variable level (see Fig. 2.10). Many such statements are possible. For the case where no diffuse radiation is incident on the upper or lower boundaries the four employed by Chandrasekhar are as follows.

(i) $I^+(\tau; \xi, \varphi)$ at any level τ can be taken to be the sum of the reflection of the reduced irradiance $f e^{-|\tau/\xi_0|}$ and the reflection of the diffuse field $I^-(\tau; \xi', \varphi')$ from a layer of optical thickness $(\tau^* - \tau)$.

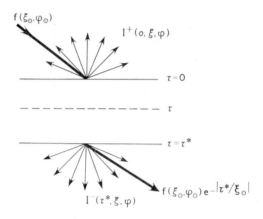

FIG. 2.10. Scattering by a finite atmosphere.

(ii) $I^-(\tau; \xi, \varphi)$ at any level τ can be taken to be the sum of the diffuse transmission through the atmosphere above τ and the reflection of the diffuse radiation, $I^+(\tau; \xi', \varphi')$, from below.

(iii) $I^+(0; \xi, \varphi)$ can be taken to be the sum of reflection from a layer of optical depth, τ, and the transmission of $I^+(\tau; \xi', \varphi')$ through this same layer.

(iv) $I^-(\tau^*; \xi, \varphi)$ can be taken to be the sum of the diffuse transmission of the reduced irradiance $f e^{-|\tau/\xi_0|}$ by the layer of optical depth $(\tau^* - \tau)$ and the transmission of the downward diffuse radiation $I^-(\tau; \xi', \varphi')$ through the same layer.

To illustrate the meanings of these four statements, we may write the fourth in mathematical terms as follows:

$$I^-(\tau^*; \xi, \varphi) = \frac{f}{4\pi\xi} T(\tau^*; \xi, \varphi; \xi_\odot, \varphi_\odot)$$

$$= \frac{f}{4\pi\xi} e^{-|\tau/\xi_\odot|} T(\tau^* - \tau; \xi, \varphi; \xi_\odot, \varphi_\odot) + e^{-|(\tau^* - \tau)/\xi|} I^-(\tau; \xi, \varphi) +$$

$$+ \frac{1}{4\pi\xi} \int T(\tau^* - \tau; \xi, \varphi; \xi', \varphi') I^-(\tau; \xi', \varphi') \, d\omega'. \quad (2.146)$$

The four equations expressing conditions (i) to (iv) can be differentiated with respect to τ and evaluated at the boundaries where the appropriate conditions are applied. The derivatives of the intensity can be eliminated with the help of the equation of transfer, which introduces the phase matrix (2.31). Four simultaneous integral equations result relating $S(\tau^*)$, $(\partial S/\partial \tau)_{\tau^*}$, $T(\tau^*)$, $(\partial T/\partial \tau)_{\tau^*}$, and $P(\xi, \varphi; \xi', \varphi')$, from which the reflection and transmission functions and hence the complete field of radiation at $\tau = 0$ and $\tau = \tau^*$ are determined. This solution, which does not allow for any upward radiation reflected from a surface at $\tau = \tau^*$, is known as the *standard problem*.

The reduction of the standard problem to manageable equations depends upon the nature of the phase matrix. If it is symmetrical the equations are simplified. We will see in Chapter 7 that the phase matrix for a spherical particle is a function of the scattering angle θ only (this is the angle between the incident ray characterized by ξ', φ', and the scattered ray characterized by ξ, φ). These angles and direction cosines are related by

$$\cos \theta = \xi \xi' + (1-\xi^2)^{1/2}(1-\xi'^2)^{1/2}\cos(\varphi'-\varphi).$$

Suppose that the phase matrix can be expressed in a finite series of Legendre polynomials $P_l(\cos \theta)$†

$$P(\cos \theta) = \sum_{l=0}^{N} \kappa_l P_l(\cos \theta). \tag{2.147}$$

Eliminating $\cos \theta$ from (2.146) and making use of the properties of the associated Legendre polynomial P_l^m, the following expression can be obtained after some manipulation:

$$P(\xi, \varphi; \xi', \varphi') = \sum_{m=0}^{N} (2-\delta_{0,m}) \left\{ \sum_{l=m}^{N} \kappa_l \frac{(l-m)!}{(l+m)!} P_l^m(\xi) P_l^m(\xi') \right\} \cos m(\varphi'-\varphi), \tag{2.148}$$

where

$$\delta_{0,m} = \begin{cases} 0 & \text{for } m \neq 0, \\ 1 & \text{for } m = 0. \end{cases}$$

For Rayleigh scattering, for example, the series (2.147) involves only three terms $\{P_0(x) = 1,\ P_1(x) = x,\ P_2(x) = \frac{3}{2}x^2 - \frac{1}{2}\}$ and $P(\xi, \varphi; \xi', \varphi')$ is represented by a sum of three terms, one independent of azimuth, one a multiple of $\cos(\varphi'-\varphi)$ and the other a multiple of $\cos 2(\varphi'-\varphi)$. We may write (2.147) in the form

$$P = \sum_m P^{(m)}.$$

---

† Here again, established conventions require that P stands for both scattering matrix and Legendre polynomial. The former appears as P and $P^{(m)}$; the latter as P_l and P_l^m.

Since the $P^{(m)}$ are orthogonal and the equation of transfer is linear, each term can be treated independently. The complete reflection and transmission functions are then

$$S = \sum_m S^{(m)}, \qquad (2.149)$$

$$T = \sum_m T^{(m)}. \qquad (2.150)$$

Let us now introduce the two new functions

$$\psi_l^m(\tau^*; \xi) = P_l^m(\xi) + \frac{(-1)^{m+l}}{2(2-\delta_{0,m})} \int_0^1 S^{(m)}(\tau^*; \xi, \xi') P_l^m(\xi') \frac{d\xi'}{\xi'}, \qquad (2.151)$$

$$\phi_l^m(\tau^*; \xi) = e^{-\tau^*/\xi} P_l^m(\xi) + \frac{1}{2(2-\delta_{0,m})} \int_0^1 T^{(m)}(\tau^*; \xi, \xi') P_l^m(\xi') \frac{d\xi'}{\xi'}. \qquad (2.152)$$

Straightforward manipulation yields expressions for $S^{(m)}$ and $T^{(m)}$ in terms of ψ_l^m and ϕ_l^m, and a pair of simultaneous, non-linear, integro-differential equations which determine them. These simultaneous equations are highly symmetrical and despite their lack of linearity can be readily solved by iterative numerical techniques. In every case in practice it has proved possible to include all the necessary iteration of these integral equations in the X- and Y-functions

$$X(\xi) = 1 + \xi \int_0^1 \frac{\Psi(\xi')}{\xi+\xi'} \{X(\xi)X(\xi') - Y(\xi)Y(\xi')\} \, d\xi', \qquad (2.153)$$

$$Y(\xi) = e^{-\tau^*/\xi} + \xi \int_0^1 \frac{\Psi(\xi')}{\xi-\xi'} \{Y(\xi)X(\xi') - X(\xi)Y(\xi')\} \, d\xi', \qquad (2.154)$$

where the characteristic function $\Psi(\xi')$ differs from problem to problem, but always has a simple algebraic form. The problem thus resolves itself into determining the appropriate X- and Y-functions (four are required for the problem of a Rayleigh phase matrix with unpolarized incident radiation), tabulating these functions for sufficiently close intervals of ξ and τ^*, and then computing the reflection and transmission matrices by relatively simple algebraic methods.

The existence of a scattering or reflecting surface at $\tau = \tau^*$ can be handled simply. An extra intensity can be defined which adds to the existing solutions, and which, in most cases, can be expressed in terms of the X- and Y-functions already computed for the standard problem.

BIBLIOGRAPHY
2.1. Definitions
The formal theory in this book is based, wherever possible, on

CHANDRASEKHAR, S., 1950, *Radiative transfer*. Oxford University Press. Reprinted 1960 by Dover Publications, Inc.

This book was written with astrophysical problems in mind. So are

CHANDRASEKHAR, S., 1949, *An introduction to the study of stellar structure*. Chicago University Press. Reprinted 1957 by Dover Publications, Inc.; and

WOOLLEY, R. v. D. R., and STIBBS, D. W. N., 1953, *The outer layers of a star*. Oxford University Press.

A more general approach is adopted by

KOURGANOFF, V., 1952, *Basic methods in transfer problems*. Oxford University Press.

The mathematical aspects of the problems are treated by

BUSBRIDGE, I. W., 1960, *The mathematics of radiative transfer*. Cambridge University Press.

2.2. Thermal emission
The treatment of Kirchhoff's laws is nowhere more explicit and readable than in

PLANCK, M., 1913, *Vorlesungen über die Theorie der Wärmestrahlung*. Leipzig: Barth. English edition 1959 by Dover Publications, Inc.

Numerical values for Planck's function are to be found in

ALLEN, C. W., 1955, *Astrophysical quantities*. London University Press; and
PIVOVONSKY, M., and NAGEL, M. R., 1961, *Tables of blackbody radiation functions*. New York: Macmillan.

The problem of the breakdown of Kirchhoff's laws for a two-level model was first treated by

MILNE, E. A., 1930, *Handbuch der Astrophysik*, **3** (i). Berlin: Springer.

The model of a vibration-rotation band, as discussed in the text, was treated by

CURTIS, A. R., and GOODY, R. M., 1956, 'Thermal radiation in the upper atmosphere', *Proc. Roy. Soc.* A, **236**, p. 193.

Equation (2.66) was first derived theoretically by

LANDAU, L., and TELLER, E., 1936, 'Zur Theorie der Schalldispersion', *Phys. Z. Sowjet.* **10**, p. 34.

Details of ultrasonic methods of measuring relaxation times are given by

VIGOUREAUX, P., 1950, *Ultrasonics*. London: Chapman & Hall.

2.3. The integral equations
Alternative derivations of these equations are given by CHANDRASEKHAR (1950) and KOURGANOFF (1952) (§ 2.1).

2.4. Approximate and numerical methods
The subject-matter of this section is treated at length by CHANDRASEKHAR (1950) and KOURGANOFF (1952) (§ 2.1). The function $H(\xi)$ is tabulated by

Chandrasekhar (1950, p. 125). There have been a number of advances in technique subsequent to 1952, but the only ones of major importance for our purposes are the series of studies of the semi-infinite case by Yamamoto and King:

YAMAMOTO, G., 1953, 'Radiative equilibrium of the earth's atmosphere. I. The grey case', *Science Rept., Tohoku Univ., Series 5*, No. 2, p. 45;

—— 1955, 'Radiative equilibrium of the earth's atmosphere. III. The exact solution for a grey, finite atmosphere', ibid. **7**, No. 1, p. 1;

KING, J. I. F., 1956, 'Radiative equilibrium of a line-absorbing atmosphere. II', *Astrophys. J.* **124**, p. 272; and

—— 1960, 'The Hopf q-function simply and precisely evaluated', ibid. **132**, p. 509.

The paper by King (1956) contains a tabulation of $Q(\tau^*)$ and $L(\tau^*)$ which disagrees with an alternative tabulation given by Yamamoto; King's figures are correct.

The exact solution mentioned in § 2.4.4 is by

MARK, C., 1947, 'The neutron density near a plane surface', *Phys. Rev.* **72**, p. 558.

The first use of the principles of invariance was by Ambarzumian but we owe their systematic exploitation to CHANDRASEKHAR (1950) (§ 2.1).

Tables of X- and Y-functions or related quantities are given by

CHANDRASEKHAR, S., and ELBERT, D., 1954, 'The illumination and polarisation of the sunlit sky on Rayleigh scattering', *Trans. Am. Phil. Soc.* **44**, p. 643,

and in unpublished papers listed by

COULSON, K. L., DAVE, J. V., and SEKARA, Z., 1960, *Tables related to radiation emerging from a planetary atmosphere with Rayleigh scattering.* Univ. of California Press.

3

THEORY OF GASEOUS ABSORPTION

3.1. Introduction

THE equations discussed in Chapter 2 involve the absorption spectrum of all atmospheric gases. The complexity of these spectra is such that a simple empirical study is of limited value and it is important to understand the underlying physical processes so that we may concentrate upon the data important to our problem.

It is, however, neither necessary nor possible to give a complete account of spectroscopy. The main concern of the following sections will be to introduce terms and expressions and to relate them to physical concepts. As an account of spectroscopy, the result is inadequate; but the details should be sufficient for the reader to understand and appreciate data as presented in the literature. For further information, the reader can consult the three valuable works of Herzberg which are discussed in the Bibliography.

If we at first neglect the interaction between modes of energy we may write for the energy of an isolated molecule

$$E = E_e + E_v + E_r + E_t, \tag{3.1}$$

where E_e = electronic energy,
E_v = vibrational energy,
E_r = rotational energy,
E_t = translational energy.

The first three terms in equation (3.1) are *quantized*, and take discrete values only, these values being specified by one or more *quantum numbers*. Any combination of quantum numbers defines an *energy state* or *quantum state*, or *term*. Radiation is absorbed or emitted when a transition takes place from one energy state to another, the frequency (ν) of the absorbed or emitted quantum being given by *Planck's relation*,

$$\Delta E = \mathbf{h}\nu, \tag{3.2}$$

where $\mathbf{h}$ is Planck's constant. The different units of frequency, wavelength, wave number, and energy used by spectroscopists are discussed in Appendix 3.

The most general transition involves simultaneous changes of electronic, vibrational, and rotational energy. The minimum energy jumps commonly observed vary considerably between the three forms, and provide a convenient preliminary method of distinguishing between them.

Minimum changes in rotational energy are usually of the order of 1 cm^{-1}, but we will also discuss lines in the solar spectrum arising from transitions of 500 cm^{-1} and more. *Rotation lines*, i.e. lines due to rotational energy changes only, therefore form part of the microwave spectrum or the far infra-red spectrum.

Vibrational energy changes are rarely less than 600 cm^{-1}. This is so much larger than the minimum for rotational energy change that vibrational transitions never occur alone, but with many simultaneous rotational changes giving a group of lines which constitute a *vibration-rotation band*, usually in the intermediate infra-red spectrum.

An electronic transition typically involves a few electron volts of energy, and the resulting absorption or emission usually occurs in the visible or ultra-violet spectrum. Atoms can exhibit *electronic line spectra*, but molecules have complex *band systems*, with simultaneous changes of all three forms of quantized energy.

For absorption or emission to take place matter must interact with the incident field of electromagnetic radiation which in practice must involve an electric or magnetic dipole or quadrupole moment. Interactions can differ widely in strength. Electric dipole interactions are greater by a factor of order 10^5 than magnetic dipole interactions; electric dipole interactions are of the order of 10^8 times stronger than electric quadrupole interactions. Electric dipole transitions are therefore responsible for the strongest spectral lines, and are called *permitted transitions*. Other transitions are loosely named *forbidden*. The nature of a transition can be specified in terms of the quantum numbers of the upper and lower states. Such a relationship is known as a *selection rule*.

The intensity of a dipole transition is proportional to the square of the matrix element of the dipole moment (**M**):

$$R_{i,j} = \int \psi_i^* \mathbf{M} \psi_j \, dV, \qquad (3.3)$$

where dV is a volume element in configuration space, and the integral is over all space. ψ_i and ψ_j are the wave functions, or solutions of Schrödinger's equation, which describe the ith and jth energy states, and the star denotes a complex conjugate. The wave functions are orthogonal so that

$$\int \psi_i^* \psi_j \, dV = 0 \quad (i \neq j), \qquad (3.4)$$

and consequently, if $\mathbf{M}$ is constant,

$$R_{i,j} = \mathbf{M} \int \psi_i^* \psi_j \, dV = 0. \tag{3.5}$$

Thus, for a dipole transition to be permitted, the electric moment must change between initial and final states.

The matrix elements (3.3) and the Einstein coefficients (§ 2.2.3) are related by

$$A_{i,j} = \frac{64\pi^4 \nu_{i,j}^3}{3hg_i c^3} |R_{i,j}|^2, \tag{3.6}$$

where g_i is the statistical weight of the upper quantum state. For a pure rotational transition, involving only a directional change of a constant dipole moment, $R_{i,j}$ is equal to a geometric factor multiplying the dipole moment. The square of this geometric factor is the *line strength*. Orders of magnitude for $A_{i,j}$ in electronic transitions are 10^8 sec^{-1} for permitted electric dipole transitions, 10^3 sec^{-1} for magnetic dipole transitions, 1 sec^{-1} for an electric quadrupole transition. For vibrational transitions the order of magnitude of $A_{i,j}$ is about 10 sec^{-1}; for pure rotational transitions it is of the order of 1 sec^{-1}.

The relationship between line intensity and Einstein coefficients given by (2.63) and (2.76) presumes that vibrational transitions superpose. This does not happen with real molecules but if we isolate a single transition and carry through the same reasoning we can show that

$$S_{i,j} = \frac{n_j}{ng_i} \frac{8\pi\nu_{i,j} |R_{i,j}|^2}{3hc} (1 - e^{-h\nu_{i,j}/k\theta}), \tag{3.7}$$

where n_j and n are the molecules per cm^3 in the lower quantum state, and in all quantum states, respectively. The term in parentheses in (3.7) owes its presence to induced emissions.

If a molecule absorbs too much energy the binding energy will be insufficient to hold the molecule together. Thus, excessive electronic excitation leads to *ionization*, and excessive vibrational excitation to *chemical decomposition*. Transitions to dissociated states (*bound-free transitions*) are unquantized and give rise to *continua*. The energy required to dissociate a molecule is normally so large that continua are commonly only found in the visible and ultra-violet spectrum. The behaviour of loosely bound, negative ions is exceptional, and H$^-$ continua in the infra-red spectrum are important features of the solar emission (Appendix 14). For this particular ion a *free-free* continuum is also observed for transitions between two parabolic orbits, neither of which is quantized. Bound-free, ionizing transitions dominate the ultra-violet absorption spectrum of the main atmospheric gases, and are of great importance to the heat economy of the upper atmosphere.

The energy levels of an atom or molecule are not strictly sharp as implied by the above discussion. If an excited state has a finite lifetime Δt, then according to Heisenberg's uncertainty principle there is an energy uncertainty

$$\Delta E \simeq \frac{h}{2\pi\,\Delta t}, \tag{3.8}$$

giving rise to a line of width

$$\Delta \nu = \frac{\Delta E}{h} \simeq \frac{1}{2\pi\,\Delta t}. \tag{3.9}$$

The lifetime cannot be longer than the natural lifetime $\phi_{i,j} = (A_{i,j})^{-1}$, which sets a lower limit to the width of a spectral line (the *natural width*). In practice, transitions are more usually induced by collisions, and line widths are of the order of $0 \cdot 1$ cm^{-1} at 1 atmosphere pressure.

3.2. Rotational energy

The characteristics of a pure rotation spectrum depend upon relationships between the three principal moments of inertia of the molecule. Four types of rotating molecule are distinguished in Table 3.1.

TABLE 3.1

Nomenclature based upon moments of inertia

Moments of inertia	Name	Atmospheric gases
$I_A = 0,\ I_B = I_C \neq 0$	Linear	CO_2, N_2O, O_2, N_2, CO
$I_A \neq 0,\ I_B = I_C \neq 0$	Symmetric top	No common atmospheric gases
$I_A = I_B = I_C$	Spherical top	CH_4
$I_A \neq I_B \neq I_C$	Asymmetric top	H_2O, O_3

Of the molecules listed, only H_2O has an important rotation spectrum, but all four groups will be considered, for the discussion is relevant to combined vibrational and rotational transitions (§ 3.5).

The nuclear angular momentum **J** is quantized according to the relation

$$|\mathbf{J}| = \frac{h}{2\pi}\sqrt{\{J(J+1)\}}, \tag{3.10}$$

where J is the angular momentum quantum number (an integer).

The component of angular momentum in the direction of the symmetry axis (**K**) is a constant of the motion for those molecules with such an axis, and is quantized according to

$$|\mathbf{K}| = \frac{Kh}{2\pi}, \tag{3.11}$$

where K is integral, positive, and $\leqslant J$. For linear molecules K is zero, and spherical tops have no figure axis; the quantization is therefore important for symmetric and asymmetric tops only. The sign of K is rarely important so that K takes only $J+1$ distinct values, which are doubly degenerate, excepting $K = 0$.

For planar molecules a reflection of all atoms at the centre of mass gives the same wave function, apart from a possible change of sign. If the sign does not change the parity of the level is said to be $(+)$, while if it does change its parity is said to be $(-)$. It is a general selection rule that only opposite parities combine in a rotational transition.

The nuclear spins of individual atoms affect the statistical weights in certain symmetry classes and, through Boltzmann's law, the population densities of the levels. Nuclear symmetry groups are classed as a or s (antisymmetric and symmetric) for linear molecules, A or E for symmetric tops, and A, E, or F for spherical tops.

The energy levels of a rigid, *linear molecule* are

$$E_\mathrm{r} = \mathbf{h}\mathbf{c}F(J), \qquad (3.12)$$

where

$$F(J) = BJ(J+1),$$

$$B = \frac{\mathbf{h}}{8\pi^2\mathbf{c}I} = \frac{2{\cdot}7994 \times 10^{-39}}{I} = \text{the } \textit{rotational constant},$$

$$I = \text{moment of inertia.}$$

The energy levels defined by (3.12) are illustrated in Fig. 3.1.

A small centrifugal stretching correction,

$$\Delta F(J) = -DJ^2(J+1)^2, \qquad (3.13)$$

is added to (3.12) for the highest precision, particularly for large J, but for many purposes the rigid rotator is an adequate model. The rotational constant D can be related to the vibrational stretching frequencies.

If a molecule has a centre of symmetry, and two nuclei not at that centre, the ratio of the nuclear statistical weights of the s and a levels is

$$\frac{w_s}{w_a} = \frac{I+1}{I} \quad \text{for Bose–Einstein particles,}$$

$$= \frac{I}{I+1} \quad \text{for Fermi–Dirac particles,} \qquad (3.14)$$

where I is the spin of one nucleus. For the carbon dioxide molecule the oxygen atoms have spin zero and the nuclear statistics are Bose–Einstein. Hence, from (3.14), the anti-symmetric levels, which are those with odd J, are missing entirely. For two hydrogen atoms the anti-symmetric levels have three times the weight of the symmetric levels.

A selection rule for a rigid, linear rotator is

$$\Delta J = 0, \pm 1 \quad (J = 0 \text{ to } J = 0 \text{ excepted}). \qquad (3.15)$$

Of these possibilities only $\Delta J = +1$ gives a positive absorption frequency in a pure rotation band. From (3.12) and (3.13) the wave numbers are

$$\nu/c = 2B(J''+1) - 4D(J''+1)^3, \qquad (3.16)$$

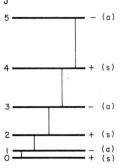

FIG. 3.1. Rotational energy levels of a linear molecule. The labelling (a) and (s) is correct for CO_2. It is not appropriate for N_2O, CO, and $CO^{16}O^{18}$, all of which have no centre of symmetry. The vertical lines indicate possible transitions. After Herzberg (1945).

where J'' is the quantum number of the initial (lower) state. If we neglect the stretching term we have an array of equally spaced lines, with interval $2B$ cm^{-1}.

A rotating molecule can only absorb or emit if it has a permanent dipole moment. Since a molecule with a centre of symmetry cannot have a dipole moment,† symmetric molecules such as CO_2 have no pure rotation spectrum. For molecules which have a dipole moment the rotational wave functions are known, and the calculation of the matrix elements (3.5) is straightforward. For the absorption process $(\Delta J = +1)$, under most circumstances

$$|R_{J'',J''+1}|^2 \propto \frac{(J''+1)}{2J''+1}. \qquad (3.17)$$

This expression has to be slightly modified for very high J-values when centrifugal stretching cannot be neglected. The line intensity (3.7) is proportional to the fraction of the molecules in the initial state,

$$\frac{n(J'')}{n} = \frac{g_{J''}}{Q_r(\theta)} e^{-E_r(J'')/k\theta}, \qquad (3.18)$$

where

$$Q_r = \sum_{J=0}^{\infty} g_J e^{-E_r(J)/k\theta}, \qquad (3.19)$$

is the *rotational partition function*. Since there is no constraint upon the angular momentum the level with a quantum number J has a directional degeneracy $(2J+1)$. g_J is the product of this and the nuclear weight factor $w_J (= I \text{ or } (I+1), \text{ see } (3.14))$; hence

$$\frac{n(J'')}{n} = \frac{w_{J''}(2J''+1)}{Q_r(\theta)} e^{-E_r(J'')/k\theta}. \qquad (3.20)$$

† Pressure induced magnetic dipole transitions have been observed, but these are too weak to influence atmospheric problems.

A graphical representation of (3.20) is shown in Fig. 3.2. If J'' is not small $|R_{J'',J''+1}|$ varies little with J'' and consequently (3.20) is a good approximation to the relative line intensity. If (3.20) is treated as a continuous function of J'', it has a maximum for

$$J'' = \sqrt{\left(\frac{k\theta}{2Bhc}\right)} - \tfrac{1}{2}. \tag{3.21}$$

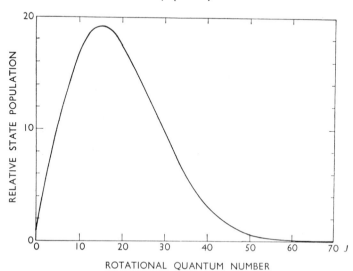

FIG. 3.2. Thermal distribution of rotational levels.
Equation (3.20) is plotted with the values

$$\frac{w_{J''}}{Q(\theta)} = 1,$$
$$B = 0\cdot418 \text{ cm}^{-1} \text{ (N}_2\text{O)},$$
$$\theta = 300° \text{ K}.$$

The foregoing discussion applies only to the linear molecule. The next important class is the *symmetric top*, which must be considered briefly as an introduction to the *asymmetric top*, although there are no important atmospheric gases with this symmetry. The moment of inertia in the direction of the symmetry axis is I_A (rotational constant, A), and the other is I_B (rotational constant, B). If $A > B$ the top is *prolate*; if $A < B$ it is *oblate*.

The rotational energy of a symmetric top is

$$E_r = hcF(J, K), \tag{3.22}$$

and, in the absence of centrifugal stretching,

$$F(J, K) = BJ(J+1)+(A-B)K^2. \tag{3.23}$$

Centrifugal stretching brings in small cross-terms and fourth-degree terms in J and K.

The statistical weights of the rotational states, apart from the nuclear contribution, depend upon the number of ways in which the angular momenta can be arranged to give the same energy. This is $(2J+1)$ if $K = 0$ and $2(2J+1)$ if $K \neq 0$; the additional factor 2 enters because the energy involves K^2 only, and the sign of K is irrelevant. As regards the nuclear spins, symmetry properties fall into two classes, A and E, which do not combine. These factors determine the population number $n(J, K)$ from (3.18).

The selection rules allow

$$\Delta K = 0, \quad \Delta J = 0, \pm 1 \ (J = 0 \text{ to } J = 0 \text{ excepted}), \qquad (3.24)$$

after restrictions imposed by the symmetry elements have been taken into consideration. For a pure rotation band, only $\Delta J = +1$ gives observable absorption frequencies and the spectrum consists of approximately equally spaced lines at wave numbers

$$\nu/\mathbf{c} = 2B(J''+1) + \text{small centrifugal terms.} \qquad (3.25)$$

The only difference between (3.16) and (3.25) is in the centrifugal term which now involves both K and J, giving multiplets of $(J+1)$ close lines, corresponding to the different values of K. Lines are distinguished by the J and K values of the initial state, written (J, K).

In the absence of centrifugal terms, the matrix elements for the transition $J'' \rightarrow J''+1$ are

$$|R_{J'',J''+1}|^2 \propto J'' \frac{(J''+1)-K^2}{(J''+1)(2J''+1)}. \qquad (3.26)$$

The *spherical top* is a special case of the symmetric top for which $A = B$. From (3.23) therefore it has the same energy levels as a linear molecule. It has, however, a much greater degree of degeneracy, since the angular momentum is not defined with respect to any direction in the molecule, and the orientation of the molecule cannot be defined with respect to any external direction. Consequently the Jth level has a weight of $(2J+1)^2$, apart from weighting factors for the nuclear spin.

If a molecule has many nuclei, the nuclear spin factors can be very complicated. For tetrahedral molecules (including CH_4) the weights have been computed. Levels fall into one of three classes, designated A, E, and F, which cannot combine in a transition.

The final symmetry type is that of the *asymmetric top*. It cannot be described in any particularly simple terms, but can be pictured as an intermediate stage between a prolate and an oblate symmetric top. Each J-level of the symmetric top is split into $(J+1)$ sub-levels with different K. For an oblate top the energy decreases as K increases while for a prolate top the order is reversed (see Fig. 3.3). Each level with $K \neq 0$ is doubly degenerate, as has already been discussed. The first deviation from a symmetric top gives rise to a splitting of these levels, and the levels for an asymmetric top can be found to a first approximation by interpolating between the prolate and oblate cases, as shown in Fig. 3.3. K now has an ambiguous role, and is no longer a useful quantum number for defining the energy levels.

When detailed expressions are worked out for the energy levels, it is found that the states with rotational quantum number J (which, as usual, defines the magnitude of the total angular momentum) have $(2J+1)$ values of $E_r(J)$. These are labelled arbitrarily in order of their energy, with an integer, τ, ranging from $-J$ to $+J$. This integer is set as a subscript to the J-value when defining the energy level and is not a quantum number in the usual sense. Reference to Fig. 3.3 shows that $\tau = K_A - K_C$, where K_A is the K-value in the related prolate symmetric top, while K_C is the K-value for the related oblate symmetric top. Some writers prefer to designate the term with the symbol $J_{K_A K_C}$ rather than J_τ.

The energy of the level J_τ for a rigid molecule is given by (3.22) with

$$F(J_\tau) = \tfrac{1}{2}(A+C)J(J+1) + \tfrac{1}{2}(A-C)E(J, \tau, \kappa), \qquad (3.27)$$

where
$$\kappa = \frac{2\{B - \tfrac{1}{2}(A+C)\}}{A-C}. \qquad (3.28)$$

Tables of $E(J, \tau, \kappa)$ are available in the literature (see Bibliography at the end of the chapter).

When evaluating statistical weights for an asymmetric top a new symmetry condition must be considered. The wave function can change its sign, or remain constant, for a half rotation about each of the principal axes. However, the behaviour of two defines the behaviour of the third and the symmetry can be specified by two symbols in four classes $++$, $+-$, $-+$, or $--$.

If a molecule has symmetry elements, nuclear spin affects the statistical weight and for a molecule such as H_2O statistical weights depend upon the nuclear spin in the same way as linear molecules. Thus the a-levels for H_2O have three times the statistical weight of the s-levels (3.14). These weight factors are additional to the directional degeneracy

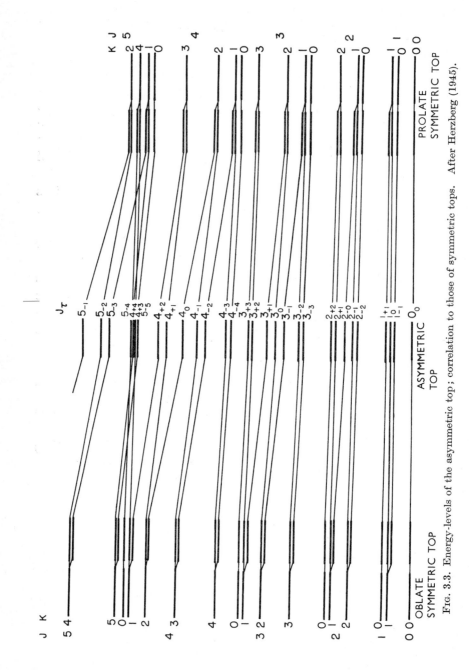

Fig. 3.3. Energy-levels of the asymmetric top; correlation to those of symmetric tops. After Herzberg (1945).

of the angular momentum with respect to the major axis of the molecule $(2J+1)$.

Selection rules for the asymmetric top are

$$\Delta J = 0, \pm 1 \quad (J = 0 \text{ to } J = 0 \text{ excepted}). \tag{3.29}$$

The rotational energy (3.27) is no longer a monotonic function of J, as for more highly symmetrical molecules. All three changes in J can therefore give a positive energy jump, and these are distinguished by the nomenclature

$$\Delta J = -1 \quad P\text{-branch,}$$
$$\Delta J = 0 \quad Q\text{-branch,}$$
$$\Delta J = +1 \quad R\text{-branch.}$$

Other things being equal, an R-branch line has a higher frequency than a Q-branch line, which in turn has a higher frequency than a P-branch line.

Selection rules between the rotational symmetry classes $++$, $-+$, etc., depend upon whether the dipole moment coincides with the largest (type C), intermediate (type B), or least (type A) moment of inertia. For water vapour it lies along the intermediate axis, and the selection rule (type B) is

$$++ \longleftrightarrow -- \quad \text{and} \quad +- \longleftrightarrow -+. \tag{3.30}$$

For type A on the other hand

$$++ \longleftrightarrow -+ \quad \text{and} \quad +- \longleftrightarrow --. \tag{3.31}$$

The matrix elements are complex, but lend themselves readily to tabulation (see Bibliography). Data for the analysis of any asymmetric top are therefore available, and they have been applied to the more important bands of O_3 and H_2O. A classical comparison between theory and observation for the water vapour spectrum is shown in Fig. 3.4. Where the three moments of inertia differ greatly from each other, as for H_2O, the net result is an apparently disordered structure.

3.3. Vibrational energy

A molecule containing N atoms has $3N$ degrees of freedom, of which three are attributable to translation of the centre of gravity and three more can be taken up by rotation of a non-linear molecule, but only two for a linear molecule. Thus for a non-linear molecule $(3N-6)$ degrees of freedom are available for vibrational modes and $(3N-5)$ for a linear molecule.

Assume for the present that the oscillations of the molecule are simple harmonic (unfortunately, this is an oversimplification and the

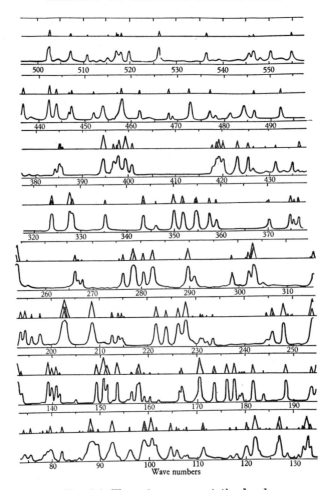

Fig. 3.4. The water-vapour rotation band.

Different amounts of water vapour were in the path for the different ranges; the strongest lines are in the region of 200 cm^{-1}. The triangles represent calculated positions and intensities. After Randall, Dennison, Ginsberg, and Weber (1937).

anharmonicities have a strong influence on intensities and selection rules). If the classical vibration frequencies are ν_1, ν_2, ν_3, etc., then the quantum theory expression for the vibrational energy is

$$E_v = (v_1 + \tfrac{1}{2})h\nu_1 + (v_2 + \tfrac{1}{2})h\nu_2 + (v_3 + \tfrac{1}{2})h\nu_3 + ..., \qquad (3.32)$$

where v_1, v_2, v_3, etc., are the quantum numbers defining the vibrational state. The term symbol simply lists the values of v_1, v_2, v_3, etc., in order e.g. (101) for $v_1 = 1$, $v_2 = 0$, $v_3 = 1$.

For triatomic molecules nomenclature is systematic. ν_1 and ν_3 both involve stretching a chemical bond, the former symmetrically, the latter asymmetrically (see Fig. 3.5). ν_2 involves bending rather than stretching bonds and consequently has the lowest frequency. Although the linear molecule has one more normal mode than the non-linear molecule, two of them (ν_{2a} and ν_{2b}) are identical. Despite this degeneracy, distin-

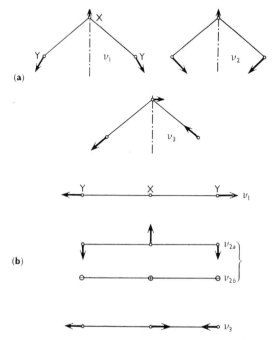

FIG. 3.5. Normal vibrations of (*a*) bent and (*b*) linear XY_2 molecules. H_2O is an example of a non-linear XY_2 molecule; CO_2 is an example of a linear molecule. After Herzberg (1945).

guishable substates of ν_2 exist, which separate when the molecule rotates. ν_{2a} and ν_{2b} can have different phases. If the phases differ by $\pi/2$, for example, the atoms perform elliptical trajectories with the consequence that there is a resultant angular momentum about the symmetry axis. This can add to or subtract from the nuclear rotational angular momentum giving rise to two energy levels. The quantum number l which defines this *vibrational angular momentum* is added as a superscript to the vibrational quantum number in the term symbol and the splitting of levels is called *l-type doubling* and occurs only if $l \neq 0$.

Anharmonicities in the fundamental modes give rise to quadratic and higher terms in the expression for the energy (3.32). Fig. 3.6 shows the effect on the energy levels of H_2O of including and excluding the anharmonic terms.

The selection rule for a harmonic oscillator permits only one quantum number at a time to change and this by unity. From (3.32) absorption or emission then takes place at frequencies

$$\nu = \frac{\Delta E_v}{h} = \nu_1, \nu_2, \text{ or } \nu_3, \text{ etc., only.} \qquad (3.33)$$

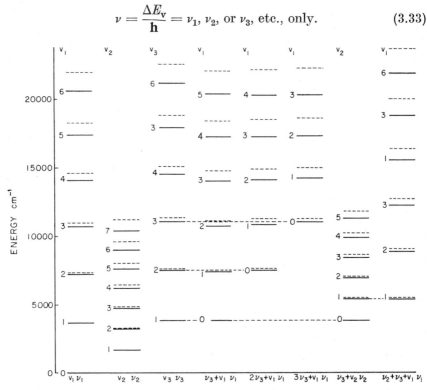

FIG. 3.6. Vibrational energy-level diagram of H_2O showing the influence of anharmonicity.

The full lines represent the actual energy levels, which are not equally spaced as (3.32) suggests. The heavy, broken lines indicate the positions of the energy levels if anharmonic terms are neglected. Not all levels below 25 000 cm^{-1} are shown. Levels that are repeated in different sets are connected by light broken lines. Since anharmonic modes do not combine linearly, a separate term scheme is required for each combination of modes. Combinations other than those shown are also possible. After Herzberg (1945).

These are the *fundamentals* and are usually the strongest bands in the infra-red spectrum. For degenerate levels there is a selection rule on the vibrational angular momentum quantum number

$$\Delta l = 0, \pm 1. \qquad (3.34)$$

All of these bands are not necessarily active, for some transitions are forbidden by symmetry conditions. We have already remarked that, unless the dipole moment differs in the initial and final state, there can be no absorption or emission. A linear molecule with a centre of symmetry, such as CO_2, can have no electric dipole moment because of the high degree of symmetry. The symmetrical stretching motion of the ν_1 vibration (see Fig. 3.5) does not disturb this symmetry, even if it is anharmonic. The dipole moment is therefore always zero and the fundamental is absent from the infra-red spectrum. For the same reason, no homonuclear, diatomic molecule has any permitted infra-red absorption.

The introduction of anharmonic terms has two main repercussions. Firstly, by destroying the equal spacing of the vibrational levels, the same frequencies do not result from all $\Delta v = 1$. For example, the absorption associated with the transition $v = 0$ to $v = 1$ will have a slightly different frequency from the transition $v = 1$ to $v = 2$. The latter is named an *upper state* band. For the CO_2 fundamental at 667·3 cm^{-1}, the difference between $(00^00) \rightarrow (01^10)$ and $(01^10) \rightarrow (02^20)$ is only 1 cm^{-1} and the ground state and upper state bands are closely interwoven. Upper state bands are always weak at atmospheric temperatures because the number of molecules above the ground state is small. By applying Boltzmann's law (2.42) to ν_2 of CO_2 ($\nu/c = 667·3$ cm^{-1}) at $\theta = 273°$ K it can be shown that 6 per cent of the molecules are in the upper state (a factor 2 arises from the effect of l-type doubling on the statistical weight). The ν_2 vibration of N_2O has a rather lower energy ($\nu/c = 588·8$ cm^{-1}) and the upper state bands are of greater importance than for CO_2. Other atmospheric gases however have no energy levels lower than the (01^10) level of CO_2 and upper state bands can generally be neglected.

A second effect of anharmonicity is to modify the selection rules, and changes $\Delta v = 2, 3, 4$ are now possible (*overtone bands*), as are simultaneous changes of two quantum numbers (*combination* or *difference bands*, depending on whether the two changes are of the same sign or not). Some transitions are still forbidden by symmetry considerations. For example, vibrational states whose wave functions change sign when inverted with respect to a centre of symmetry only combine with those whose wave functions do not change sign, and the selection rule (3.34) on the vibrational angular momentum cannot be violated. The main result of anharmonicity is that every molecule shows many more bands than simply the fundamentals which, however, are usually the strongest.

Anharmonicity can also cause strong interactions between modes which have accidental resonance. For example, CO_2 has close resonance between the first overtone $2\nu_2$ ($2 \times 667 \cdot 3$ cm^{-1} = $1334 \cdot 6$ cm^{-1}) and the lowest state of ν_1 at 1337 cm^{-1}. The effect of this resonance (*Fermi resonance*) is to increase the separation between the ν_1 and $2\nu_2$ energy levels, and also to make it impossible to specify whether the molecule is in one state or the other; we have instead to think of simultaneous occupancy of the two states. Symmetry conditions restrict the occurrence of Fermi resonance. For example, the (020) level of CO_2 is degenerate, having the two possible vibrational angular momentum quantum numbers $l = 0$ and $l = 2$. The ($10^0$0) level can only resonate with the ($02^0$0) level and not with the ($02^2$0) level for this would violate the selection rule (3.34). The former is therefore perturbed but not the latter.

Isotopic molecules can have markedly different infra-red spectra. The exchange of an atom for its isotope can change the symmetry character of the molecule and alter its selection rules, and it can also alter the frequencies of the normal modes. The latter depend upon the way in which the particular isotope is involved in the vibration. The difference between H_2O^{16} and H_2O^{18} only amounts to shifts varying between $6 \cdot 5$ and 18 cm^{-1}. The change from H_2O to HDO, however, gives rise to what is essentially a new molecule. Since isotopic ratios are nearly constant in nature (see § 1.3) it is usually possible to treat isotopic bands as extra, weak bands of the predominant species.

Evaluation of the intensity of a vibration band on a theoretical basis requires a knowledge of the wave functions. This demands a detailed knowledge of the force field between the atoms of the molecule, which is not generally known with the required precision. Attempts to evaluate matrix elements theoretically meet, in general, with little success and it is more usual to measure *band intensities* ($\int k_\nu \, d\nu$) in the laboratory or alternatively to form the sum over a band on a partly-theoretical basis using a few measured line intensities and predicted relative intensities.

3.4. Electronic energy

The energy levels of an atom with a single electron are mainly determined by the *principal quantum number* n. There is also a dependence of the energy upon the angular momentum (**l**), which is quantized according to the rule

$$|\mathbf{l}| = \frac{h}{2\pi} \sqrt{\{l(l+1)\}}, \tag{3.35}$$

where l is an integer (the *azimuthal quantum number*). For dipole transitions there is a strict selection rule

$$\Delta l = \pm 1. \tag{3.36}$$

In the absence of external electric or magnetic fields the energy levels are

$$E_e(n, l) = -\frac{\mathrm{hc}\,RZ^2}{\{n+a(l)\}^2}, \tag{3.37}$$

where $a(l)$ is a function of l, R is *Rydberg's constant*, and Z the *atomic number*. $a(l)$ is zero for the hydrogen atom, but not for other atoms with a single valence electron. From Planck's relation (3.2), lines will be formed at wave numbers

$$\frac{\nu_{1,2}}{c} = RZ^2\left(\frac{1}{(n_2+a_2)^2} - \frac{1}{(n_1+a_1)^2}\right). \tag{3.38}$$

Any set of spectral lines which can be described by such a formula, with integral values for n_1 and n_2, is a *Rydberg Series*. If $n_1 > n_2$, and n_2 is held constant, then (3.38) converges to a limit as $n_1 \to \infty$. This is the *series limit* and corresponds to the lowest energy state of the ionized atom.

The *magnetic quantum number* m_l defines the quantized component of the angular momentum in the direction of an imposed magnetic field. The electron also has its own angular momentum or *spin* (**s**) and associated quantum number s related together analogously to (3.35). This brings in a second magnetic quantum number m_s which has only two values for a single electron.

For one-electron atoms the designation of spectroscopic terms is simple. The principal quantum number is written first, and then the azimuthal quantum number in the code

$$
\begin{array}{lll}
S & \text{for} & l = 0,\\
P & \text{for} & l = 1,\\
D & \text{for} & l = 2,\\
F & \text{for} & l = 3.
\end{array}
$$

Finally, the *multiplicity*, $(2s+1)$, corresponding to the number of possible values of m_s, is written as a left-hand superscript. Thus $2\,^2S$ designates an energy level with $n = 2$, $l = 0$, and $s = \frac{1}{2}$. Fig. 3.7 shows an energy level diagram for sodium, with the designations of the different terms.

A description of the state of electrons in an atom can be made by listing the spectrographic term values for each separate electron. In doing so the multiplicity is omitted but the number of electrons in each multiplet state is written as a right superscript, and the symbols S, P,

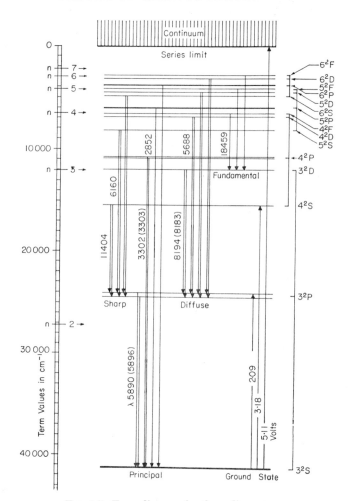

FIG. 3.7. Term diagram for the sodium atom.

The Rydberg series are named Sharp, Principal, Diffuse, Fundamental. Note
the doublet structure of the P-terms. Wavelengths are in Å. The symbols $n2$,
$n3 \ldots n7$ on the left-hand scale show, for comparison, the energy levels of the
H atom. After White (1934).

D, F are written in the lower case. The electron configuration with
lowest energy is determined by *Pauli's principle* which requires that
no two electrons can have the same set of quantum numbers n, l, m_l,
and m_s. As the atomic number increases successive shells specified by
different values of n and l, but all possible m_s and m_l, are filled, giving
stable configurations, and only the electrons in the incomplete outer

shell are of interest spectrographically or as valence electrons. Groups of electrons with one value of n are conventionally referred to as K, L, M, or N shells for $n = 1$, 2, 3, or 4 respectively. For example, atomic oxygen in its lowest or *ground state* of energy has the electron configuration $1s^2 2s^2 2p^4$. The four $2p$ electrons form an incomplete sub-shell, two more electrons being required to complete the $l = 1$ sub-shell of the L-shell; the K-shell however is complete. Nitrogen has one less electron and its ground state configuration is $1s^2 2s^2 2p^3$; singly ionized oxygen† has the same configuration.

Atoms with more than one electron in an incomplete shell combine their orbital angular momenta vectorially to give the angular momentum

$$\mathbf{L} = \sum \mathbf{l}. \tag{3.39}$$

The spin momenta also combine

$$\mathbf{S} = \sum \mathbf{s}, \tag{3.40}$$

and the total angular momentum is

$$\mathbf{J} = \mathbf{S} + \mathbf{L}. \tag{3.41}$$

$\mathbf{J}$, $\mathbf{S}$, and $\mathbf{L}$ are each separately quantized, and specified by quantum numbers J, S, and L by relations analogous to (3.35).

The designation of terms follows that for a single electron, the letters S, P, D, F now representing $L = 0$, 1, 2, 3. The multiplicity $(2S+1)$ is placed as a left-superscript, and the value of J is added as a right-subscript. Sometimes a superscript (o) is added to denote that $\sum l$ is odd (a symmetry condition of the wave function which is involved in the selection rules, see below).

Two selection rules are rigorously obeyed for electric dipole transitions

$$\left.\begin{array}{l} \Delta J = 0, \pm 1 \text{ (excepting } J = 0 \text{ to } J = 0) \\ \sum l \text{ odd} \leftrightarrow \sum l \text{ even} \end{array}\right\}. \tag{3.42}$$

Other restrictions on ΔL, ΔS, and on the individual Δl depend upon the strength of coupling between the various angular momenta.

A diatomic molecule differs importantly from an atom in having an axis of symmetry. The total angular momentum is no longer a conservative quantity and the resolved momentum along the axis of symmetry decides the important properties of the spectroscopic state. The total orbital spin momenta $\mathbf{L}$ and $\mathbf{S}$ are formed by vectorial addition, as for atoms. The component of orbital momentum along the symmetry

---

† An unionized atom is sometimes indicated by the Roman numeral I, e.g. OI. OII, OIII, OIV then represent singly-, doubly-, and triply-ionized atoms.

axis is specified by the quantum number M_L which can vary from $-L$ to $+L$. The sign of M_L does not affect the energy of the state which is therefore specified by a quantum number

$$\Lambda = |M_L|. \tag{3.43}$$

The term is now designated by the symbols Σ, Π, Δ, Φ,..., representing $\Lambda = 0, 1, 2, 3,....$ The component of the spin momentum along the axis of symmetry is represented by the quantum number M_S, which can have $(2S+1)$ values, both positive and negative. $(2S+1)$ is given as a left-hand superscript to the term symbol. Corresponding to the quantum number J is now a new number, formed from components along the symmetry axis, and sometimes added to the term symbol as a right-subscript

$$\Omega = |\Lambda + M_S|. \tag{3.44}$$

Two symmetry conditions are often added to the term symbol. Firstly u or g are used as right subscripts to denote an odd or even character of the sign of the wave function for reflection at the origin, but only for molecules whose two atoms have the same charge. Secondly, for molecules in a Σ state, the sign of the electronic wave function can either change or not change when reflection takes place at a plane passing through both nuclei. $(-)$ or $(+)$ as right superscripts are used to denote these symmetry properties. Thus $^3\Sigma_g^-$ denotes a state with $\Lambda = 0$, $S = 1$ and the symmetry properties $(-)$ and g. This is the term symbol for the ground state of the oxygen molecule. A shorthand for this complex notation identifies the ground state with the letter X, and the various states above the ground state, in order, with the letters A, B, C, D, etc., for states with the same multiplicity, and a, b, c, d, etc., for states with multiplicities differing from that of the ground state. For the oxygen molecule, for example,

$$^3\Sigma_g^- \equiv X,$$
$$^1\Delta_g \equiv a,$$
$$^3\Sigma_u^+ \equiv A, \text{ etc.}$$

Selection rules are exceedingly complicated. The rules (3.42) generally apply with J replaced by the total momentum, and certain relationships between the symmetry of upper and lower states must be observed. However, all other selection rules—and there are many—depend upon the nature of the coupling between the various kinds of angular momentum.

Two ionization continua are of importance. If the ejected electron is from an outer shell an ionization energy between 5 and 20 eV will probably be involved, corresponding to a wavelength between 2480 Å and 620 Å. Since the translational energy is unquantized, absorption is continuous for wavelengths short of the ionization edge. The ion has, however, quantized energy states and the continuum may show corresponding features.

The second type of ionization continuum corresponds to the ejection of an electron from a completed inner shell. Since these electrons are very strongly bound the ionization energy is high, and the absorption edges appear in the X-ray region (1 to 300 Å).

3.5. Interaction of electronic, vibrational, and rotational energies

To a first approximation the three forms of internal energy are additive. A typical term scheme is shown in Fig. 3.8. If the energies were strictly independent the selection rules for a combined transition could be regarded as a combination of the three separate selection rules, except that where symmetry properties are involved (e.g. the parity rule) the symmetry of the complete wave function must be considered.

A vibration-rotation transition will give a band consisting of P-, Q-, and R-branches spaced about the vibrational frequency ν_0. All three branches can occur, negative jumps in rotational energy giving lines at frequencies lower than ν_0 and vice versa. An electronic band of a molecule with one vibrational mode ν_0 will consist of an array of vibration-rotation bands spaced by multiples of ν_0 from the electronic band centre. Transitions with a lower vibrational state in common are known as *progressions*, while the group of transitions involving the change of a given number of vibrational quanta is a *sequence*.

In practice bands resulting from combined transitions differ markedly from expectation on a non-interacting model. This is partly because of the need to consider the total wave function where symmetry rules are involved, e.g. the total angular momentum $\mathbf{J}$ includes all types of angular momentum, and it is to this that the selection rule (3.15) must be applied. There are, however, also energy interactions between electronic, vibrational, and rotational states which strongly influence the resulting spectra.

The classification of these interactions is more a historical matter than anything else. Differences from the simple non-interacting model were discovered experimentally and a separate theoretical treatment was developed for each. All forms of interaction, including anharmonicity in vibrations and centrifugal stretching of rotating molecules, should logically be treated together. Following normal convention, however, we distinguish first a group of interactions which causes the rotational constants to depend slightly upon the vibrational quantum number. This has the effect of separating slightly lines in Q branches and distorting the structure of P- and R-branches. This will be discussed

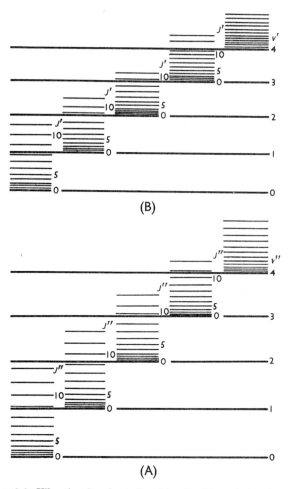

Fig. 3.8. Vibrational and rotational levels of two electronic states A and B of a molecule (schematic).

After Herzberg (1950).

in relation to a vibration-rotation band, involving no change of electronic energy.

We may write the rotational constant formally as a function of the vibrational energy in the following way

$$B_\mathrm{v} = B_\mathrm{e} - \sum_i \alpha_i(v_i + \tfrac{1}{2}d_i), \tag{3.45}$$

where the i suffixes denote different vibrational modes, v_i are the vibrational quantum numbers, d_i are the degeneracies, and B_e is the rotational constant for the hypothetical case of no vibrational energy. The expression in parentheses is proportional to the vibrational energy. We can now distinguish three reasons why the α_i are not zero. Firstly, the rotational constant is not a linear function of the nuclear separation,

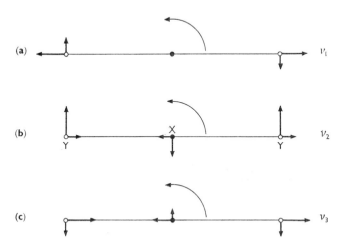

FIG. 3.9. Coriolis interactions in a linear XY_2 molecule.

The curved arrow indicates the direction of rotation. The heavy arrows show the nuclear motions, and the small arrows give the resultant Coriolis forces (orthogonal to the translational and rotational vectors). The Coriolis forces on ν_2 tend to excite ν_3 and vice versa. If ν_2 and ν_3 have similar, or close multiple frequencies, the interaction can be large. After Herzberg (1945).

and its value will vary with the amplitude even for a harmonic vibration. Secondly, anharmonicities will cause a shift in the mid-point of the vibration. Thirdly, there is the *Coriolis interaction* between vibrational states, which gives a coupling depending upon the rotation. The principle of the Coriolis interaction between ν_2 and ν_3 for a linear molecule is illustrated in Fig. 3.9. The Coriolis force (as for winds on the surface of the earth) is orthogonal both to the linear motion of the nuclei and to the rotation. The figure shows how this can couple one normal mode of vibration with another.

A second group of interactions give rise to what are known as *perturbations*. These are irregular disturbances indicating a close resonance. For example, the combination of vibrational and rotational energy may lead to very close resonance between two particular rotational sublevels of two different vibrational modes. If these levels can interact, either through the anharmonic cross terms in the vibration (*Fermi interactions*), or Coriolis forces, the result will be an anomalous shift of the two levels. Alternatively two entire vibrational modes may resonate because of anharmonic terms; this is the Fermi resonance already mentioned in § 3.3.

For a molecule with the energy levels (3.12), the selection rule (3.15), and ignoring for the moment other selection rules associated with symmetry, we arrive at the following line positions in wave numbers when the rotational constants differ in upper and lower states,

$$\left.\begin{aligned}P(J'') &= \nu_0/c-(B'+B'')J''+(B'-B'')J''^2 \\ Q(J'') &= \nu_0/c+(B'-B'')J''+(B'-B'')J''^2 \\ R(J'') &= \nu_0/c+2B'+(3B'-B'')J''+(B'-B'')J''^2\end{aligned}\right\}. \qquad (3.46)$$

In (3.46) the lines are labelled according to an accepted convention, with the letter P, Q, or R to define the branch and the J-value of the initial state in parentheses. This nomenclature obscures the fact that the lines $P(J+1)$ and $R(J)$ involve the same pair of rotational levels. We will see in § 3.6 that some properties of a line depend mainly on the two rotational levels involved; it is therefore convenient to define a parameter

$$\begin{aligned}m &= -J \quad \text{for } P\text{-branches,} \\ &= J+1 \quad \text{for } R\text{-branches.}\end{aligned} \qquad (3.47)$$

The modulus $|m|$ then defines the pair of levels and m itself provides a continuous numbering system for lines throughout the band.

We will now consider individually bands arising from the different types of rotator discussed in § 3.2. For linear polyatomic molecules the vibrational angular momentum quantum number plays an important role, imposing the selection rule (3.34) on the vibrational transitions, but additionally forbidding transitions $\Delta J = 0$ when $l = 0$ in both upper and lower states. Some bands therefore have Q-branches and some have not, markedly affecting the general appearance of the band. Two categories are distinguished:

(i) Σ–Σ transitions have $l = 0$ in both upper and lower states. These are sometimes called parallel bands and have no Q-branches. The P- and R-branches are regular arrays of lines with nearly constant spacing, and a gap between them at $\nu = \nu_0$. Intensities are not quite symmetrical about the band centre because the first lines on either side ($R(0)$ and $P(1)$) have initial states with different populations. The ν_1 band of N_2O is a typical Σ–Σ band; a tracing of a section of the band is shown in Fig. 3.10. Nitrous oxide has no centre of symmetry, and spin factors

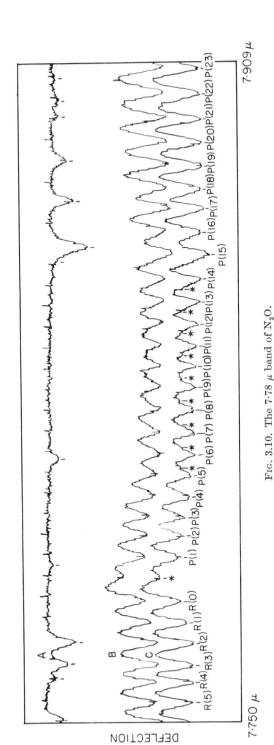

DEFLECTION

7·750 μ

7·909 μ

R(5) R(4) R(3) R(2) R(1) R(0)

P(1) P(2) P(3) P(4) P(5)

P(6) P(7) P(8) P(9) P(10) P(11) P(12) P(13) P(14)

P(15)

P(16) P(17) P(18) P(19) P(20) P(21) P(22) P(23)

A

B

C

Fig. 3.10. The 7·78 μ band of N_2O.

This is the ν_1 band. The amount of nitrous oxide in runs B and C is equivalent to a 1·6 cm path of pure N_2O at s.t.p.; A is recorded with no N_2O and shows unwanted H_2O lines. The lines marked with an asterisk (∗) are upper state lines with an excited ν_2 level. One such line falls fortuitously in the central gap between $R(0)$ and $P(1)$. After Migeotte, Neven, and Swensson (1957).

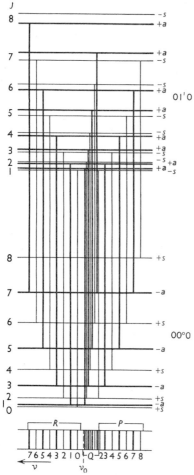

FIG. 3.11. The spectrum of an $(01^10) \rightarrow$ (00^00) transition of a linear molecule. The levels are drawn as if $B' = B''$ but the spectrum below is not based upon this assumption. In the upper state the l-type doubling is exaggerated. J, the total angular momentum quantum number, cannot be less than l; hence $J = 0$ is missing in the upper state, for which $l = 1$. Parity and nuclear symmetry are shown on the right; a possible intensity alternation caused by nuclear spin factors is indicated by the thickness of the levels and spectral lines. After Herzberg (1945).

do not enter into the intensity. Σ–Σ carbon dioxide bands, however, have alternate lines missing because the two, symmetric, oxygen atoms have zero nuclear spin. If $\Delta l = 0$ but $l \neq 0$, there results a parallel band which is not a Σ–Σ band, and it will show a very weak Q branch. No important fundamentals are of this type although some difference bands are.

(ii) $\Delta l = \pm 1$ leads to the perpendicular bands. A Q-branch is permitted and it is stronger than the P- and R-branches combined. The build-up of a schematic spectrum of a perpendicular band is illustrated in Fig. 3.11. An important example is the ν_2 band of carbon dioxide at $15\,\mu$.

We will not consider further the class of symmetric tops since there are no important atmospheric examples. A few words about the spherical top are appropriate, however, since methane falls into this class.

Discussion of pure rotation spectra indicated similarities between linear and spherical molecules, only one angular momentum being quantized in both cases. Thus, at moderate resolution, the ν_3 band of methane looks like a perpendicular band of a linear molecule, except that the statistical weight of the J-level is $(2J+1)^2$ rather than $(2J+1)$ and the J-dependence of

the intensity differs. This apparent simplicity results from a high degree of degeneracy, and is deceptive. At high resolution the lines in the ν_3

band are seen to be split by Coriolis interactions. For the ν_4 band of methane, at $7.65\,\mu$, Fermi resonance and Coriolis interactions give rise to a band with no obvious regularities to its structure at all.

The question as to whether a transition for an asymmetric top is of type A, B, or C is now no longer a matter of the direction of the permanent dipole moment as for a rotation band but of the alternating dipole moment induced by the vibration. The intermediate moment of inertia for water vapour lies along the axis of symmetry so that the ν_1 and ν_2 fundamentals are type B bands, while ν_3 is a type A band (see Fig. 3.12). The appropriate selection rules have been given in equations (3.30) and (3.31). Selection rules governing parity and a- and s-symmetry refer to the symmetry of the complete wave function and the statistical weight of a given J-level varies from band to band.

Type A bands, viewed under low resolution, show an increased intensity in the centre which is mainly caused by Q-branch lines, although such lines also occur in other parts of the band. Type B bands, while possessing P-, Q-, and R-branches, show no intensification near the centre. Thus the important ν_2 fundamental of H_2O has, under low resolution, a simple doublet structure. Despite the complexity of the asymmetric top the agreement between theory and observation can be very good. Fig. 3.13 shows observed and predicted spectra of the H_2O ν_2 band at $6.26\,\mu$.

The combination of electronic with vibrational and rotational changes introduces additional complications. Vibrational frequencies and rotational constants depend strongly upon the electronic state. This destroys the regularity of simple sequences and progressions and of P- and R-branches. Non-linear terms in the energy may become so important that rotation lines double back forming *band heads*, with all the lines either at longer or shorter wavelengths. The greatest complication relates to the degree of coupling between the different angular momenta. While the selection rules remain the same, the angular momenta to which they refer depend a great deal on the nature of the coupling. If the electronic orbital and spin momenta are strongly coupled then the total angular momentum $\mathbf{J}$ is of major importance. This follows the usual selection rules for the total momentum. If spin and orbital momenta are weakly coupled, however, nuclear and the orbital momenta combine to form a quantized resultant $\mathbf{K}$ (quantum number K). The selection rules for K are then of major importance. These are similar to those for J except that $\Delta K = 0$ is forbidden for $\Sigma \to \Sigma$ transitions. Other angular momenta are defined in special cases,

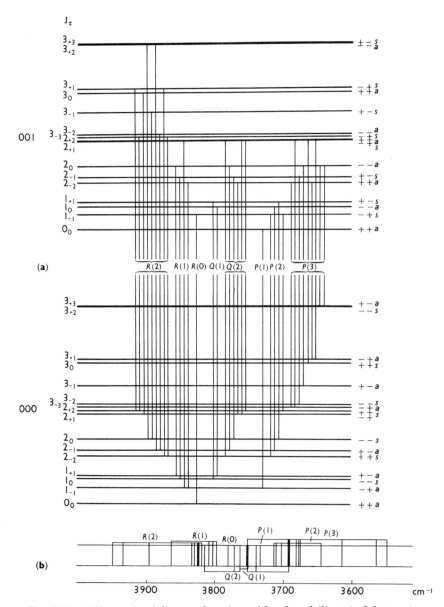

FIG. 3.12. (a) Energy level diagram for a type A band, and (b) part of the spectrum. The spacing of the rotational levels simulates the ν_3 fundamental of H_2O. Note the great complication arising from four J-values only. Intensities are not indicated in (b), and there is a very large variation among the particular lines shown. After Herzberg (1945).

and there are appropriate selection rules. In addition, there are all the
normal symmetry restrictions on transitions.

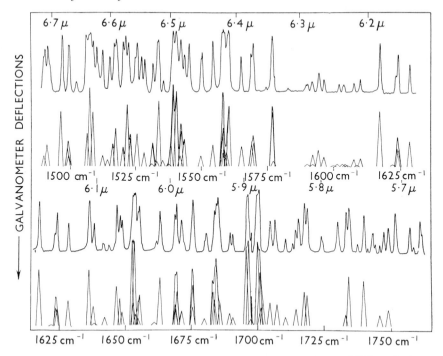

FIG. 3.13. Observed and predicted fine structure of the ν_2 band of H_2O at 6·26 μ.

The upper, smooth curves are the observed spectrum; the triangles represent computed
lines. Even more complete agreement with spectra of higher resolution has been obtained
by Dalby and Nielsen (1956). After Nielsen (1941).

The final result is too complex for brief discussion but many of the
properties of a mixed electronic, vibrational, and rotational transition
are typified by the N_2 band in Fig. 3.14.

The interaction of vibrational and electronic energies can best be
described in terms of the potential curves of the molecule. Fig. 3.15
shows potential curves for the known states of the $O^{16}O^{16}$ molecule.
Each curve shows the work required to produce a given nuclear separation for one electronic state. A minimum in such a curve indicates
a stable equilibrium separation (r_e). A vibrational level is represented
by a horizontal straight line, with the correct energy excess over the
minimum of the potential curve, the intersections of this line with the
potential curve giving the amplitude of the vibration.

As the internuclear distance becomes large the potential energy tends
to a limit, representing the energy of two nuclei or groups of nuclei as

independent entities, or in other words, the energy of a dissociated molecule. The dissociation products may be in excited states and there are several dissociation energies; two are shown in the lower half of

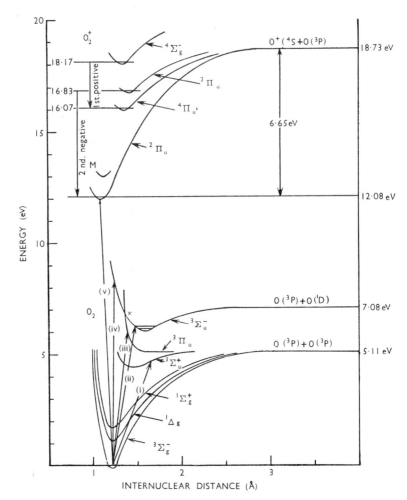

Fig. 3.15. Potential curves for observed states of the $O^{16}O^{16}$ molecule.

Some observed transitions are indicated by arrows; (i) gives the Herzberg bands, (ii) the Schumann–Runge bands. At the nuclear separation indicated by X a resonant transition $^3\Sigma_u^- \to {}^3\Pi_u$ can take place which would be immediately followed by dissociation to $2O(^3P)$ (*predissociation*). After Watanabe (1958).

Fig. 3.15. The upper half of Fig. 3.15 shows the potential curves for ionized oxygen, with a minimum energy 12·08 eV higher than the minimum for the unionized molecule. Transitions to a number of

FIG. 3.14. The fine structure of the $^3\Pi_u \rightarrow {}^3\Pi_g$ ($v = 0 \rightarrow v = 2$) band of N_2 (2nd positive group). The head of the ($v = 1 \rightarrow v = 3$) band is shown on the right. The short leading lines show the triplets of the P-branch, while the long lines show the R-branch. The numbers are the K-values. After Herzberg (1950).

ionized states are possible, and internal transitions of the ionized molecule take place in the same way as for the unionized molecule.

3.6. The shape of a spectral line

3.6.1. *Doppler broadening*

The preceding sections have treated transitions between discrete energy levels, involving strictly monochromatic quanta. Emission or absorption is not concentrated at a single frequency, however, and this deviation from monochromatism is important in atmospheric computations. Let us write for one line

$$k(\nu) = Sf(\nu - \nu_0), \tag{3.48}$$

where

$$\int_{-\infty}^{+\infty} f(\nu - \nu_0) \, d(\nu - \nu_0) = 1.$$

ν_0 is the frequency of an ideal, monochromatic line, and S is the line intensity. Our problem is to determine the *shape factor* $f(\nu - \nu_0)$.

The most ubiquitous source of line broadening is the Doppler effect, associated with molecular thermal motions. If an absorbing or emitting molecule has a component of velocity u in the line of sight (the line joining the molecule and the observer) then, provided $u \ll \mathbf{c}$, the Doppler shift can be written

$$\Delta\nu = \frac{u}{\mathbf{c}}\nu_0. \tag{3.49}$$

If a stationary molecule has a shape factor $f'(\nu - \nu_0)$ and all the molecules have a line-of-sight velocity u, then the observed $f(\nu - \nu_0)$ will be $f'\{\nu - \nu_0 - (u/\mathbf{c})\nu_0\}$. Let the probability that the velocity component lies between u and $u + du$ be $p(u)\,du$. The contribution to the shape factor at ν from all Doppler-shifted components is

$$f(\nu - \nu_0) = \int_{-\infty}^{+\infty} p(u) f'\left(\nu - \nu_0 - \frac{u}{\mathbf{c}}\nu_0\right) du. \tag{3.50}$$

If the translational states are in thermodynamic equilibrium, $p(u)$ is given by the Maxwell distribution

$$p(u) = \left(\frac{m}{2\pi \mathbf{k}\theta}\right)^{\frac{1}{2}} \exp\left(-\frac{mu^2}{2\mathbf{k}\theta}\right), \tag{3.51}$$

where m is the mass of the molecule.

Both $p(u)$ and $f'(u)$ have a single maximum. The effective width of (3.51) is $\Delta u = \sqrt{(2k\theta/m)}$ (giving a decrease of $p(u)$ to e^{-1} of its peak value). The width of f' is characterized by the line width α of the stationary molecule. From (3.49) we can interpret this as an effective velocity dispersion $\Delta u = \alpha c/\nu_0$. The general behaviour of the integrand

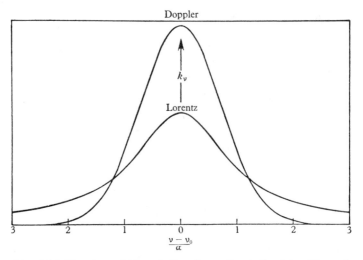

FIG. 3.16. Doppler and Lorentz line shapes for similar intensities and line widths.

of (3.50) can therefore be expected to be governed by the non-dimensional parameter

$$d = 2\frac{\alpha}{(\nu_0/\mathbf{c})(2k\theta/m)^{1/2}}, \tag{3.52}$$

the factor 2 being included by some authors and not by others for reasons of convenience, which are irrelevant here.

If $d \ll 1$, $f'(u)$ is a narrow function compared to $p(u)$, and in the limit can be treated as a delta function which replaces u by $\{(\nu - \nu_0)/\nu_0\}\mathbf{c}$. There results

$$f(\nu - \nu_0) = \frac{1}{\alpha_D\sqrt{\pi}}\exp\left\{-\left(\frac{\nu - \nu_0}{\alpha_D}\right)^2\right\}, \tag{3.53}$$

where

$$\alpha_D = \frac{\nu_0}{\mathbf{c}}\left(\frac{2k\theta}{m}\right)^{\frac{1}{2}}, \tag{3.54}$$

is the *Doppler width* of the line. The multiplying factor in (3.53) has been chosen to satisfy the normalization condition (3.48) on f. A graphical representation of the Doppler line shape is shown in Fig. 3.16.

Doppler widths vary with m and ν_0 and two extreme values for atmospheric lines are:

(i) the 5577 Å forbidden line of atomic oxygen, at 300° K,

$$\alpha_D = 3\cdot3 \times 10^{-2}\ \text{cm}^{-1};$$

(ii) a line near 200 cm^{-1} in the water-vapour rotation band, at 300° K,

$$\alpha_D = 3\cdot5 \times 10^{-4}\ \text{cm}^{-1}.$$

These figures may be compared with typical collision line widths (see § 3.6.4) of 0·08 cm^{-1} at s.t.p. which are proportional to the air pressure. Doppler and collision widths are equal at a pressure of about 0·41 atmospheres in case (i) and $4\cdot4 \times 10^{-3}$ atmospheres in case (ii), or at altitudes in the earth's atmosphere of about 7 km and 37 km respectively. Above these altitudes (3.53) can, under certain circumstances, provide a useful approximation to the line shape, but the criterion is not a simple one as will be discussed in § 4.2.3.

If $d \gg 1$, $p(u)$ is a very narrow function compared with $f'(u)$, and acts as a delta-function making the identical substitution $u \to u$. The effect of molecular motions can then be neglected, and

$$f(\nu - \nu_0) = f'(\nu - \nu_0). \tag{3.55}$$

3.6.2. *Natural broadening*

A molecule remains in an excited state until a downward transition is induced by interactions with other molecules or quanta, or until spontaneous decay takes place. We can be certain that the actual lifetime will not be longer than the natural lifetime, and according to Heisenberg's uncertainty principle (3.9) the width of the observed spectral line cannot be less than the *natural width*

$$\alpha_N = \frac{\Delta\nu}{2} \simeq \frac{1}{4\pi\phi}. \tag{3.56}$$

In practice, α_N is negligible, but it is worth a brief discussion in its role as a fundamental limit.

On the basis of quantum theory, the probability that an atom or molecule in the ith energy level has an energy between E and $E+dE$ has been shown to be

$$p_i(E)dE = \frac{(1/\mathbf{h}\phi_i)dE}{(2\pi/\mathbf{h})^2(E-E_i)^2+(1/2\phi_i)^2}. \tag{3.57}$$

The ground state of an atom or molecule, has by definition, an infinite lifetime ($\phi = \infty$). When a transition takes place between the excited state and the ground state, there is a probability $p_i(E)$ that a quantum of frequency $\nu = E/\mathbf{h}$ is emitted or absorbed. Writing $\nu_0 = E_i/\mathbf{h}$ in (3.57),

$$f(\nu-\nu_0) = \frac{1}{\pi}\frac{\alpha_{\mathrm{N}}}{(\nu-\nu_0)^2+\alpha_{\mathrm{N}}^2}, \qquad (3.58)$$

where

$$\alpha_{\mathrm{N}} = \frac{1}{4\pi\phi_i}. \qquad (3.59)$$

(3.58) is known variously as the Natural, Dispersion, or Lorentz profile. In Fig. 3.16 this line shape is compared with the Doppler profile. The important difference between the two lies in the wings, which cut off sharply for the Doppler line as opposed to the slower variation as $(\nu-\nu_0)^{-2}$ for the Lorentz profile. Different names are used for α_{N} in the literature, and the reader is warned to be sure what is meant by the term *line width* in any paper whose numerical results are used. We will consistently use the words as applied to the Lorentz profile to mean half the frequency width to half peak intensity.

In § 2.2.4 we found a typical value of ϕ for a vibrational transition to be of the order of 0·1 s. From (3.59) α_{N} is therefore of the order of $1 \ \mathrm{s}^{-1}$ in frequency units or $3\times10^{-11} \ \mathrm{cm}^{-1}$ in wave number units. This is trivial compared with the Doppler widths mentioned in the last paragraph, and is negligible in atmospheric problems.

3.6.3. *Combined Doppler and natural broadening*

While natural broadening is unimportant in our problems, its combination with Doppler broadening has been studied extensively for astrophysical purposes, and we will briefly examine this work because it happens also to have applications to pressure-broadened lines.

Substituting (3.58) for $f'(\nu)$ in equation (3.50) we find

$$f(\nu-\nu_0) = \int_{-\infty}^{+\infty} \frac{\alpha}{\pi}\frac{du}{(\nu-\nu_0-u\nu_0/\mathbf{c})^2+\alpha^2}\left(\frac{m}{2\pi\mathbf{k}\theta}\right)^{\frac{1}{2}}\exp\left(\frac{-mu^2}{2\mathbf{k}\theta}\right). \qquad (3.60)$$

This expression has been evaluated numerically by a number of writers. The two asymptotic forms which we discussed in § 3.6.1 do not have quite the simple significance that the discussion might have indicated. The size of d only defines the relative sharpness of $p(u)$ and $f'(u)$ near to their maxima. Even for $d \ll 1$ there is always a point in the far wings of a Lorentz line beyond which variations of intensity are small over one Doppler line width. Thus the line can have a Doppler core and Lorentz wings. However, for $d \gg 1$ the Lorentz form is correct at all frequencies.

Various expansions and computations of (3.60) have been made. Tabulations add little to the discussion which has been given above, however, and the original literature as indicated in Table 3.2 can be consulted, if required.

TABLE 3.2

Computations of the combined Doppler and Lorentz line shape

Author	Date	Nature of contribution
Zemansky . . .	1930	Tables for $d = 1, 2$, and 3
van der Held. . .	1931	Table for $d = 1$ and complete numerical solution for $d \leqslant 0.1$
Born	1932	Expansions for $d \gg 1$. Tables for $d = 1, 2, 4$, and 20
Mitchell and Zemansky .	1934	Expansions for $d < 0.01$
Hjerting . . .	1938	Tables for $d = 0 \ (0.02) \ 0.4$, $\nu/\alpha_D = 0 \ (0.25) \ 5 \ (0.5) \ 20$
Harris	1948	Expansions in ascending powers of d, for $d < 1.0$

3.6.4. *Pressure broadening*

3.6.4.1. *The phase-shift approximation*

The effect of mutual interference by neighbouring air molecules on absorption and emission increases with the pressure, and the resulting phenomenon is usually called *pressure broadening* or *collision broadening*. Before going into details, we will develop a general mathematical framework, which is conveniently stated in terms of a Fourier integral, and we will make one assumption which is common to all modern theories. In doing so, we will use a mixture of quantal and classical ideas which can be justified without difficulty if so desired. The final result can be stated in either classical or quantal terms.

If we neglect Doppler effects and natural broadening, an isolated molecule will emit and absorb one frequency only. In classical terms this means that the absorbed or emitted wave train is simple-harmonic, and that it is infinite in duration. Suppose the amplitude of the wave train to be $a(t)$, where t is the time. This monochromatic wave can be represented by

$$a(t) = a(0)e^{2\pi i \nu_0 t}, \tag{3.61}$$

where ν_0 is the frequency of the undisturbed oscillation.

If the absorbing or emitting molecule interacts with other molecules, all of which are in motion, the energy levels become time-dependent, and (3.61) is no longer correct. Instead, we are concerned with a wave train which is *frequency modulated* and *amplitude modulated*, i.e. ν_0 and $a(0)$ are both functions of time.

Subject to certain reasonable conditions, $a(t)$ can be described by a spectrum

$$a(\nu) = \frac{1}{\pi} \int_{-\infty}^{+\infty} a(t) e^{-2\pi i \nu t} \, dt. \tag{3.62}$$

An hypothesis common to most modern treatments is that we are concerned with frequency rather than amplitude modulation. This is the *phase-shift* approximation.

With this approximation

$$a(t) = a \exp\left(2\pi i \int_{0}^{t} \nu(t') \, dt'\right), \tag{3.63}$$

where a is a constant. It is worth keeping in mind how such a time-dependent frequency can arise. From Planck's relation

$$\nu_{ij} = \frac{E_i - E_j}{h} = \frac{\Delta E_{ij}}{h}. \tag{3.64}$$

Thus if ν_{ij} is to vary with time, so must the energy gap between the two levels involved, which means in general that both E_i and E_j must vary, but by different amounts.

Now write
$$\nu(t) = \nu_0 + \nu'(t), \tag{3.65}$$

and
$$\eta(t) = \int_{0}^{t} \nu'(t') \, dt'. \tag{3.66}$$

The intensity of a wave train is proportional to the square of the amplitude, and the spectral distribution of intensity is from (3.62), (3.63), and (3.66)

$$I(\nu) = a(\nu) a^*(\nu) \propto \int_{-\infty}^{+\infty} dt \int_{-\infty}^{+\infty} dt' \exp[2\pi i (t - t')(\nu_0 - \nu) + 2\pi i \{\eta(t) - \eta(t')\}], \tag{3.67}$$

where a^* is the complex conjugate of a.

After some straightforward manipulation (3.67) can be reduced to the following form which is a formal statement of the phase-shift approximation:

$$I(\nu) \propto \mathscr{R} \int_{-\infty}^{+\infty} dt \exp 2\pi i (\nu_0 - \nu) t \, \overline{\exp 2\pi i \eta(t)}, \tag{3.68}$$

where $\mathscr{R}$ denotes the operation of taking the real part and the overbar denotes an average over all starting times.

No complete solution of (3.68) has been made which takes into account every feature of an encounter between two molecules. All existing treatments involve approximations, which fall into a number of general classes, depending upon the nature of the collision. One approximation common to all treatments is to assume that collisions are binary, which is a good approximation for all pressures less than 1 atmosphere.

As two molecules approach one another, they interact through the electrons in their outer shells. The interactions are always complex although models involving dipole-dipole, dipole-quadrupole, or dipole-induced dipole are employed. Whatever the nature of the interaction it will depend upon the angle for all but spherically symmetric molecules, with the result that perturbations will depend upon rotational quantum numbers. Thus line broadening, insofar as it is induced by such interactions, will necessarily differ from line to line. When we consider Anderson's work (see § 3.6.4.2), we will see that there are additional reasons for widths to vary from line to line.

When two molecules interact they are said to have made an *encounter*. The encounter is *strong* or *weak*, according to whether the total change of phase is large or small. From (3.68), $I_\nu \to 0$ as $\eta \to \infty$, and the Fourier integral may simply be broken off when a strong encounter takes place; not so when the encounter is weak.

In a weak encounter, a certain phase change η_i takes place during the whole of the ith encounter. Since the number of encounters increases linearly with time it follows that $\eta(t)$ must contain a term which is real, and linear in t. As far as the integral (3.68) is concerned such a term has precisely the same effect as changing ν_0. Thus weak encounters necessarily give rise to a shift of the central frequency (*line shift*).

When a *collision* in the kinetic theory sense occurs we have a strong encounter. The theory of line broadening, as originally developed, envisaged kinetic collisions alone as important. Early work in this area is usually associated with the name of Lorentz (1906), although Michelson (1895) can more properly be claimed to be the originator. In a framework of classical theory, used by these authors, it is, however, difficult to see how strong encounters can take place unless accompanied by a complete range of weaker encounters. On a classical basis therefore it is difficult to find justification for the theory of strong encounters, although there is experimental evidence to suggest that unshifted lines often occur in the infra-red spectrum as predicted by the theory.

Quantum mechanical considerations are required to explain this apparent contradiction. If the interaction energy is introduced into the Schrödinger wave equation there is a calculable probability that transitions will take place in both emitting and colliding molecules. If the energy jumps in the two molecules are of opposite sign, and happen to be of nearly the same magnitude, we have near-resonance and transitions are highly probable, even though the energy of interaction is slight. When a transition takes place, the emission or absorption process is terminated and we have an analogy to a strong encounter, there having been almost no interaction and phase change up to this point, and consequently negligible shift of the line centre.

An encounter in which transitions are induced is called *non-adiabatic*, as opposed to *adiabatic* encounters, which only distort the energy levels. Weak encounters are essentially adiabatic, and the theory of weak encounters becomes increasingly applicable as energy jumps become larger and further from resonance, as for example in optical atomic spectra.

The results for adiabatic and non-adiabatic theories tend to differ considerably. In the latter class we will discuss Anderson's theory. Non-adiabatic theories are the most pertinent to atmospheric studies, for line shapes are important primarily in the infra-red spectrum where energy gaps are small and transitions take place readily. Most work in the field of line shapes, like Lindholm's (§ 3.6.4.3), attempts to explain optical spectra, where energy gaps are large and transitions difficult. We only consider Lindholm's theory because it is complete in some respects where Anderson's has not been so fully explored.

The final two classifications of theory cut across the above classes. These are the *statistical theories* and *impact theories*. The former assume that the molecules are stationary, and take account only of the static distribution of energy levels during encounters. A complete theory of strong or weak encounters should take all time into account and will necessarily include a statistical theory as a special case. However, it is more often assumed that the duration of the encounter is negligible compared with the time between encounters. This defines an *impact theory*.

Let us consider (3.68) as it applies to a statistical theory. Since it is a static theory $\nu'(t)$ is constant and $\eta(t) = \nu't$. From (3.64) and (3.65) we can now write (3.68) in the form

$$I(\nu) \propto \mathscr{R} \int\limits_0^\infty dt \exp 2\pi i \left(\frac{\Delta E}{h} - \nu \right) t$$

$$= \delta\left(\frac{\Delta E}{h}, \nu \right). \tag{3.69}$$

We must now average (3.69) over all ΔE, which depends upon the spacing between the absorbing or emitting molecule and its nearest neighbour. If $p(\Delta E)\, d\,\Delta E$ is the probability that ΔE lies between ΔE and $\Delta E + d\Delta E$ then (3.69) shows that $I(\nu)$ is proportional to $p(\mathbf{h}\nu)$. $p(\Delta E)$ is normally a highly asymmetric function of ΔE, and the line shape predicted from a statistical theory is shifted and asymmetric.

Consider, for example, a statistical theory involving a van der Waals interaction. The static frequency shift is

$$\Delta\nu = \frac{\Delta E}{\mathbf{h}} = \frac{-\beta}{r^6}, \tag{3.70}$$

where β is generally unknown. The corresponding statistical line shape is

$$f(\nu-\nu_0) \propto \beta^{1/2} n (\nu_0-\nu)^{-3/2} e^{-(4/9)\pi^3\beta n^2/(\nu_0-\nu)} \quad \text{for } \nu < \nu_0,$$

$$f(\nu-\nu_0) = 0 \quad \text{for } \nu > \nu_0, \tag{3.71}$$

where n is the molecular number density.

Note the $(\nu_0-\nu)^{-3/2}$ dependence in the red (long-wave) wing and the sharp cut-off in the blue wing. The maximum of $f(\nu)$ is shifted to the red by frequency $\Delta\nu = \frac{4}{9}\pi^2\beta n^2$. A more general result can be found in the red wing alone. If (3.70) is replaced by an inverse rth power law, we find in the red wing,

$$f(\nu-\nu_0) \sim (\nu_0-\nu)^{-(r+3)/r}. \tag{3.72}$$

It is interesting to note that for $r = 3$ (3.72) gives, fortuitously, the same line-wing shape as the Lorentz profile.

3.6.4.2. *The theory of strong encounters*

In (3.68) we consider the situation in which there is negligible phase shift before t, at which time a strong encounter occurs, and the integral terminates.

$$I(\nu) \propto \mathscr{R} \int_0^t dt' \exp 2\pi i(\nu_0-\nu)t'$$

$$= \mathscr{R} \frac{1}{2\pi i(\nu_0-\nu)} \big[\exp 2\pi i\{(\nu_0-\nu)t\} - 1\big]$$

$$= \frac{\sin 2\pi(\nu_0-\nu)t}{2\pi(\nu_0-\nu)}. \tag{3.73}$$

This result must now be averaged over all possible values of t. In doing so we will neglect the time spent in collision, and the contribution to (3.68) from this period; the theory thus becomes an impact theory.

According to the kinetic theory of gases, the distance (l) travelled between collisions by a molecule of velocity (u) follows the distribution function

$$p(l)\,dl = \frac{dl}{\bar{l}_u}\,e^{-l/l_u}, \tag{3.74}$$

where $\bar{l}_u$ is the mean-free-path for molecules of velocity u. The distribution function for the time between collisions follows from the substitution $u\,dl = dt$:

$$p(t)\,dt = \frac{u}{\bar{l}_u}\,dt\,e^{-tu/l_u}$$

$$= \frac{dt}{\tau_u}\,e^{-t/\tau_u}, \tag{3.75}$$

where τ_u is the mean time between collisions. Taking the product of (3.73) and (3.75), and integrating over all t we obtain

$$I(\nu, u) \propto \frac{1}{(\nu-\nu_0)^2+(1/2\pi\tau_u)^2}. \tag{3.76}$$

If we normalize this expression in accordance with (3.48),

$$f(\nu, u) = \frac{1}{\pi}\,\frac{\alpha_{\mathrm{L}}(u)}{(\nu-\nu_0)^2+\alpha_{\mathrm{L}}^2(u)}, \tag{3.77}$$

where

$$\alpha_{\mathrm{L}}(u) = \frac{1}{2\pi\tau_u}. \tag{3.78}$$

We will later return to the point that τ_u in (3.78) depends upon the velocity of the emitting or absorbing molecule and that an integration must be made over all velocities. Some writers identify τ_u with $\bar{\tau}$, the mean time between collisions; let us denote by $\bar{\alpha}_{\mathrm{L}}$ the line width resulting from this substitution. Let σ_i be the distance of approach required to bring about a strong encounter between an emitter and a broadener belonging to a class distinguished by the subscript i. From the kinetic theory of gases

$$\bar{\alpha}_{\mathrm{L}} = \frac{1}{2\pi\bar{\tau}} = \sum_i n_i\sigma_i^2\left\{\frac{2\mathbf{k}\theta}{\pi}\left(\frac{1}{m}+\frac{1}{m_i}\right)\right\}^{\frac{1}{2}}, \tag{3.79}$$

where n_i is the number density of the ith class of molecule, and m_i is its molecular mass.

In any mixture of approximately constant composition

$$n_i \propto \frac{p}{\theta}, \tag{3.80}$$

where p is the pressure. Collision line widths from impact theory are therefore proportional to the pressure. In atmospheric problems we are concerned with large ranges of pressure, and this relation is fundamental to our studies.

A molecular diameter, or distance of closest approach, can be derived from viscosity and conductivity measurements. The best value for water-vapour–nitrogen collisions is $\sigma = 3 \cdot 2$ Å, from which $\tilde{\alpha}_L = 0 \cdot 033$ cm^{-1} at s.t.p. A collision capable of transferring kinetic energy and

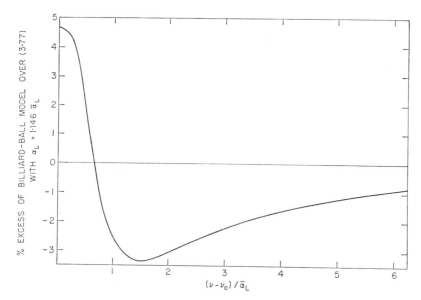

FIG. 3.17. Comparison between a Lorentz line of width $1 \cdot 146 \tilde{\alpha}_L$ and the profile for a billiard-ball model.

momentum should be a strong encounter in the optical sense, and we may expect such 'kinetic theory' widths to be lower limits to the observed widths.

We must now briefly return to the problem of averaging (3.77) over all τ_u. There are conditions under which τ_u is constant, and then (3.77), i.e. the Lorentz profile, should be the actual line shape. In general, τ_u is a function of u, although only a billiard-ball model has been investigated in detail. With tabulated data for this model (3.77) can be averaged with the results shown in Fig. 3.17. The main effect of the averaging is to increase (3.79) by a factor of about $1 \cdot 146$.

A second modification to the impact theory attempts to resolve a contradiction between the Lorentz theory for a line at zero frequency and a theory of static polarization due to Debye. The classical analogue to the foregoing derivation of the Lorentz profile is an oscillatory charge, starting with zero displacement and velocity, absorbing energy from the radiation field. Infinite absorption is prevented by a collision after a time t. If this model is analysed consistently an expression is obtained which differs slightly from (3.77), namely

$$f(\nu) = \frac{1}{\pi} \frac{\nu}{\nu_0} \left(\frac{\alpha_{\mathrm{L}}}{(\nu - \nu_0)^2 + \alpha_{\mathrm{L}}^2} + \frac{\alpha_{\mathrm{L}}}{(\nu + \nu_0)^2 + \alpha_{\mathrm{L}}^2} \right). \qquad (3.81)$$

The difference arises because the phase-shift approximation implicitly assumes α_{L} and $(\nu - \nu_0)$ to be much smaller than ν_0. Under these conditions (3.77) and (3.81) are identical. The difference is only significant in the microwave region of the spectrum.

The derivation can be criticized on another count. If ν_0 is close to zero the initial velocity and displacement of the oscillating charge become important, and a Maxwell distribution should be used. This leads to

$$f(\nu) = \frac{1}{\pi} \left(\frac{\nu}{\nu_0} \right)^2 \left(\frac{\alpha_{\mathrm{L}}}{(\nu - \nu_0)^2 + \alpha_{\mathrm{L}}^2} + \frac{\alpha_{\mathrm{L}}}{(\nu + \nu_0)^2 + \alpha_{\mathrm{L}}^2} \right), \qquad (3.82)$$

sometimes called the Van Vleck and Weisskopf profile.

Microwave evidence points to (3.82) as a good approximation if $(\nu - \nu_0)$ is comparable to ν_0, but it must be at fault for $(\nu - \nu_0) \gg \alpha_{\mathrm{L}}$ because $\int_0^\infty f(\nu) \, d\nu$ diverges, implying infinite energy absorption at high frequencies. The situation is relieved when statistical effects are considered, but this seems unlikely to be the key to this particular contradiction, for we may expect a properly-executed impact theory to give plausible results without appeal to other considerations.

It was pointed out in the previous section that an adiabatic theory of strong encounters is conceptually inconsistent, for we cannot have strong encounters alone without weak encounters. However, the foregoing theory is relevant if a small interaction energy can lead to a large change, e.g. a non-adiabatic transition between quantum states. Anderson (1949) has developed a non-adiabatic theory for molecular spectra, which appears to fit the observations well in the microwave and infra-red regions of the spectrum. It is not of necessity an impact theory, but we will present it in this form.

Let us designate the initial and final states of the emitting or absorbing molecule by i and j; one quantum of radiation then appears or disappears when the transition $i \leftrightarrow j$ takes place. Under the influence of another molecule, other transitions are possible, namely $i \to i'$ and $j \to j'$. When any transition occurs, a strong encounter has taken place.

Following through the consequences of this model, it is found, as expected from the analogy to impact theory, that (3.77), (3.78), and (3.79) still hold, and that the appropriate collision diameter is that which gives an average transition probability of one-half. The collision diameter for a transition from the initial or final state can be written $\sigma(i, i'; j)$ or $\sigma(i; j, j')$ respectively, and the total collision diameter is

$$\sigma(i; j) = \sum_{i'} \sigma(i, i'; j) + \sum_{j'} \sigma(i; j, j'). \tag{3.83}$$

However the transitions $i \rightarrow i'$ or $j \rightarrow j'$ will generally be difficult to bring about and if the perturbing molecule has no quantized levels with similar energy jumps the collision diameters given by (3.83) are negligible in comparison with the 'kinetic theory' collision diameter. If the perturber is a molecule it also has quantized rotational and vibrational energy levels whose totality of quantum numbers will be designated by the symbol J_2. While the encounter is taking place a transition can take place in the perturber $J_2 \nrightarrow J_2'$. If it should happen that either

$$\Delta E_i = (E_i - E_{i'}) + (E_{J_2} - E_{J_2'})$$

or

$$\Delta E_j = (E_j - E_{j'}) + (E_{J_2} - E_{J_2'})$$

is small for any particular value of J_2, the transition is close to resonance, and the probability that it will take place is high. For collisions between molecules, whose rotational levels provide a wide variety of values of $E_{J_2} - E_{J_2'}$, it is almost always possible to find a near-resonant transition. It is fortunate from the point of view of computation that the number of near-resonant transitions is restricted and that some are forbidden by selection rules. Although tedious, it is not a difficult task to compute

$$\sigma(i, j, J_2) = \sum_{J_2', i'} \sigma(i, i'; J_2, J_2'; j) + \sum_{J_2', j'} \sigma(i; J_2, J_2'; j, j'). \tag{3.84}$$

Finally, we must sum over J_2, taking account of the variable number of colliding molecules, $n(J_2)$, in the J_2 states, and from (3.79)

$$\alpha_{i,j} = \sum_{J_2} \left\{ \frac{2\mathbf{k}\theta}{\pi} \left(\frac{1}{m} + \frac{1}{m_0} \right) \right\}^{\frac{1}{2}} n(J_2) \sigma^2(i, j, J_2), \tag{3.85}$$

where m and m_0 are the molecular masses of absorber and perturber. The computation of the approach distance for a transition probability of a half is a straightforward application of quantum mechanical perturbation theory based upon the particular interaction energies involved.

3.6.4.3. *Lindholm's theory*

We will not consider as a separate class the theory of weak encounters. In its simplest form the theory gives a Lorentz line with a line shift proportional to the line width (i.e. proportional to the pressure). Nor will we consider statistical theories as a class by themselves. Instead we will discuss the results of a theory by Lindholm (1945) which is not an impact theory, nor is it restricted to strong encounters. It is an attempt to provide a complete adiabatic theory for a van der Waals interaction, which is not directly applicable to atmospheric problems, but shows how the approximations discussed in § 3.6.4.1 can fit into a single theory.

We can, on general grounds, predict some features of the solution. Frequencies near the line centre are the result of weak interactions; we may therefore expect the line centre to approximate to that resulting from a theory of weak encounters, i.e. a shifted Lorentz shape. The line wings, on the other hand, are formed by the large energy interactions taking place during the encounter; here the line shape should be more closely related to that derived from a statistical theory.

On Lindholm's theory the transition from an 'impact' line centre to a 'statistical' wing takes place near the frequency

$$\nu_p = 0 \cdot 0361 \beta^{-1/5} (\bar{u})^{6/5}, \tag{3.86}$$

where

$$\bar{u} = \left\{ \frac{8k\theta}{\pi} \left(\frac{1}{m} + \frac{1}{m_0} \right) \right\}^{\frac{1}{2}}, \tag{3.87}$$

and β is defined in (3.70). The line width is

$$\alpha = 1 \cdot 34 \beta^{2/5} (\bar{u})^{3/5} n, \tag{3.88}$$

and, like the Lorentz line width, is directly proportional to the pressure. To gain an idea of the orders of magnitude involved, we may substitute observed line widths in (3.88) and hence compute β. Substituting this value of β in (3.86) yields values of ν_p for atmospheric molecules between 1 and 3 cm^{-1}. This, then, is the frequency displacement out to which we may expect impact theories to hold; we will see that experimental evidence tends to confirm this estimate.

We now introduce the dimensionless parameters

$$\epsilon = \alpha/\nu_p, \tag{3.89}$$

and

$$\mu = \frac{\nu - \nu_0 + 0 \cdot 726}{\nu_p}. \tag{3.90}$$

In terms of these parameters the expression

$$f_{\mathrm{L}}(\nu)\,d\nu = \frac{\epsilon\,d\mu}{\pi(\mu^2+\epsilon^2)}, \tag{3.91}$$

corresponds to a Lorentz line which is shifted by $(\nu-\nu_0) = -0\cdot7260\alpha$. We can express Lindholm's results in terms of a modified Lorentz shape

$$f(\mu) = f_{\mathrm{L}}(\mu)F(\mu,\epsilon), \tag{3.92}$$

where $F(\mu,\epsilon)$ can be determined by numerical quadrature. If ϵ is small F becomes a function of μ only. The condition is fulfilled in the atmosphere and a computation for this degenerate case is available in the range $-4 \leqslant \mu \leqslant 4$. The results correspond quite closely to the following approximate algebraic forms:

(i) centre approximation $-2\cdot5 \leqslant \mu \leqslant +1\cdot5$

$$F(\mu) = 1-0\cdot286\mu; \tag{3.93}$$

(ii) red wing approximation $\mu < -2\cdot5$

$$f(\mu) = 0\cdot932|\mu|^{1/2}+0\cdot319|\mu|^{-1/3}; \tag{3.94}$$

(iii) blue wing approximation $\mu > 1\cdot5$

$$F(\mu) = 0\cdot638\mu^{-1/3}. \tag{3.95}$$

The importance of Lindholm's theory lies in this general picture of a line centre given by impact theory and wings dominated by statistical effects.

3.6.5. *Measurements of the widths and shapes of spectral lines in the infra-red and microwave spectrum*

In the visible and ultra-violet spectrum, much of the energy is absorbed in continua; the problem of line absorption is therefore most important in the infra-red spectrum. In this spectral region, the best resolution obtainable with a modern spectrometer is of the order of magnitude of an average line width at s.t.p., and this technical limitation together with difficulties caused by the overlapping of rotation lines precludes any simple and direct test of the foregoing theory. Conditions are more favourable in the microwave region, since almost unlimited resolution is available, and by making measurements at very low pressures the overlap of spectral lines can be minimized. However, microwave spectroscopists have shown little interest in line shapes and the information available is scanty; it is, moreover, not exactly what we require for atmospheric problems, which are rarely concerned with such long wavelengths.

Although detailed confirmation of the theory is lacking, nevertheless many special features have been investigated. The overall picture is encouraging and there seems little reason to doubt that the theory is essentially correct. The following is a brief outline of the evidence supporting this statement.

(i) *Direct measurements of line shape.* Fig. 3.18 shows one of the few detailed tests that have been made of the theory of strong encounters.

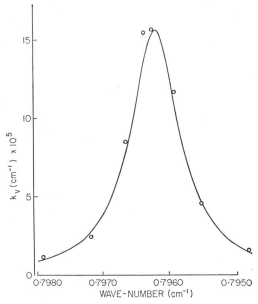

FIG. 3.18. The line $(J, K) = (3, 3)$ of NH_3 at a pressure of 0·5 mm Hg. The full line follows (3.82). After Bleaney and Penrose (1948a).

The points are measured on the $(3, 3)$ rotation line of ammonia, at 0·7962 cm^{-1}, self-broadened at a pressure of 0·5 mm Hg. These measurements were made by microwave techniques and are compared with an unshifted Van Vleck and Weisskopf profile. Collisions with identical molecules are necessarily resonant and the conditions for an unshifted Lorentz line shape are most favourable. Similar agreement between theory and measurement has been obtained on a microwave line of oxygen. In neither of these cases does the confirmation extend to frequencies displaced by more than one wave number from the centre of the line.

(ii) *Tests of special features.* All impact theories predict that at a line centre the absorption coefficient is proportional to α^{-1}, that is to p^{-1}. A sample of pure gas in an absorption tube of fixed length has an optical path at the line centre proportional to the amount multiplied by the absorption coefficient: the optical path is therefore independent of pressure. This simple consequence of the impact theory can be tested with great precision in the microwave spectrum. All experimenters who have reported investigations of this type agree that, in the range of pressures in which impact theory can be expected to hold, the prediction is precisely fulfilled.

Another special feature of impact theory which can be readily tested is the proportionality between line width and pressure. Careful investigations in the microwave spectrum of ammonia have confirmed this feature, excepting at very low and very high pressures, when departures are observed for reasons which are irrelevant to atmospheric problems.

(iii) *Absence of line shift*. According to Anderson's theory there should be no shift of the spectral lines for resonant self broadening, and even for most cases of non-resonant broadening shifts should be small. Measurements on the microwave spectrum of ammonia with a wide variety of broadening gases have not revealed any measurable line shift. Similar measurements on the $0 \cdot 741$ cm^{-1} line of water vapour broadened by air lead to the same result. In the infra-red region of the spectrum an investigation on the 3μ ammonia band has also failed to reveal any line shift. This prediction of the theory is therefore fully confirmed.

(iv) *The shape of the central region*. The *curve of growth* of a single spectral line or the average absorption near to the centre of a band are functions of the profile within one or two wave numbers of the line centre (see Chapter 4). Agreement between measurement and theoretical prediction based upon the Lorentz line shape tends to confirm the validity of the theory of strong encounters. This evidence, which will be discussed at great length in the next chapter, is not decisive. One of the most convincing investigations is that of Benedict, Hermann, Moore, and Silverman (1956) on self-broadening in the vibration-rotation bands of HCl. Although this type of verification is of disappointing precision, it nevertheless provides a very direct test of the type of information which we require for application to atmospheric problems.

(v) *The effect of temperature*. The expression (3.85) for the line width involves the temperature in a number of ways. It appears in the molecular velocity and in the state populations $n(J_2)$. For resonant self-broadening, the theory predicts that the average cross-section varies in such a way that the temperature factors in (3.85) cancel and, in a closed tube where the number of molecules is constant, there should be no temperature-dependence of the line width. This has been confirmed on the $(3, 3)$ microwave line of ammonia. In the infra-red spectrum, the problems of measurement are severe and the only reported results have been on the $7\cdot8\,\mu$ band of nitrous oxide broadened by air. These results suggested a possible increase of collision cross-section with temperature.

(vi) *Doppler effects.* The combination of Doppler and natural broadening has received much attention in the astrophysical literature and agreement between theory and observation is, on the whole, satisfactory. There have, however, been very few tests in the laboratory of combined Doppler and pressure broadening. The theory has been worked out in certain cases (§ 4.2.3) and laboratory results at low pressures where Doppler effects are important have been recorded. The only numerical comparison of theory and experiment, however, appears to be on the ν_3 band of ozone (§ 5.6), and in this case the agreement was satisfactory. The theory of the Doppler line shape is based on such elementary and well-established physical principles that there can be little reasonable doubt as to its correctness, despite the small amount of direct evidence.

(vii) *Calculated line widths.* The work described in the above paragraphs all confirms that, when applicable, the general features of Anderson's theory of pressure broadening are correct, at least as regards the line profile within about 1 cm^{-1} of the centre. The other important consequence of the theory is the possibility of predicting actual line widths. Here the conclusions of the theory can be examined in a critical manner, for, if the Lorentz shape is presumed, it is possible to derive precise figures for the line width from laboratory data.

Fig. 3.19 shows a test of Anderson's theory for P- and R-branch lines of HCl bands. The points show the measurements, and the full line shows what the authors choose to call the resonant-dipole-billiard-ball approximation. This refers to the approximation used for very close collisions, and introduces as a parameter the asymptotic value of the line width for large values of $|m|$. The results for small $|m|$, upon which this approximation has little effect, show excellent agreement with theory; the agreement for large $|m|$ could be greatly improved if the asymptotic collision diameter were chosen for best fit, and not taken from viscosity and conductivity data, as has been done here.

A less exhaustive but probably more important test of Anderson's theory as applied to atmospheric problems is shown in Table 3.3. Only for the $13_5 12_{-1}$ line at $914\cdot03$ cm^{-1} is a separate comparison available

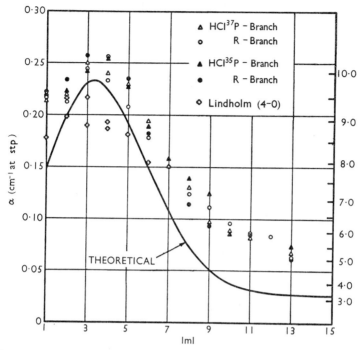

FIG. 3.19. Experimental values of line widths against $|m|$ for self-broadening in HCl (1–0) band.

The full curve is that calculated for the resonant-dipole-billiard-ball approximation. The corresponding collision diameters are given on the right-hand side. After Benedict *et al.* (1956b).

TABLE 3.3

Measured and theoretical parameters for water-vapour rotation lines

Measurements were made in the free atmosphere, i.e. with air as the broadening gas. After Saiedy (1961) and Benedict and Kaplan (1959)

ν_0/c	J'_τ	J''_τ	Measured			Theoretical		
			S	$\alpha(s.t.p.)$	$S\alpha$	S	$\alpha(s.t.p.)$	$S\alpha$
(cm^{-1})			(cm g^{-1})	(cm^{-1})	(g^{-1})	(cm g^{-1})	(cm^{-1})	(g^{-1})
871·32	11_{-2}	10_{-10}	—	—	0·0231($\pm$0·0016)	0·444	0·0469	0·0208
878·61	12_{-4}	11_{-10}	—	—	0·0127($\pm$0·0007)	0·177	0·0674	0·0119
914·03	13_5	12_{-1}	0·031($\pm$0·001)	0·047($\pm$0·002)	0·00147($\pm$0·00007)	0·036	0·0445	0·00160

for both intensity and width. The agreement is good although just outside the limits of the probable error. For the other lines in Table 3.3, only the product of line intensity and width can be compared. Again the agreement is reasonably good, when one considers that the lines arise from transitions with very high J-values for which the theoretical intensities are likely to be in error, and further that a single molecular velocity is assumed in the line-width calculations.

(viii) *The far wings of a rotation line.* The theoretical discussion in § 3.6.4.3 led to the conclusion that at displacements of more than about 2 cm^{-1} from the centre of a rotation line, statistical effects would

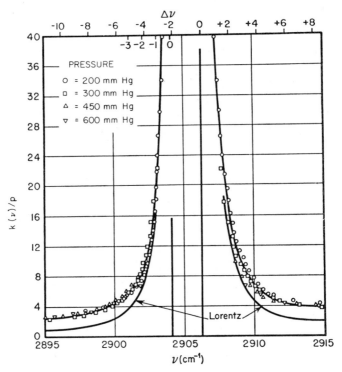

FIG. 3.20. $k(\nu)/p$ for self-broadening as a function of ν near the lines of $R(0)$ (HCl^{35} and HCl^{37}) at high optical density.

The absorption expected from Lorentz lines is also shown. The two vertical lines indicate positions and relative intensities of the two components. After Benedict *et al.* (1956*b*).

become important. The actual prediction made by Lindholm for a van der Waals type interaction should not be given serious consideration; we are completely dependent upon observation for information about far wings of lines.

In Fig. 3.20 are shown measurements on a blend of two HCl lines. Plotting the parameter $k(\nu)/p$ causes all of the points to lie on a single curve. The significance of this observation is that in the far wings of the line, where statistical effects are important, the absorption coefficient is proportional to the pressure (or molecular number density). This is predicted by all statistical theories (see (3.71)) and is a result of great importance both for interpretation of laboratory data and for application to atmospheric problems. In Fig. 3.20 we see that for self-broadening of HCl the wing absorption coefficient is considerably greater than predicted by the Lorentz profile and that the absorption coefficient varies as $(\nu-\nu_0)^{-1.73}$, which from (3.72) could be taken to suggest an inverse fourth-power interaction.

This profile is not valid for all gases to judge by results for self-broadening in the $4\cdot3\,\mu$ band of carbon dioxide. The lines of this band have an exponential cut off in the wings:

$$f(\nu) = \frac{\alpha \exp(-0\cdot135|\nu-\nu_0|^{0\cdot7})}{\pi\{(\nu-\nu_0)^2+\alpha^2\}}. \tag{3.96}$$

For water vapour broadened by air, a few measurements are available of the absorption in line wings as much as 800 cm^{-1} from line centres (see § 5.4.8). These show much more absorption than is indicated by (3.96) but less than is indicated by the observed profile in Fig. 3.20.

BIBLIOGRAPHY

3.1. Introduction
3.2. Rotational energy
3.3. Vibrational energy
3.4. Electronic energy
3.5. Interaction of electronic, vibrational, and rotational energies

There are many good general books on spectroscopy, but the three books of Herzberg are unique in giving a uniform treatment of a large body of fact. They are not easy reading, but are essential to anyone who wishes to go further than the brief remarks of this chapter. The titles below are in order of production and increasing complexity; the dates refer to the latest editions.

HERZBERG, G., 1944, *Atomic spectra and atomic structure.* New York: Dover Publications;
—— 1950, *Molecular spectra and molecular structure. I. Spectra of diatomic molecules.* New York: Van Nostrand; and
—— 1945, *Molecular spectra and molecular structure. II. Infra-red and Raman spectra of polyatomic molecules.* New York: Van Nostrand.

An exhaustive study of rotational spectra, with particular reference to the microwave region is by

TOWNES, C. H., and SCHAWLOW, A., 1955, *Microwave spectroscopy.* New York: McGraw-Hill.

This has a more restricted scope than Herzberg's books and the authors are able to cover the field in much greater detail. For example, there is a good chapter on line shapes, including an account of Anderson's theory, and tables of energy levels and line strengths for the rigid rotator are included in appendixes.

The following book has much in common with Chapters 3, 4, and 5 of the present volume. The main difference is in the point of view, which is concerned with constant-pressure paths at higher temperatures, i.e. the view of a combustion engineer. However, the book makes valuable supplementary reading:

PENNER, S. S., 1959, *Quantitative molecular spectroscopy and gas emissivities.* Reading, Mass.: Addison-Wesley.

The following additional references have provided certain of the figures in the text:

MIGEOTTE, M., NEVEN, L., and SWENSSON, J., 1957, *An atlas of nitrous oxide, methane and ozone infrared absorption bands. Part I. The photometric records.* Institute d'Astrophysique de l'Université de Liège;

NIELSEN, H. H., 1941, 'The near infra-red spectrum of water vapour. I. The perpendicular bands ν_2 and $2\nu_2$', *Phys. Rev.* **59**, p. 515;

NIELSEN, A. H., and NIELSEN, H. H., 1935, 'The infrared absorption bands of methane', *Phys. Rev.* **48**, p. 864;

DALBY, F. W., and NIELSEN, H. H., 1956, 'Infrared spectrum of water vapour. Part I. The 6·26 μ region', *J. Chem. Phys.* **25**, p. 934;

WATANABE, K., 1958, 'Ultra-violet absorption processes in the upper atmosphere', *Advances in Geophysics No. 5.* New York: Academic Press;

WHITE, H. E., 1934, *Introduction to Atomic Spectra.* New York: McGraw-Hill; and

RANDALL, H. M., DENNISON, D. M., GINSBERG, N., and WEBER, L. R., 1937, 'The far infrared spectrum of water vapour', *Phys. Rev.* **52**, p. 160.

Tables of the energy levels of the asymmetric top are given by

KING, G. W., HAINER, R. M., and CROSS, P. C., 1943, 'The asymmetric rotor. I. Calculation and symmetry classification of energy levels', *J. Chem. Phys.* **11**, p. 27.

Matrix elements are given by

CROSS, P. C., HAINER, R. M., and KING, G. W., 1944, 'The asymmetric rotor. II. Calculation of dipole intensities and line classification', ibid. **12**, p. 210; and

SCHWENDEMAN, R. H., and LAURIE, V. W., 1958, *Tables of line strengths for rotational transitions of asymmetrical rotor molecules.* New York: Pergamon.

3.6. The shape of a spectral line

3.6.1. *Doppler broadening*

3.6.2. *Natural broadening*

3.6.3. *Combined Doppler and natural broadening*

Maxwell's theory is treated by (for example)

FOWLER, R. H., and GUGGENHEIM, E. A., 1939, *Statistical thermodynamics.* Cambridge University Press.

The quantum theory of natural broadening is discussed by

WOOLLEY, R. v. d. R., and STIBBS, D. W. N., 1953, *The outer layers of a star.* Oxford University Press.

References to Table 3.1 are as follows:

BORN, M., 1932, *Optik.* Berlin: Springer, § 93;

HARRIS, D. L., 1948. 'On the line-absorption coefficient due to Doppler effect and damping', *Astrophys. J.* **108**, p. 112;

VAN DER HELD, E. F. M., 1931, 'Intensität und natürliche Breite von Spektrallinien', *Z. Phys.* **70**, p. 508;

HJERTING, F., 1938, 'Tables facilitating the calculation of line absorption coefficients', *Astrophys. J.* **88**, p. 508;

MITCHELL, A. C. G., and ZEMANSKY, M. W., 1934, *Resonance radiation and excited atoms.* Cambridge University Press, p. 319; and

ZEMANSKY, M. W., 1930, 'Absorption and collision broadening of the mercury resonance line', *Phys. Rev.* **36**, p. 219.

3.6.4. *Pressure broadening*

A literature of great complexity is reviewed by

BREENE, R. G., 1957, 'Line shape', *Rev. Mod. Phys.* **29**, p. 94;

―― 1961, *The shift and shape of spectral lines*, London: Pergamon;

LINDHOLM, E., 1942, *Über die Verbreiterung und Verschiebung von Spektrallinien.* Uppsala: Almqvist and Wiksells;

MARGENAU, H., 1939, 'Van der Waals' forces', *Rev. Mod. Phys.* **11**, p. 1; and

―― and WATSON, W. W., 1936, 'Pressure effects on spectral lines', ibid. **8**, p. 22.

The theory of strong encounters for a single collision time was initiated by

MICHELSON, A. A., 1895, 'On the broadening of spectral lines', *Astrophys. J.* **2**, p. 251.

Nearly 30 years had to pass before the extension to a realistic distribution of collision times by

LENZ, W., 1924, 'Einige korrespondenzmässige Betrachtungen', *Z. Phys.* **25**, p. 299.

In the meantime a complete theory was given for another purpose by

LORENTZ, H. A., 1906, 'The absorption and emission lines of gaseous bodies', *Proc. R. Acad. Sci.* (Amsterdam), **8**, p. 591.

The correct distribution function of free paths for a billiard-ball model is given by

JEANS, J. H., 1921, *The dynamical theory of gases.* Cambridge University Press.

The modified theory leading to (3.82) is by

VAN VLECK, J. H., and WEISSKOPF, V. F., 1945, 'On the shape of collision-broadened lines', *Rev. Mod. Phys.* **17**, p. 227.

Anderson's theory was originally described by

ANDERSON, P. W., 1949, 'Pressure broadening in the microwave and infra-red regions', *Phys. Rev.* **76**, p. 647.

Further work has been done by many authors; among these are

BENEDICT, W. S., and KAPLAN, L. D., 1959, 'Calculation of line widths in H_2O–N_2 collisions', *J. Chem. Phys.* **30**, p. 388;

SMITH, W. V., and HOWARD, R., 1950, 'Microwave collision diameters. II. Theory and correlation with molecular quadrupole moments', *Phys. Rev.* **79**, p. 132; and

―― LACKNER, H. A., and VOLKOV, A. B., 1955, 'Pressure broadening of linear molecules II', *J. Chem. Phys.* **23**, p. 389.

Lindholm's theory is given in

LINDHOLM, E., 1945, 'Pressure broadening of spectral lines', *Ark. Mat. Astr. Fys. Stockholm* **32**a, No. 17;

and further discussed by

CURTIS, A. R., and GOODY, R. M., 1954, 'Spectral line shape and its effect on atmospheric transmissions', *Quart. J. R. Met. Soc.* **80**, p. 58.

It may be compared with the statistical theory described by MARGENAU (1939).

3.6.5. *Measurements of the widths and shapes of spectral lines*

The following series of papers is concerned with the (3, 3) ammonia line in the microwave spectrum:

BLEANEY, B., and PENROSE, R. P., 1947a, 'The inversion spectrum of ammonia at centimetre wave-lengths', *Proc. Roy. Soc.* A, **189**, p. 358;

—— —— 1947b, 'Collision broadening of the inversion spectrum of ammonia at centimetre wave-lengths. I. Self-broadening at high pressure', *Proc. Phys. Soc.* **59**, p. 418;

—— —— 1948a, 'Pressure broadening of the inversion spectrum of ammonia. Part II. Disturbance of thermal equilibrium at low pressures', ibid. **60**, p. 83;

—— —— 1948b, 'Collision broadening of the inversion spectrum of ammonia. III. The collision cross-sections for self-broadening and for mixtures with non-polar gases', ibid. **60**, p. 540;

—— and LOUBSER, J. H. N., 1950, 'The inversion spectra of NH_3, CH_3Cl and CH_3Br at high pressures', *Proc. Roy. Soc.* A, **63**, p. 483;

HOWARD, R., and SMITH, W. V., 1950a, 'Microwave collision diameters. I. Experimental', *Phys. Rev.* **79**, p. 128; and

—— —— 1950b, 'Temperature dependence of microwave line widths', ibid. **77**, p. 840.

Microwave spectroscopy on atmospheric gases is restricted to one paper on oxygen:

BERINGER, R., 1946, 'The absorption of one-half centimetre electromagnetic waves in oxygen', ibid. **70**, p. 53;

and two on water vapour:

TOWNES, C. H., and MERRITT, F. R., 1946, 'Water spectrum near one-centimetre wave-length', ibid. **70**, p. 558; and

BECKER, G. E., and AUTLER, S. H., 1946, 'Water vapour absorption of electromagnetic radiation in the centimetre wave-length range', ibid., p. 300.

A general review of microwave spectroscopy is provided by

GORDY, W., SMITH, W. V., and TRAMBARULO, R. F., 1953, *Microwave spectroscopy*. New York: John Wiley.

Two important papers on the infra-red spectrum of HCl are:

BENEDICT, W. S., HERMAN, R., MOORE, S. E., and SILVERMAN, S., 1956a, 'The strengths, widths and shapes of infrared lines I', *Canad. J. Phys.* **34**, p. 830; and

—— —— —— —— 1956b, 'The strengths, widths and shapes of spectral lines II', ibid., p. 851.

Line shifts in the ammonia spectrum are discussed by

NETHERCOT, A. H., and PETERS, C. W., 1950, 'The pressure shift of the inversion frequency of ammonia', *Phys. Rev.* **79**, p. 225.

Temperature effects in the 7·8 μ band of nitrous oxide are discussed by

GOODY, R. M., and WORMELL, T. W., 1951, 'The quantitative determination of atmospheric gases by infra-red spectroscopic methods', *Proc. Roy. Soc.* A, **209**, p. 178.

The transition from pressure broadening to Doppler broadening for the 9·6 μ band of ozone is treated by

WALSHAW, C. D., 1955, 'Line widths in the 9·6 μ band of ozone', *Proc. Phys. Soc.* A, **68**, p. 530.

Measurements in the water-vapour window near 1000 cm^{-1} are by

ROACH, W. T., and GOODY, R. M., 1958, 'Absorption and emission in the atmospheric window from 770 to 1250 cm^{-1}', *Quart. J. R. Met. Soc.* **84**, p. 319; and

SAIEDY, F., 1961, 'Atmospheric observations of line intensity and half-width in the rotation band and ν_2 vibration-rotation bands of water vapour', ibid. **87**, p. 578.

Measurements in the far wings of the 4·3 μ carbon dioxide band were made by

BENEDICT, W. S., and SILVERMAN, S., 1954, 'Line shapes in the infrared', *Phys. Rev.* **94**, p. 752:

but the results have been more fully reported by

KAPLAN, L. D., 1954, 'A quasi-statistical approach to the calculation of atmospheric transmission', *Proceedings of the Toronto Meteorological Conference, 1953*, p. 49.

4

BAND MODELS

4.1. Introduction

THE problem of radiative heating in the atmosphere involves four distinguishable scales of frequency-dependence. Firstly, there is the comparatively slow variation with frequency of the Planck function and its derivative with respect to temperature (see Fig. 2.3). About one half of the radiation from a black body at terrestrial temperatures lies in a wave-number range of 500 cm^{-1}.

The second scale is that of the unresolved contour of a band. The water-vapour bands are by far the widest of importance to atmospheric studies and these have a width to half intensity of about 300 cm^{-1}. For molecules other than water vapour the Planck function can usually be treated as a constant for each band; water-vapour bands must be divided into independent sections of the order of 50 cm^{-1} wide before this can be done.

For a rotating molecule, the next relevant scale of frequency is the spacing between rotation lines, approximately 1–5 cm^{-1}. Finally, there is the scale on which Lambert's law of absorption is obeyed; of the order of one-fifth of a line width. This varies from about 2×10^{-2} cm^{-1} for a gas at atmospheric pressure to about 2×10^{-4} cm^{-1} for Doppler broadening. This step takes us simultaneously into great experimental difficulties and to a division of the frequency scale which, when taken together with other features of the problem, presents a formidable computational task even for the fastest electronic computers.

Both for interpreting measurements and computing numerical problems it has therefore proved desirable to develop techniques for averaging over the finest structure. A few line spacings is a sufficient range of averaging to smooth out lines without distorting the band contour. This leads to the idea of a *band model*.

The abstraction employed is that of an infinite array of absorption lines of uniform statistical properties. An interval of this array (containing several lines) is taken to have properties similar to those of an interval of the band under consideration. In the model each interval is flanked by statistically similar intervals, but in a real band this will not be so. Here lies the major source of discrepancy between abstraction and actuality; possible errors on this account will be discussed in § 4.7.

In this chapter we will consider only the laboratory problem in which the state of the gas does not vary along the absorption path. This is a special case, not immediately applicable to atmospheric problems, but appropriate to the analysis of laboratory results and the derivation of band parameters, such as line strengths, widths, and spacings; extension to atmospheric problems will be discussed in Chapter 6.

Excepting the trivial case of grey absorption, the transmission of energy by a band will not follow Lambert's law. This was first established experimentally for vibration-rotation bands and was called 'the breakdown of Lambert's law'. This is an unfortunate misnomer, for Lambert's law fails only if there are non-linear radiative effects, as discussed in § 2.1.2, and it is a case of misapplication rather than breakdown of the law. The term *transmission function* is used to describe the mean of many exponential terms involved in band transmissions.

Before treating band models in detail, we must first consider an important property of observed transmission, which may be called the *multiplication property*. This is illustrated in Fig. 4.1, which shows low-resolution absorption spectra for the ν_1 and ν_3 fundamentals of water vapour and the resonating combination bands of carbon dioxide near $2 \cdot 7\,\mu$, individually for the two gases and also when they are mixed in one absorption tube. The transmission of the mixture, averaged over a spectrometer width embracing many lines, is shown to be the product of the mean transmissions of the two components. It is not obvious that this should be so. Suppose that the two absorbing molecules do not interact,† then for strictly monochromatic radiation the transmission of the mixture must be the product of the transmissions (T_ν) of the two components. If we distinguish the two components by the numbers (1) and (2) and the mixture by (1, 2) we have

$$T_\nu(1, 2) = T_\nu(1) \times T_\nu(2). \tag{4.1}$$

The multiplication property now implies that, for certain averages over spectral intervals,

$$\bar{T}_\nu(1, 2) = \bar{T}_\nu(1) \times \bar{T}_\nu(2). \tag{4.2}$$

We wish to know the circumstances under which (4.1) and (4.2) are

---

† This possibility could have been avoided in the experiment described if absorption tubes containing the two gases had been placed in series, rather than mixing the two in one tube. Fortunately, it happens that the dilution in nitrogen causes N_2–CO_2 and N_2–H_2O collisions to dominate the CO_2–H_2O collisions.

compatible. Taken together, they imply (as a matter of definition) that $T_\nu(1)$ and $T_\nu(2)$ are uncorrelated, a condition which cannot be satisfied for any limited frequency range. However, if the lines in either array are randomly arranged (*random* bands), and if the width of the array

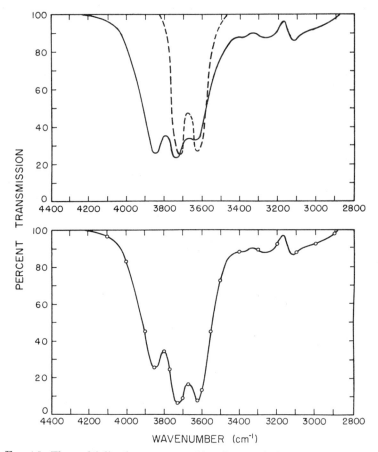

Fig. 4.1. The multiplication property of band transmission at low resolution. The upper spectra are those for CO_2 and H_2O individually. The lower spectrum is for the mixture; the full line is observed and the points are obtained by multiplying the two transmissions in the upper panel. Path length 88 m, H_2O pressure 5 mm Hg. CO_2 pressure 4 mm Hg. The total pressure is made up to 140 mm Hg with nitrogen. The dilution is sufficient for interaction between absorbing molecules to be neglected. After Burch, Howard, and Williams (1956).

is large enough, we can expect the condition to be met within any desired limits. It will be shown that some important bands can be treated as if they are random arrays, and this is therefore a relevant situation.

Another circumstance under which the correlation coefficient is zero, is for two arrays with regular line-spacings (*regular* bands) but with the two individual spacings non-commensurate. This condition is realized in practice in arrays which are almost regular, but with the line spacing varying slowly over the band.

It is difficult to conceive of a state of affairs in which one or the other of these conditions does not hold, and in view of the experimental evidence supporting (4.2) we can safely employ the multiplication property whenever absorption by two different gases is involved.

The two models—*regular* and *random*—mentioned above are the only known two-parameter models which are physically distinct.

4.2. Single line models

4.2.1. *Single line of Lorentz shape*

Before considering regular and random models we will discuss a condition common to both, where the lines are so narrow that there is effectively no overlapping. The necessary and sufficient conditions for this to be so cannot be stated simply. Clearly, there must be large gaps between the lines, so that the average absorption over many lines must necessarily be small,

$$\bar{A} \ll 1. \tag{4.3}$$

The question of a sufficient condition cannot be further discussed until models involving overlap have been evolved.

The absorption can be written as a function of the absorption coefficient and the amount of material (a),

$$A_\nu = 1 - T_\nu = 1 - \exp(-k_\nu a). \tag{4.4}$$

If k_ν is contributed by one line only, then the integral of (4.4) over all frequencies (the *absorption area*) is finite. For historical reasons, connected with astronomical spectroscopy, it is alternatively known as the *equivalent width* of the line,†

$$W(a) = \int_{-\infty}^{+\infty} A_\nu \, d\nu = \int_{-\infty}^{+\infty} (1 - \exp[-k_\nu a]) \, d\nu. \tag{4.5}$$

The relation between $W(a)$ and a is called the *curve of growth*. These curious terms are now so deeply embedded in the literature that we will not try to avoid them.

† The name refers to the width of a rectangular line, whose centre is totally absorbed, having the same absorption area.

If the average spacing between lines in a band is δ, the average absorption on the single line model is[†]

$$\bar{A} = \frac{1}{\delta} \int\limits_{-\infty}^{+\infty} \{1 - \exp(-k_\nu a)\}\, d\nu. \qquad (4.6)$$

For the Lorentz line shape (3.77)

$$k_\nu a = \frac{Sa\alpha_{\rm L}}{\pi(\nu^2 + \alpha_{\rm L}^2)}. \qquad (4.7)$$

Introducing the non-dimensional quantities

$$\left.\begin{aligned} x &= \nu/\delta \\ y &= \alpha_{\rm L}/\delta \\ u &= Sa/2\pi\alpha_{\rm L} \end{aligned}\right\}, \qquad (4.8)$$

we can write

$$\bar{A} = \int\limits_{-\infty}^{+\infty} dx\left\{1 - \exp\left(-\frac{2uy^2}{(x^2 + y^2)}\right)\right\}. \qquad (4.9)$$

This result may be conveniently expressed in terms of Bessel functions of the first kind with imaginary arguments (see Whittaker and Watson, 1915, chapter xvii). There results

$$\begin{aligned} \bar{A}(10) &= 2\pi y u e^{-u}\{I_0(u) + I_1(u)\} \\ &= 2\pi y L(u). \end{aligned} \qquad (4.10)$$

Some values of the function $L(u)$ are given in Appendix 9. This solution was discovered by Ladenberg and Reiche (1913).

$L(u)$ is a comparatively simple function of its argument. For small values of u it is linear, while for large values it varies as $u^{1/2}$. The following series expansion is valid for small u,

$$L(u) = u\left[1 - \sum_{n=1}^{\infty} (-1)^{n+1} \frac{(2n-1)(2n-3)\ldots 5.3.1}{n!\,(n+1)!} u^n\right]. \qquad (4.11)$$

For large u there is an asymptotic expansion

$$L(u) = \sqrt{\left(\frac{2u}{\pi}\right)}\left[1 - \sum_{n=1}^{\infty} \frac{(2n-1)^2(2n-3)^2(2n-5)^2\ldots 3^2.1^2}{n!\,(8u)^n}\right]. \qquad (4.12)$$

As u increases (4.12) tends rapidly to its limit $\sqrt{(2u/\pi)}$ and for all practical purposes it may be so approximated for $u > 3$. This is called

[†] Except where specifically stated, the zero of frequency will be placed at ν_0, in the centre of a line or band. The limits of ν are therefore $-\nu_0$ and ∞, but in all examples of band models considered by us there is no contribution to $\bar{A}$ from the lower limit, and it can be replaced by $-\infty$; if the microwave spectrum were under consideration a finite lower limit would have to be employed.

the *square-root law* régime, while $u \ll 1$ defines the *linear law* régime. The rapid changeover from *linear* to *square-root* law is illustrated in Fig. 4.2 in which $\log\{L(u)/u\}$ is plotted as a function of $\log u$. Such a plot, in which the slope changes from zero for small u to $-\frac{1}{2}$ for large u, is known as a *Matheson diagram* after the first user of this device.

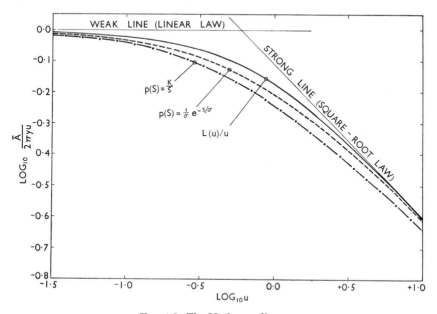

FIG. 4.2. The Matheson diagram.

The two curves marked $p(S) = K/S$ and $p(S) = (1/\sigma)e^{-S/\sigma}$ are for arrays of unequal but independent lines adjusted to correspond asymptotically in both linear law and square-root law régimes (see § 4.2.2).

The physical significance of the two asymptotic régimes is clearer if we note from (4.7) and (4.8) that u is half the optical path at the centre of the line. Thus if $u \ll 1$ the optical path is small at all frequencies, and the exponential in (4.6) can be replaced by the first two terms in its expansion:

$$\bar{A} = \frac{1}{\delta} \int_{-\infty}^{+\infty} k_\nu a \, d\nu = Sa/\delta = 2\pi y u. \tag{4.13}$$

This equation has been derived without reference to the line shape and is therefore valid for all line shapes. It is also valid for overlapping lines provided $k_\nu a \ll 1$ for all ν, and provided that some means is available of isolating the contribution from a single line. Thus (4.13) is a valid asymptotic form under all circumstances; only the limits of validity differ from one case to another.

If $u \gg 1$ the centre of the line is strongly absorbed. In this case the integral in (4.9) is insensitive to changes in the integrand near $x = 0$, provided that these changes increase the exponent. In other words, if the centre of a line is completely blacked out, an approximation which assumes increased absorption in the blacked-out region is a good approximation. If we neglect y^2 in the denominator of the exponent

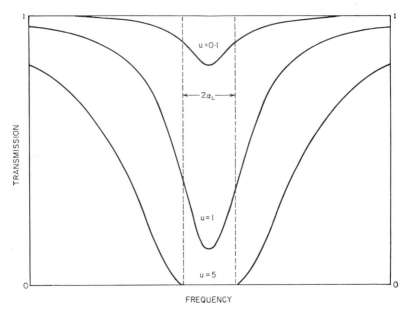

FIG. 4.3. The physical significance of the linear and square-root régimes.

in (4.9) absorption is increased for all x. However for $x \gg y$ the increase is negligible and for small x it has no effect. For sufficiently large u, we may therefore write (with $\xi = 2uy^2/x^2$)

$$\bar{A}(10)/y = 2\surd(2u) \int\limits_0^\infty (1-e^{-\xi}) \frac{d\xi}{\xi^3} \qquad (4.14)$$

$$= 2\surd(2\pi u),$$

in agreement with (4.12).

This discussion suggests a more significant nomenclature for the two asymptotic absorption régimes, viz. the *weak line* and *strong line* approximations, and these terms will be used in subsequent sections (see Fig. 4.3).

The nature of the absorption law for strong lines explains two phenomena which perplexed the earlier investigators. It was observed that band absorption tended to vary as a low power of amount of absorbing material rather than according to Lambert's law, and that the presence of neutral dilutant gas appeared to be as important in determining absorption as the amount of the absorbing gas. From (4.14), we find

$$\bar{A}(10) = \frac{1}{\delta}\sqrt{(4Sa\alpha_{\mathrm{L}})}. \tag{4.15}$$

From (3.79) α_{L} is, in dilute mixtures, proportional to the pressure of the dilutant gas. (4.15) therefore predicts that for strong lines the partial pressure of the absorbing gas and the total pressure of the dilutant stand on the same footing.

Equation (4.10) can be tested by measuring the equivalent width of an isolated line using a high-resolution spectrometer. A spectrometer is a device for measuring energy as a function of apparent frequency ν'. On account of its finite resolution, however, it mistakenly interprets a fraction $f(\nu-\nu')$ of energy of frequency ν as being of frequency ν'. $f(\nu-\nu')$ is a sharply-peaked function with a maximum at $(\nu-\nu') = 0$. It often has a triangular form, and is approximately symmetrical; we will assume symmetry and normalize so that

$$\int_{-\infty}^{+\infty} f(\nu-\nu')\,d\nu = \int_{-\infty}^{+\infty} f(\nu-\nu')\,d\nu' = 1. \tag{4.16}$$

The width of a line is small and the intensity of a laboratory source (I_ν) can be assumed constant over the line. However, we will include the first term in a Taylor expansion of I_ν so that the result may also be applicable to some complete bands. The apparent transmission measured by the spectrometer is

$$T'_{\nu'} = \frac{\displaystyle\int_{-\infty}^{+\infty} \{I_\nu+(\nu-\nu')(\partial I_\nu/\partial\nu)\}\{\exp(-k_\nu a)\}f(\nu-\nu')\,d\nu}{\displaystyle\int_{-\infty}^{+\infty} \{I_\nu+(\nu-\nu')(\partial I_\nu/\partial\nu)\}f(\nu-\nu')\,d\nu}. \tag{4.17}$$

Owing to the symmetry of $f(\nu-\nu')$

$$\int_{-\infty}^{+\infty} (\nu-\nu')f(\nu-\nu')\,d\nu = 0, \tag{4.18}$$

and the denominator of (4.17) is I_ν. The apparent equivalent width is then

$$W'(a) = \int_{-\infty}^{+\infty} d\nu' \int_{-\infty}^{+\infty} d\nu \left[1+\left\{1-(\nu-\nu')\frac{1}{I_\nu}\!\left(\frac{\partial I_\nu}{\partial\nu}\right)\right\}\exp(-k_\nu a)\right]f(\nu-\nu'). \tag{4.19}$$

Performing the integration over ν', with the aid of (4.16) and (4.18)

$$W'(a) = \int\limits_{-\infty}^{+\infty} d\nu \, \{1 - \exp(-k_\nu a)\}$$

$$= W(a). \tag{4.20}$$

The equivalent width is therefore independent of the measuring apparatus to the degree of approximation used in this derivation.

Table 4.1 gives experimental evidence confirming the relation (4.20). It can also be applied to complete band areas provided that the background intensity over the whole band can be adequately represented by two terms of a Taylor expansion, and that the band is approximately symmetrical about its central frequency.

T A B L E 4.1

The invariance of equivalent width

Measurements of equivalent width of some HCl lines on the same sample, with varying slit width. After Benedict *et al.* (1956)

Spectrometer slits (cm^{-1}) .	0·24	0·29	0·34	0·41	0·47	0·71	1·02	1·78
W' for HCl^{35} $R(5)$ . .	0·525	0·546	0·564	0·560	0·566	0·538	0·536	0·522
W' for HCl^{37} $R(5)$ . .	0·322	0·306	0·327	0·319	0·322	0·314	0·308	0·297

Having established that a spectrometer measures true equivalent widths, we can now use observed data to construct a Matheson diagram. This is done for the $P(6)$ line of the CO fundamental in Fig. 4.4. The points fit the theoretical curve well, and the band intensity derived agrees well with independent measurements.

Another test of equation (4.10) is illustrated in Fig. 4.5. Here a pure gas is contained in a tube of fixed length, and the pressure is varied. The amount of gas is now proportional to the pressure, and from (4.8) u is a constant; therefore W/y or W/p should also remain constant according to (4.10). The range of validity of the statement is illustrated in Fig. 4.5. Between 6 and 80 mm Hg the relationship is obeyed; below 6 mm Hg Doppler effects invalidate the use of the Lorentz line shape; above 80 mm the isotopic lines begin to overlap. The full line is based upon a theoretical analysis of both these factors.

4.2.2. *Independent lines of unequal intensity*

Consider a region, of width $N\delta$, containing N lines of mean spacing δ which do not mutually interfere. The relevance of this model will

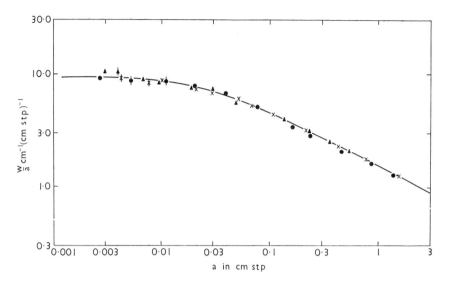

FIG. 4.4. Matheson diagram for the $P(6)$ lines of CO.

The total pressure was held constant at 700 mm Hg. The intercepts give
$\alpha_L(760 \text{ mm Hg}) = 0\cdot069 \text{ cm}^{-1}$ and an intensity per cm s.t.p. for the whole band
of 237 cm^{-2}. The full line is from equation (4.10). After Shaw and France
(1956).

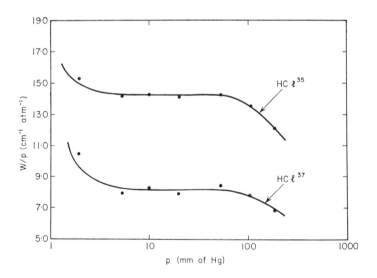

FIG. 4.5. Equivalent width of pure HCl^{35} and HCl^{37}.

The full line is theoretical, allowing for Doppler effects at low pressures (§ 4.2.3)
and line overlap at high pressure. After Benedict et al. (1956).

become clear in § 4.5. If the equivalent width of the ith line is $W(i)$ the total absorption of all the lines is $\sum_{i=1}^{N} W(i)$, and the mean absorption

$$\bar{A} = \frac{1}{N\delta} \sum_{i=1}^{N} W(i)$$

$$= \frac{1}{\delta}\, \bar{W}, \tag{4.21}$$

where $\bar{W}$ is the arithmetic mean of the $W(i)$.

By fitting an approximate mathematical form to the observed distributions of shape, intensity, and width, it is possible to obtain expressions for $\bar{A}$ in terms of mean values of the statistics involved. Then both laboratory and atmospheric problems can be treated in terms of these mean values without reference to the detailed data.

Two distribution functions have been investigated; both refer to line intensities alone, all line widths being assumed equal. The validity of the latter assumption has been tested by Godson (1955a) for sections of the water-vapour spectrum, who found the errors in computed heating rates to be small. Godson (1955b) also showed that errors between two and three times as great can result from an incorrect choice of the distribution of line intensities.

Let $p(S)\, dS$ be the probability that a line has an intensity between S and $S+dS$. The two representations which have been proposed are:

$$p(S) = \frac{1}{\sigma}\, e^{-S/\sigma}, \tag{4.22}$$

where σ is the mean line intensity; and an inverse first power law

$$p(S) = \frac{K}{S} \quad \text{for } S \leqslant S', \tag{4.23}$$

where K and S' are, at present, undefined.

The main physical difference between (4.22), (4.23), and a delta-function, which corresponds to a single line intensity, lies in the number of weak lines which the representations permit. The presence of weak lines means that heat exchange can occur over very long paths.

From (4.6), (4.21), and (4.22) we have for a large number of lines

$$\bar{A}(22) = \frac{1}{\delta} \int_{0}^{\infty} \frac{dS}{\sigma}\, e^{-S/\sigma} \int_{-\infty}^{+\infty} \{1-\exp(-k_\nu a)\}\, d\nu. \tag{4.24}$$

Following (3.48) let us write

$$k_\nu = Sf_\nu. \tag{4.25}$$

The integral in (4.24) may now be performed, yielding

$$\bar{A}(22) = \frac{1}{\delta} \int_{-\infty}^{+\infty} \frac{a\sigma f_\nu}{1+a\sigma f_\nu}\, d\nu. \tag{4.26}$$

For the Lorentz shape,

$$f_\nu = \frac{\alpha_L}{\pi(\nu^2+\alpha_L^2)}, \tag{4.27}$$

and

$$\bar{A}(22) = \frac{2\pi u y}{\sqrt{(2u+1)}}, \tag{4.28}$$

where u is defined according to (4.8) with σ in place of S.

The exponential distribution (4.22) is at best an approximation to a real absorption band; σ and δ are therefore not precisely defined and may be chosen to give minimum errors. A number of different procedures may be followed; one is to make an exact fit in the strong and weak line regions. For $u \gg 1$, (4.22) gives

$$\bar{A}(22) \to \pi y(2u)^{1/2}, \tag{4.29}$$

while from (4.15) and (4.21)

$$\bar{A} \to \frac{1}{N\delta} \sum_{i=1}^{N} 2\sqrt{\{S(i)\alpha_L(i)a\}}. \tag{4.30}$$

For $u \ll 1$, (4.22) gives $\quad \bar{A}(22) \to 2\pi y u, \tag{4.31}$

while from (4.13) and (4.21)

$$\bar{A} \to \frac{1}{N\delta} \sum_{i=1}^{N} aS(i). \tag{4.32}$$

Equating (4.29) to (4.30) and (4.31) to (4.32), and writing $\Delta\nu (= N\delta)$ for the width of the range, we find

$$u(22) = \frac{a}{8}\left(\frac{\sum_{i=1}^{N} S(i)}{\sum_{i=1}^{N} \sqrt{\{S(i)\alpha_L(i)\}}}\right)^2, \tag{4.33}$$

and

$$y(22) = \frac{4}{\pi\,\Delta\nu} \frac{\left[\sum_{i=1}^{N} \sqrt{\{S(i)\alpha_L(i)\}}\right]^2}{\sum_{i=1}^{N} S(i)}. \tag{4.34}$$

Since α_L is proportional to the pressure, these expressions vary with pressure in a known way and need only be evaluated for one pressure. However, when neighbouring lines have different Boltzmann factors (e.g. for asymmetric top molecules), y and u are not predictable functions of temperature and have to be recomputed for each temperature.

According to (4.23), the cumulative probability of lines with intensity greater than S is

$$\int_{S}^{S'} p(S)\,dS = K\log\frac{S'}{S}. \tag{4.35}$$

Fig. 4.6 shows the number of lines with intensity greater than S for some ranges of the water-vapour spectrum, plotted as a function of

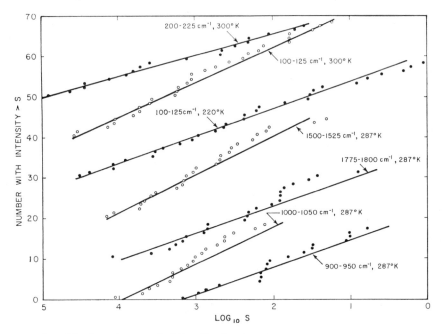

FIG. 4.6. Cumulative probability distributions for water-vapour line intensities. Note the comparative computations for two different temperatures for the range 100–125 cm^{-1}. After Godson (1954).

$\log S$. The relation (4.35) is closely obeyed, indicating many more weak lines than are required by (4.22).

From (4.23), we have

$$\bar{A}(23) = \frac{1}{\delta}\int_{0}^{S'}\frac{K\,dS}{S}\int_{-\infty}^{+\infty}\{1-\exp(-k_\nu a)\}\,d\nu. \tag{4.36}$$

For the Lorentz line shape (4.27), this becomes

$$\bar{A}(23) = 2\pi y[e^{-u}I_0(u)+2ue^{-u}\{I_0(u)+I_1(u)\}-1], \tag{4.37}$$

where $y = K\alpha_L/\delta$, $u = S'a/2\pi\alpha_L$ and the I's are Bessel functions with imaginary arguments.

Best fits for y and u can again be obtained from strong and weak line limits. For $u \gg 1$,

$$\bar{A}(23) \to 4y\sqrt{}/(2\pi u), \qquad (4.38)$$

while for $u \ll 1$,

$$\bar{A}(23) \to 2\pi yu. \qquad (4.39)$$

Hence, for a frequency range of width $\Delta\nu$

$$u(23) = \frac{2a}{\pi} \left(\frac{\sum\limits_{i=1}^{N} S(i)}{\sum\limits_{i=1}^{N} \sqrt{\{S(i)\alpha_L(i)\}}} \right)^2, \qquad (4.40)$$

$$y(23) = \frac{1}{4\,\Delta\nu} \frac{\left[\sum\limits_{i=1}^{N} \sqrt{\{S(i)\alpha_L(i)\}} \right]^2}{\sum\limits_{i=1}^{N} S(i)}. \qquad (4.41)$$

To compare different expressions, let us write

$$u_0 = \frac{a}{2\pi} \left(\frac{\sum\limits_{i=1}^{N} S(i)}{\sum\limits_{i=1}^{N} \sqrt{\{S(i)\alpha_L(i)\}}} \right)^2, \qquad (4.42)$$

$$y_0 = \frac{1}{\Delta\nu} \frac{\left[\sum\limits_{i=1}^{N} \sqrt{\{S(i)\alpha_L(i)\}} \right]^2}{\sum\limits_{i=1}^{N} S(i)}. \qquad (4.43)$$

Then

$$\left. \begin{aligned} u(22) &= \frac{\pi}{4}\, u_0 \\[2mm] y(22) &= \frac{4}{\pi}\, y_0 \end{aligned} \right\}, \qquad (4.44)$$

$$\left. \begin{aligned} u(23) &= 4u_0 \\ y(23) &= \tfrac{1}{4}y_0 \end{aligned} \right\}. \qquad (4.45)$$

In Fig. 4.2, expressions (4.28) and (4.37) are compared with the expression for a single line (4.10), using the same y_0 and u_0 to ensure correspondence in strong and weak line regions. Differences are considerable. In particular, (4.37) requires large values of u_0 before the strong line region is reached on account of the many weak lines allowed by the model.

Although (4.28) and (4.37) have been derived on the assumption that only intensities vary from line to line, it should be noted that (4.33),

(4.34), (4.42), and (4.43) give conditions for a best fit which permit line widths also to vary. If u_0 and y_0 are evaluated from the complete data the results are therefore necessarily correct in the two asymptotic cases, and it is difficult to imagine circumstances under which the errors might be serious.

4.2.3. Doppler effects

Substituting the Doppler profile (3.53) in (4.7) we find

$$\bar{A}(46) = \frac{\alpha_D}{\delta} \int_{-\infty}^{+\infty} \{1 - \exp(-we^{-x^2})\} \, dx, \qquad (4.46)$$

where $x = \nu/\alpha_D$ and $w = Sa/\alpha_D \sqrt{\pi}$ is the optical path at the line centre. (4.46) can be expanded in ascending powers of w and integrated term by term, but the expression converges slowly for large w:

$$\bar{A}(46) = \sqrt{\pi} \frac{w\alpha_D}{\delta} \Big\{ 1 + \sum_{n=1}^{\infty} (-1)^n \frac{w^n}{n! \sqrt{n}} \Big\}. \qquad (4.47)$$

If $w \to 0$
$$\bar{A}(46) \to Sa/\delta, \qquad (4.48)$$

which agrees (as it should) with (4.13).

For large w an asymptotic expansion for (4.46) can be developed,

$$\bar{A}(46) = \frac{2\alpha_D}{\delta} \{(\ln w)^{1/2} + 0 \cdot 2886(\ln w)^{-1/2} - 0 \cdot 1335(\ln w)^{-3/2} +$$
$$+ 0 \cdot 0070(\ln w)^{-5/2} + ...\}. \qquad (4.49)$$

Equation (4.49) gives a very slow increase of $\bar{A}$ with w. The reason lies in the sharply cut-off wings of a Doppler line. A strongly-absorbed Doppler line resembles a square-sided line, whose width is

$$\Delta\nu = \alpha_D(\ln w)^{1/2},$$

(see (3.53)). Then, in agreement with the first term of (4.49),

$$\bar{A}(46) \simeq \frac{2\,\Delta\nu}{\delta} = \frac{2}{\delta} \alpha_D(\ln w)^{1/2}. \qquad (4.50)$$

In § 3.6.3 we saw that the far wings of a line are always influenced by molecular collisions. If the absorbing path is long enough the absorption must ultimately be dominated by the line-wing profile. For mixed Lorentz and Doppler broadening therefore we may predict that for sufficiently large w the absorption will be given by (4.10).

The mixed Doppler and Lorentz line shape is too complex for analytical solution, and Fig. 4.7 is based upon numerical quadrature. The parameter d was introduced in (3.52) and is equal to $2\alpha_L/\alpha_D$. The ordinate and abscissa in Fig. 4.7 are chosen so that the weak line region

is represented by a single straight line of unit slope. The inclusion of
the factor d in both axes means that the strong line limit for Lorentz
lines (4.14) is represented by a series of parallel lines of slope $\frac{1}{2}$. The
curve for Doppler broadening alone is labelled $d = 0$.

In Fig. 4.7 let us follow the curve marked $d = 0\cdot001$. For small u
(i.e. small amounts) the absorption varies linearly with amount. When

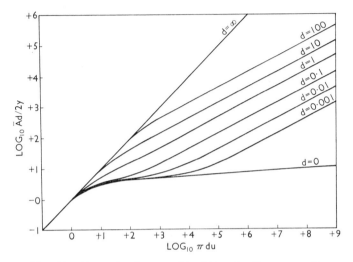

FIG. 4.7. Absorption for a combined Doppler–Lorentz profile.
The ordinate and abscissa are chosen for convenience rather than physical
significance. After van der Held (1931).

$\pi u d = \sqrt{\pi}w \simeq 1$, the centre of the Doppler core is strongly absorbed
and, following (4.50), the absorption varies very slowly with amount.
When $\pi u d \simeq 10^3$ the entire Doppler core is absorbed and the Lorentz
wings begin to be important. For $\pi u d > 10^5$ the curve of growth
behaves as if there were no Doppler core at all. For $d \geqslant 1$ the Doppler
core has negligible influence under all conditions.

If the equivalent width of the line is considerably greater than either
α_L or α_D, then it necessarily follows that $u \gg 1$, and the following
expansion can be shown to be valid:

$$\bar{A} = 2\pi y L(u)\{1-(1-6d^{-2})(8u)^{-1}+O(u^{-2})\}. \qquad (4.51)$$

The integral (4.26) is no more difficult to evaluate numerically than
the equivalent width of a single line, and, by chance, the same expression
happens to be relevant to the problem of line formation in a scattering
atmosphere using the Schwarzschild–Schuster approximation (§ 2.4.1).

Numerical solutions are therefore available and a few values are given in Table 4.2.

TABLE 4.2

$$\frac{d}{uy}\,\bar{A}(22) \text{ as a function of } d \text{ and } ud\sqrt{\pi}$$

After Walshaw (1954)

	d			
$ud\sqrt{\pi}$	0	0·6	0·20	1·0
0·1	0·166	—	—	—
0·316	0·461	—	—	9·504
1	1·07	—	1·14	1·33
3·16	1·95	2·02	2·20	2·94
10	2·82	3·06	3·54	5·48
31·6	3·56	4·16	5·34	9·77
100	4·18	5·70	8·26	17·0
316	4·73	8·30	13·8	—
1000	5·16	11·9	—	—

4.2.4. *Other line shapes*

Plass and Warner (1952) have evaluated the equivalent width of a Lindholm line (§ 3.6.4.3) by numerical quadrature. The differences from the Lorentz shape in the wings give rise to differences from the Lorentz expression (4.10) which increase indefinitely with u. These differences do not become important however until the equivalent width is comparable to ν_p (3.86).

Calculations have also been made of the equivalent width for a line which has a Lorentz profile for $\nu \leqslant \nu_p$ and $k_\nu \propto \nu^{-n}$ outside the range (Fig. 3.20). If W_L is the equivalent width for a Lorentz profile, and W that for the modified line shape, the strong line approximation gives

$$\frac{W-W_L}{W_L} = \frac{1}{2\pi}\left(\frac{2-n}{n-1}\right)\left(\frac{W_L-2\alpha_L}{\nu_p}\right) - \frac{\alpha_L}{\nu_p\,\pi^2}\left(\frac{2-n}{n-1}\right)^2\left(\frac{W_L-2\alpha_L}{\nu_p}\right) + \dots .$$

$$(4.52)$$

Since $\alpha_L \ll \nu_p$ in all practical cases it follows from (4.52) that, unless W_L is comparable to ν_p, the change from a Lorentz line is small. The implication of this as regards real bands is treated in § 4.7.3.

4.3. Modification of single-line models to allow for overlap

Attempts to modify single-line models to include some effect of line overlap have not on the whole proved fruitful. They are superseded in practice by models which regard an array of lines as a statistical entity, rather than a group of individuals, but the techniques are sufficiently interesting to warrant a brief account.

4.3.1. *Schnaidt's model*

Schnaidt assumed that the effect of overlap was to cut off each line at displacements $\pm\frac{1}{2}\delta$ from its centre. Modifying (4.6) we have

$$\bar{A} = \frac{1}{\delta} \int_{-\frac{1}{2}\delta}^{+\frac{1}{2}\delta} \{1-\exp(-k_\nu a)\}\, dv. \tag{4.53}$$

This takes no account of the contribution to the absorption in the range $-\frac{1}{2}\delta$ to $+\frac{1}{2}\delta$ from lines outside the range. One effect of this is that the correct form (4.13) for the weak line limit is no longer obtained.

For a symmetrical line (4.53) can be rewritten in the form

$$\bar{A}(53) = \bar{A}(10) - \frac{2}{\delta} \int_{\frac{1}{2}\delta}^{\infty} \{1-\exp(-k_\nu a)\}\, dv. \tag{4.54}$$

For the Lorentz shape we have

$$\bar{A}(53) = \bar{A}(10) - 2 \int_{1/2}^{\infty} \left\{1-\exp\left(\frac{2uy^2}{x^2+y^2}\right)\right\} dx. \tag{4.55}$$

For all atmospheric problems the line spacing greatly exceeds the line width and therefore $y \ll 1$. Since $x \geqslant \frac{1}{2}$ in the integral in (4.55) we may neglect y^2 in the denominator of the exponent. After some re-arrangement there results

$$\bar{A}(55) = \bar{A}(10) + \{1-\exp(-8uy^2)\} - \sqrt{(8\pi uy^2)}\,\mathrm{erf}\{2y\sqrt{(2u)}\}, \tag{4.56}$$

where

$$\mathrm{erf}\, x = \frac{2}{\sqrt{\pi}} \int_0^x e^{-t^2}\, dt, \tag{4.57}$$

(see Appendix 10 for values of the error function).

In Fig. 4.8 the absorption according to (4.56) with $y = 0{\cdot}05$ has been plotted as a function of u and is compared with the absorption for regular and random models with uniform line intensities (see below).

4.3.2. *A doublet model*

Reiche (1956) has investigated the absorption of two equally intense Lorentz lines of equal width, spaced apart by frequency Δ. The analysis required to solve this problem provides no particular insight into the physics of the absorption process, and need not be reproduced. Some

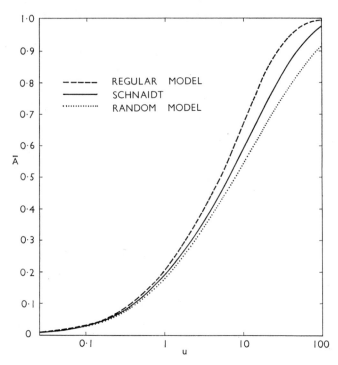

Fig. 4.8. Comparison of Schnaidt's model with random and regular models
($y = 0.05$).

results are obvious. As $\Delta \to 0$ the lines coincide and the total absorption is that of a line with double intensity. The absorption area will
then vary between twice and $\sqrt{2}$ times the single-line case, depending
upon whether the absorption is weak or strong.

As $\Delta \to \infty$, on the other hand, the two lines are effectively independent of each other and the absorption is simply doubled. If we
denote the mean absorption by $\bar{A}$, then $(2\bar{A}(10)-\bar{A})/2\bar{A}(10)$ varies between 0 (for $\Delta \gg 1$ or $u \ll 1$ independently) and $1-1/\sqrt{2} = 0.2928$
(for $\Delta \ll 1$ and $u \gg 1$ simultaneously).

4.3.3. *The method of Matossi, Meyer, and Rauscher*

The method of Matossi, Meyer, and Rauscher (1949) is an attempt
to increase indefinitely the number of overlapping Lorentz lines. The
equations are ingeniously reduced to a series of approximate, simultaneous, partial differential equations which are solved using the Laden-
burg and Reiche solution (4.10) as a boundary condition.

The analysis is complicated, and the results will only be stated. The mean absorption over a frequency range of width $\Delta\nu$, containing N lines is

$$\bar{A}(58) = \frac{\pi y}{N}\left\{ \sum_{i=1}^{N} L\left(u_i - \sum_{j=i+1}^{N} \frac{2}{\pi} u_j X_{ij}^+\right) + \sum_{i=1}^{N-1} L\left(\sum_{j=i+1}^{N} \frac{2}{\pi} u_j X_{ij}^+\right) + \right.$$

$$\left. + \sum_{i=1}^{N} L\left(u_i - \sum_{j=i+1}^{N} \frac{2}{\pi} u_j X_{ij}^-\right) - \sum_{i=1}^{N-1} L\left(\sum_{j=i+1}^{N} \frac{2}{\pi} u_j X_{ij}^-\right)\right\}, \quad (4.58)$$

where L is the function defined in (4.10), i and j are indices denoting individual lines, and

$$X_{ij}^{\pm} = \tan^{-1}\left\{\frac{\Delta\nu}{2\alpha_L}\left(1 \pm \frac{2(\nu_j - \nu_i)\Delta\nu}{(\nu_j - \nu_i)^2 + \alpha_L^2}\right)\right\}. \quad (4.59)$$

One restriction on the validity of this equation is that y must be small. Other restrictions also exist, but they are not explicitly stated. The effect of overlap occurs in the X factors. If these are all zero then (4.58) reduces, as expected, to

$$\bar{A}(58) = \frac{1}{N}\sum_{i=1}^{N} \bar{A}_i(10). \quad (4.60)$$

According to the authors (4.58) can be used to compute the mean absorption for water vapour. However, the double sums are tedious and the method cumbersome for general use; it has not in fact been exploited although, in principle, it is more general than the models which will be discussed in the following sections.

4.4. Regular models

4.4.1. *The Elsasser model*

This model (Fig. 4.9) consists of an infinite array of lines of equal intensity, spaced at equal intervals, and is most applicable to bands of linear molecules. Even for these bands such a simple array is uncommon, since upper state and isotopic lines are usually present. However, by the multiplication property discussed in § 4.1, each sub-band can be represented by a separate Elsasser band and the resulting transmissions multiplied. Fig. 4.10 shows a schematic representation of a part of the carbon dioxide spectrum which can be treated in this way.

The Elsasser model has been solved only for the Lorentz line shape. The absorption coefficient at a frequency displacement ν from the centre of one particular line is then

$$k_\nu = \sum_{i=-\infty}^{i=+\infty} \frac{S}{\pi} \frac{\alpha_L}{(\nu - i\delta)^2 + \alpha_L^2}, \quad (4.61)$$

where δ is the line spacing, or, from (4.8),

$$ak_x = \sum_{i=-\infty}^{i=+\infty} \frac{2uy^2}{(x-i)^2+y^2}. \qquad (4.62)$$

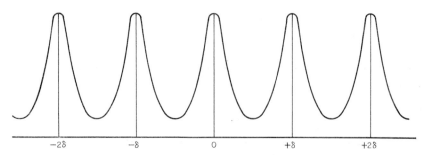

FIG. 4.9. The Elsasser model.

The Mittag–Leffler theorem permits this infinite sum to be expressed in terms of periodic and hyperbolic functions,

$$ak_x = 2\pi uy \, \frac{\sinh 2\pi y}{\cosh 2\pi y - \cos 2\pi x}. \qquad (4.63)$$

The average absorption is now evaluated by integrating $(1-T_x)$ with respect to x from $-\tfrac{1}{2}$ to $+\tfrac{1}{2}$,

$$\bar{A}(64) = 1 - \int_{-1/2}^{+1/2} \exp\!\left(-2\pi uy \, \frac{\sinh 2\pi y}{\cosh 2\pi y - \cos 2\pi x}\right) dx \qquad (4.64)$$

$$= 1 - E(y,u).$$

This integral cannot be expressed in terms of known functions and must be integrated numerically; some results are shown in Appendix 11 which can be extended with the help of one of the three following asymptotic forms.

Consider the limit $y \to \infty$. Then $\sinh 2\pi y \to \cosh 2\pi y \to \infty$ and,

$$E(y,u) = 1 - \bar{A}(64) \to \exp(-2\pi yu) = \exp(-Sa/\delta). \qquad (4.65)$$

In this limit, lines overlap and there is no fine structure. The transmission is now independent of pressure (Beer's law of absorption). This is a feature of all band models when absorption lines overlap and it contains the weak line approximation (4.13) as a special case, for u small.

The second important limit is that of non-overlapping lines, when the mean absorption is given by (4.10). The conditions for validity of this approximation have no simple form. $\bar{A} \ll 1$ is, as discussed in § 4.2.1, necessary, but not always sufficient. In practice the ranges of

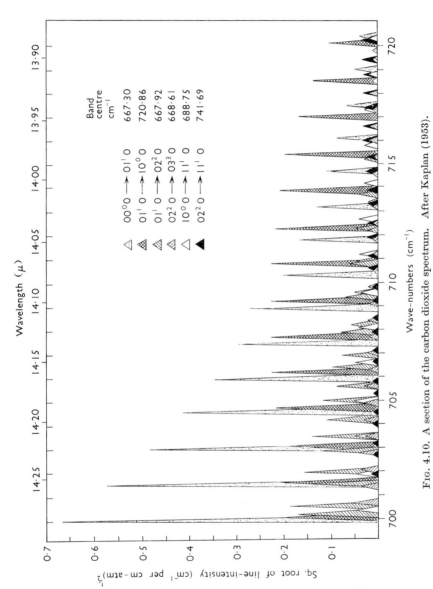

Fig. 4.10. A section of the carbon dioxide spectrum. After Kaplan (1953).

validity of different approximations overlap considerably, and this uncertainty does not give rise to serious difficulties.

Finally, there is the strong line asymptotic limit, originally derived by Elsasser. If, in (4.64), we write

$$\begin{aligned} \sinh 2\pi y &= 2\pi y \\ \cosh 2\pi y &= 1 \end{aligned} \Bigg\},$$

the integral can be transformed into

$$E(y, u) = 1 - \frac{2}{\sqrt{\pi}} \int_{0}^{\pi y \sqrt{(2u)}} e^{-\xi^2} \, d\xi, \tag{4.66}$$

and

$$\bar{A}(64) = \mathrm{erf}\{\pi y \sqrt{(2u)}\}. \tag{4.67}$$

According to Elsasser the condition for (4.65) to be valid is $y \ll 1$, which, indeed, seems reasonable. However the term $\cos 2\pi x$ in the denominator of the exponent in (4.64) approaches unity for x small and it is not correct to approximate one only of two nearly-equal terms. Returning to (4.62), it can be seen that by neglecting y^2 in comparison with $(x-n)^2$ we are led to (4.66) and (4.67). In other words (4.67) is the strong line approximation to (4.64) and the required condition for its validity is $u \gg 1$, rather than $y \ll 1$.

An interesting feature of these three asymptotic forms is that they all provide envelopes to experimental curves of $\bar{A}$ as a function of a or p provided that the correct axes are chosen. If $\bar{A}$ is plotted against ap ($\propto y^2 u$) then (4.67) is the envelope curve. If $\bar{A}$ is plotted against a ($\propto yu$) then (4.65) is the envelope. Lastly if $\bar{A}p$ or $\bar{A}/a$ is plotted against a/p ($\propto u$) we have the single-line approximation (4.10) as an envelope. These statements are illustrated in Fig. 4.11.

4.4.2. *The Curtis model*

This model is not realized in practice, but has the characteristic line spacings of a regular model, and is conveniently expressed in terms of tabulated functions.

Consider an array of Lorentz lines with the same widths, but arbitrary intensities. Equation (4.61) becomes

$$k_\nu a = \sum_{i=-\infty}^{i=+\infty} \frac{aS(i)}{\pi} \frac{\alpha_\mathrm{L}}{(\nu - i\delta)^2 + \alpha_\mathrm{L}^2}. \tag{4.68}$$

The absorption at frequency ν is

$$A_\nu = 1 - \exp\left(-\sum_{i=-\infty}^{i=+\infty} \frac{aS(i)}{\pi} \frac{\alpha_\mathrm{L}}{(\nu - i\delta)^2 + \alpha_\mathrm{L}^2}\right), \tag{4.69}$$

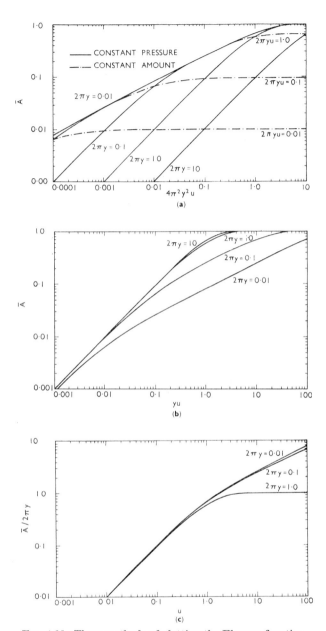

Fig. 4.11. Three methods of plotting the Elsasser function.
(a) $\bar{A}$ as a function of $4\pi^2y^2u$, with curves at constant pressure and amount.
The left-hand curve is the strong-line approximation. (b) $\bar{A}$ as a function of yu,
with constant-pressure curves. The left-hand curve is the Beer's law approximation. (c) $\bar{A}/2\pi y$ as a function of u. The left-hand curve is the single line
approximation. After Plass (1960).

and its average value is

$$A_\nu^* = \int_0^\infty dS(1)p\{S(1)\} \int_0^\infty dS(2)p\{S(2)\}... \int_0^\infty dS(i)p\{S(i)\}... \times$$

$$\times A_\nu\{S(1), S(2)... S(i)...\}$$

$$= 1 - \prod_{i=-\infty}^{i=+\infty} \int_0^\infty p\{S(i)\}\exp\left(-\frac{aS(i)\alpha_L}{\pi\{(\nu-i\delta)^2+\alpha_L^2\}}\right)dS(i), \qquad (4.70)$$

where $p(S)$ is the intensity probability function. Using the exponential distribution (4.22) to describe the distribution of line intensity and replacing S by σ in the definition of u (4.8) we find

$$A_x^* = 1 - \prod_{i=-\infty}^{i=+\infty} \frac{(x-i)^2+y^2}{(x-i)^2+y^2(1+2u)}. \qquad (4.71)$$

The Mittag–Leffler theorem can be used to transform the infinite product into an expression involving periodic and hyperbolic functions

$$A_x^* = \frac{\cosh 2\pi y(1+2u)^{1/2} - \cosh 2\pi y}{\cosh 2\pi y(1+2u)^{1/2} - \cos 2\pi x}. \qquad (4.72)$$

The average absorption is now obtained by integrating this expression with respect to x, from $-\frac{1}{2}$ to $+\frac{1}{2}$, with the result

$$\bar{A}(70) = \frac{\cosh 2\pi y(1+2u)^{1/2} - \cosh 2\pi y}{\sinh 2\pi y(1+2u)^{1/2}}. \qquad (4.73)$$

It is interesting to compare asymptotic forms of (4.73) with those for the Elsasser model.

Firstly, if y becomes large

$$\bar{A}(70) \to 1 - \exp 2\pi y\{1-(1+2u)^{1/2}\}. \qquad (4.74)$$

In the laboratory this limit is achieved by increasing the pressure. If, at the same time, the path length is not increased indefinitely then u must go to zero and
$$\bar{A}(70) \to 1 - \exp(-2\pi yu), \qquad (4.75)$$

in agreement with (4.65).

Now consider the limit of (4.73) as $2\pi y(1+2u)^{1/2} \to 0$. It follows necessarily that $2\pi y \to 0$ and

$$\bar{A}(70) \to \frac{2\pi yu}{(1+2u)^{1/2}} = \bar{A}(22)$$

$$= 2\pi y(1+2u)^{1/2}\left(\frac{u}{1+2u}\right)$$

$$\ll 1.$$

This is the single-line limit, and it differs from that for the Elsasser model, which leads to the Ladenberg and Reiche solution (4.10).

Finally, if $u \to \infty$ we have the strong-line case, which is equivalent to Elsasser's error function approximation. If the transmission is not to be vanishingly small, the limit must be approached keeping $2\pi y(1+2u)^{1/2}$ finite. Thus we must also let $y \to 0$. Then

$$\bar{A}(70) \to \frac{\cosh 2\pi y(2u)^{1/2}-1}{\sinh 2\pi y(2u)^{1/2}}. \tag{4.76}$$

This expression differs from (4.67) when the absorption is close to unity, giving a smaller absorption.

4.4.3. A regular array of doublets

Consider an array of doublets, with doublet spacing Δ, and δ cm^{-1} between doublets. Let one component have an intensity S_1, and if it were present alone let the mean absorption be $\bar{A}(1)$. Let the second array have intensity S_2. Writing $r = \Delta/\delta$, we have from (4.62), for the doublet absorption

$$\bar{A}(1,2) = 1 - \int_{-1/2}^{+1/2} \exp\left(-2\pi y u_1 \frac{\sinh 2\pi y}{\cosh 2\pi y - \cos 2\pi x} - \right.$$
$$\left. -2\pi y u_2 \frac{\sinh 2\pi y}{\cosh 2\pi y - \cos 2\pi(x+r)}\right) dx,$$

or

$$\Delta\bar{A} = \bar{A}(1,2) - \bar{A}(1)$$

$$= \int_{-1/2}^{+1/2} \left\{\exp\left(-2\pi y u_1 \frac{\sinh 2\pi y}{\cosh 2\pi y - \cos 2\pi x}\right)\right\} \times$$

$$\times \left\{1 - \exp\left(-2\pi y u_2 \frac{\sinh 2\pi y}{\cosh 2\pi y - \cos 2\pi(x+r)}\right)\right\} dx. \tag{4.77}$$

Table 4.3 shows computations of $\Delta\bar{A}$ for $y = 0.0413$ and $y = 0.00826$ (representative of the $15\,\mu$ CO$_2$ band at 1.0 and 0.2 atmosphere pressure).

In their discussion of the $15\,\mu$ carbon dioxide band, Yamamoto and Sasamori (1958) assume that, for multiplets, one line can be treated as the principal line and the others as perturbing factors. If two weak lines are on one side of the strong line then they are treated as a combined line, using the results of Table 4.3. If they occur on opposite sides then the absorption increments are derived separately from Table 4.3, and added. After the authors had superposed a number of subbands in this way, they came to the interesting conclusion that the resulting absorption follows more closely the random model (see below) than the original regular model. A similar result will be obtained in § 4.6.1.

T A B L E 4.3

Increment of absorption due to addition of a doublet line

Extracts from tables of Yamamoto and Sasamori (1958): $g = \log_{10}\left(\dfrac{2\pi y u}{\sinh 2\pi y}\right)$;

$h = \log_{10}(u_1/u_2)$; $r = \Delta/\delta$; $\delta = 1\cdot55$ cm^{-1}; $\alpha_L/p = 0\cdot064$ cm^{-1} atm^{-1}

		1·0 *atm*					0·2 *atm*		
g \ r	0	0·1	0·3	0·5	g \ r	0	0·1	0·3	0·5
		$h = 0$					$h = 0$		
−1	0·0226	0·0238	0·0247	0·0248	−1	0·0045	0·0054	0·0055	0·0055
0	0·0924	0·1226	0·1581	0·1634	0	0·0194	0·0340	0·0350	0·0350
1	0·1646	0·1854	0·2667	0·3116	1	0·0536	0·1006	0·1239	0·1250
2	0·0089	0·0092	0·0093	0·0093	2	0·1407	0·1710	0·2730	0·3043
					3	0·0808	0·0828	0·0940	0·1002
		$h = -1$					$h = -1$		
−1	0·0025	0·0026	0·0026	0·0027					
0	0·0123	0·0175	0·0226	0·0265	0	0·0022	0·0035	0·0040	0·0040
1	0·0226	0·0336	0·0904	0·1130	1	0·0064	0·0292	0·0345	0·0345
2	0·0028	0·0032	0·0057	0·0087	2	0·0175	0·0399	0·0993	0·1082
					3	0·0157	0·0184	0·0464	0·0670
		$h = -2$					$h = -2$		
−1	0·0003	0·0003	0·0003	0·0003					
0	0·0010	0·0018	0·0024	0·0025	1	0·0006	0·0042	0·0048	0·0049
1	0·0023	0·0040	0·0130	0·0163	2	0·0017	0·0095	0·0280	0·0322
2	0·0003	0·0004	0·0016	0·0041	3	0·0018	0·0021	0·0159	0·0294

4.4.4. *Godson's approximation to the Elsasser function*

Godson has proposed an approximate form of the Elsasser function which can be shown numerically to be the asymptotic form for $y \ll 1$. Since y is always small for atmospheric gases, this approximation has wide applicability. The error-function approximation to the Elsasser function (4.67) can be written

$$\mathrm{erf}^{-1}\bar{A} = \pi y \sqrt{(2u)},$$

or
$$\frac{1}{yu}\mathrm{erf}^{-1}\bar{A} = \pi \sqrt{(2/u)}. \qquad (4.78)$$

If the argument of the error function is small

$$\mathrm{erf}(x) \to \frac{\sqrt{\pi}}{2}\, x,$$

$$\mathrm{erf}^{-1}(x) \to \frac{2x}{\sqrt{\pi}}. \qquad (4.79)$$

Operating upon the *weak-line* approximation (4.13) with erf^{-1} we find

$$\text{erf}^{-1}\bar{A} = 4\sqrt{\pi}\,yu,$$

$$\frac{1}{yu}\text{erf}^{-1}\bar{A} = 4\sqrt{\pi}. \qquad (4.80)$$

Both (4.78) and (4.80) have the form

$$\frac{1}{uy}\text{erf}^{-1}\bar{A} = F(u). \qquad (4.81)$$

Godson suggested that since this is correct in the *strong-* and *weak-line* limits it may always be approximately true. Hence a plot of $\log\{(1/yu)\,\text{erf}^{-1}\bar{A}\}$ against $\log u$ should give a single line, like the Matheson diagram (Fig. 4.2).

If the Elsasser absorption $\bar{A}(64)$ is plotted in this way, it immediately becomes apparent that for $y \leqslant 0.1$ a single, universal curve exists. Now, if $y \to 0$, holding u constant, the lines must eventually cease to overlap and the single line approximation must become valid. At the same time $\bar{A} \to 0$, and

$$\text{erf}^{-1}\bar{A}(64) \to \text{erf}^{-1}\bar{A}(10) \to \frac{\sqrt{\pi}\bar{A}(10)}{2}. \qquad (4.82)$$

We note from (4.10) that (4.82) has the same form as (4.81). It has only been justified for y much less than some function of u (unspecified since there is no necessary and sufficient condition for the applicability of the single-line approximation), but since (4.81) is always correct for $y \ll 1$, the same must be true for (4.82). Thus we conclude that for $y \ll 1$ and *all* u, $\qquad \bar{A}(64) \simeq \text{erf}\tfrac{1}{2}\sqrt{\pi}\,\bar{A}(10). \qquad (4.83)$

The accuracy of this approximation and its range of validity are demonstrated in Table 4.4. For $y \leqslant 0.1$ the accuracy is high, and the table brings out its relationship to the single-line approximation. Clearly (4.83) contains the single-line approximation as a special case for $\bar{A}$ small; hence for $u \leqslant 1$ the two approximations are equally valid. For $u > 1$, however, (4.83) continues to be applicable, provided only that y is small. The extension is important for carbon dioxide since u may be as high as 10^3 in the ν_2 band.

4.4.5. *Experimental verification of the theory*

Although Elsasser's theory is more than 20 years old, accurate tests are few. It has been compared with laboratory data on the $15\,\mu$ CO_2 band, the $7.8\,\mu$, and $8.4\,\mu$ N_2O bands, and (less appropriately) the $9.6\,\mu$ O_3 band. In none of these cases has there been a sufficiently detailed

TABLE 4.4

Comparison of exact and approximate forms of the Elsasser model

$\bar{A}_{\mathrm{app}} = \mathrm{erf}\,\frac{1}{2}\sqrt{\pi}\bar{A}(10)$. Below the line the single line approximation is valid to better than 0·01

$\log_{10} y$	$u = 0.1$		$u = 1$		$u = 10$	
	$\bar{A}_{\mathrm{app}}$	$\bar{A}(64)$	$\bar{A}_{\mathrm{app}}$	$\bar{A}(64)$	$\bar{A}_{\mathrm{app}}$	$\bar{A}(64)$
0	0·5464	0·4665	1·0000	0·9981	—	—
−0·2	0·3636	0·3273	0·9992	0·9809	—	—
−0·4	0·2344	0·2210	0·9652	0·9146	—	—
−0·6	0·1492	0·1451	0·8172	0·7718	—	—
−0·8	0·0946	0·0934	0·5992	0·5788	0·9984	0·9972
−1·0	0·0598	0·0595	0·4040	0·3975	0·9512	0·9464
−1·2	0·0378	0·0377	0·2622	0·2603	0·7838	0·7811
−1·4	0·0238	0·0238	0·1672	0·1668	0·5648	0·5639
−1·6	0·0150	0·0150	0·1060	0·1059	0·3776	0·3774
−1·8	0·0095	0·0095	0·0670	0·0670	0·2440	0·2440
−2·0	0·0060	0·0060	0·0423	0·0423	0·1555	0·1555
−2·2	0·0038	0·0038	0·0267	0·0267	0·0984	0·0984
−2·4	0·0024	0·0024	0·0169	0·0169	0·0623	0·0623

treatment of upper state and isotopic bands combined with sufficiently precise data to provide a critical test of the theory. This has only been done with the R-branch of the CO fundamental, for which some data are plotted in Fig. 4.12. This branch offers a simple test of the model since there are no superposed upper-state or isotopic bands. The data are plotted on the basis of the Godson approximation and points for all amounts and pressures mingle in a random manner around the theoretical curve. Not only is the agreement with theory qualitatively good, but the derived widths and intensities agree excellently with those obtained from high-resolution measurements (see § 4.2.1 and Fig. 4.4). It should be noted that with $\alpha(760\ \mathrm{mm\ Hg}) = 0\cdot069\ \mathrm{cm}^{-1}$ and $\delta = 3\cdot6$ cm^{-1}, the value of y at 3000 mm Hg pressure is 0·08, and the use of the Godson approximation is therefore justified for the entire pressure range of the experimental data.

4.5. Random models

4.5.1. *Introduction*

In 1950 Cowling computed the average absorption in the rotation band of water vapour, using theoretical line positions and intensities and the Lorentz profile. His computations were for six independent frequency ranges, each covering 25 cm^{-1}, and after he had performed them, he was led to the conclusion that: 'in atmospheric work complication is avoided, and remarkably little error is involved, if a single absorption curve is used at all wave lengths.' Inspection of the spectrum in Fig. 3.4 suggests that the only common feature of 25 cm^{-1} ranges is the apparently random line positions, and therefore that we should inquire into the absorption of a band with certain random properties.

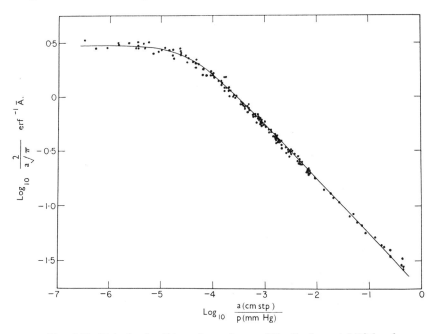

FIG. 4.12. Data for the R-branch maximum of the fundamental CO band.
The amount of CO ranged from 0·000 96 to 45·6 cm s.t.p., the total pressure of nitrogen from 5 to 3000 mm Hg, and the path lengths from 1·55 cm to 400 cm. The full line is given by (4.83).

Two approaches have been made to the problem. The first considers the average absorption at the centre of a finite array of N lines and examines the limit as N goes to infinity.

The second takes as a starting-point any infinite array, such as the Elsasser model. A number of differing arrays are combined by multiplication, and in the limit of a large number of arrays the same result as the first approach is achieved.

The main difference between the two methods is that, with the second approach, the intermediary stages, when finite numbers of bands combine together, are real physical situations while, with the first approach, a limited number of lines is a mathematical fiction. The essential feature common to both is the existence of all possible phase relations between lines, which implies that the lines or arrays are placed at random with respect to frequency. The probability of a line (or one line of an array) lying between ν and $\nu+d\nu$ is then proportional to $d\nu$ and the inverse of the proportionality factor is, by definition, the mean line spacing δ.

This implies a Poisson distribution of the spacing between neighbouring lines. Consider a band of frequencies from $\nu = -\frac{1}{2}N\delta$ to $\nu = +\frac{1}{2}N\delta$ containing N lines. Let there be a sub-range of frequencies, inside this larger range, with width Δ. If one line is placed randomly in the larger range, then the probability that it will fall in the sub-range is $\Delta/N\delta$. The probability that a line will not fall in Δ is therefore $\{1-(\Delta/N\delta)\}$, and the probability that N lines will not fall in Δ is

$$\pi(\Delta) = (1-\Delta/N\delta)^N. \tag{4.84}$$

But
$$\lim_{N\to\infty} (1+x/N)^N = \exp x, \tag{4.85}$$

and therefore as $N \to \infty$, the probability that there is a gap or spacing of width Δ is
$$\pi(\Delta) = \exp(-\Delta/\delta). \tag{4.86}$$

When a limited number of lines are considered the expression (4.86) is naturally not closely obeyed, and even for 200 water-vapour lines significant deviations from (4.86) have been found. In one case investigated, for example, there was a significant lack of very large gaps, and for very small transmissions such large gaps can be of great importance. Since line positions are determined by quantum-mechanical formulae, some order is to be expected, even in the most complex band, and the hypothesis of randomness must be judged by the results achieved.

4.5.2. *Constant line intensity*

Consider an array of identical lines, whose shapes are described by the absorption coefficient k_ν, ν being the frequency displacement from the line centre. Let N lines be distributed randomly between $-\frac{1}{2}N\delta$ and $+\frac{1}{2}N\delta$. The absorption coefficient at $\nu = 0$ contributed by a line centred at $\nu = \nu_i$ is k_{ν_i} and the total absorption coefficient from all lines in the interval is
$$\sum_{i=1}^{N} k_{\nu_i}.$$

The resultant transmission is
$$T = \exp\left(-a\sum_{i=1}^{N} k_{\nu_i}\right)$$
$$= \prod_{i=1}^{N} \exp(-ak_{\nu_i}). \tag{4.87}$$

Since the lines are randomly placed, the probability that a line lies in the interval v_i to v_i+dv_i is dv_i/δ, and the joint probability that there are lines in the intervals v_1 to v_1+dv_1, v_2 to v_2+dv_2, ..., v_i to v_i+dv_i, ..., v_N to v_N+dv_N, is

$$\prod_{i=1}^{N} \frac{dv_i}{\delta}.$$

If we now consider all possible arrangements of lines we must permit each line to lie anywhere in the range $-\frac{1}{2}N\delta$ to $+\frac{1}{2}N\delta$, and the appropriate average value of (4.87) is

$$\overline{T} = \frac{\prod_{i=1}^{N} \int_{-\frac{1}{2}N\delta}^{+\frac{1}{2}N\delta} (dv_i/\delta)\exp(-ak_{v_i})}{\prod_{i=1}^{N} \int_{-\frac{1}{2}N\delta}^{+\frac{1}{2}N\delta} (dv_i/\delta)}. \tag{4.88}$$

The N integrations in both numerator and denominator of (4.88) are identical and

$$\overline{T} = \left\{ \frac{1}{N} \int_{-\frac{1}{2}N\delta}^{+\frac{1}{2}N\delta} \frac{dv}{\delta} \exp(-ak_v) \right\}^N$$

$$= \left[1 - \frac{1}{N} \int_{-\frac{1}{2}N\delta}^{+\frac{1}{2}N\delta} \frac{dv}{\delta} \{1 - \exp(-ak_v)\} \right]^N. \tag{4.89}$$

As $N \to \infty$, $\quad \overline{T}(88) \to \exp\left[-\frac{1}{\delta} \int_{-\infty}^{+\infty} \{1 - \exp(-ak_v)\} dv \right]$

$$= \exp(-W/\delta). \tag{4.90}$$

It is tempting not to proceed to the limit of large N, as we have done here, but to adopt (4.89) as a generalized transmission function, which includes (4.90) as a special case. The integral in the exponent of (4.89) is available for the Lorentz profile and has been discussed in § 4.3.1. However, the integration performed in (4.89) assumes the coexistence of all possible line positions and this can only approximately be realized if the range of integration is large, or if an average is taken over a great many ranges such as that discussed here.

No particular line profile has been used to derive (4.90), so that it is correct for all line shapes. Thus the theory of single line models, as described in § 4.2, can be used in a very simple way to construct a realistic band model with overlapping lines.

Let us compare the properties of the random and regular models. Both have single line and Beer's law approximations as asymptotic limits, and a little consideration shows that the two models are identical in these limits, provided that the line intensities and widths are the same. Interest therefore lies mainly in the case where y is small but $\bar{A}$ is not. If we let $\bar{A}(10) \to \infty$ in the regular model expression (4.83), we find

$$\bar{A} \text{ (regular)} \to 1 - \frac{1}{2\bar{A}(10)} e^{-4\bar{A}^2(10)/\pi}, \tag{4.91}$$

which may be compared with

$$\bar{A} \text{ (random)} = 1 - e^{-\bar{A}(10)}. \tag{4.92}$$

The main difference between (4.91) and (4.92) is that the former represents a more rapid approach to total absorption as the path length or pressure is increased. The reason for this is that the random model permits a few large gaps between lines and long path lengths are necessary before these are blacked out.

There are no known gases with random line spacings but constant line intensities, and the only use which will be made of (4.90) is to derive the more general case of an arbitrary distribution of line intensities.

4.5.3. *The general random model*

Suppose a certain frequency range to contain N lines, and let us regard each of these N lines as one member of an infinite array of randomly spaced lines of equal intensity and mean spacing $N\delta$. The range under consideration therefore contains, on the average, one line from each of N arrays. According to (4.90), the mean transmission of one array is

$$\bar{T}_i = \exp\left(-\frac{W_i}{N\delta}\right), \tag{4.93}$$

where W_i is the equivalent width of one line.

The conditions for application of the multiplication property are met for random arrays, and therefore the mean transmission averaged over the N superposed arrays is

$$\bar{T}(94) = \prod_{i=1}^{N} \bar{T}_i = \exp\left(-\frac{1}{N\delta} \sum_{i=1}^{N} W_i\right)$$

$$= \exp\left(-\frac{1}{\delta}\bar{W}\right). \tag{4.94}$$

The profiles of lines have not been assumed identical in this derivation, and (4.94) is therefore valid without other restrictions than that lines have random spacing and that integration is over a sufficiently wide spectral region for the multiplication property to be valid.

Evaluation of $\overline{W}/\delta$ was the purpose of § 4.2.2. Solutions are available for the Doppler–Lorentz profile, both for a single line intensity and for an exponential distribution of line intensities. For the Lorentz shape we have in addition Godson's analysis for an intensity distribution according to the inverse first power of the intensity. All the tabular results and comparisons made in that section can be taken over directly, and no purpose is served in re-discussing the results here. If, in any range, line intensities and widths are known, and if they do not correspond closely to any of the three distributions discussed, then the summation in (4.94) can be performed numerically, using the actual spectrographic data and computed equivalent widths. With the help of an electronic computer this is not a difficult procedure; considerably simpler than performing the detailed frequency integration.

4.5.4. *Verification of the theory*

The hypothesis of random line positions can be tested theoretically by comparing model transmissions with detailed computations based on actual line positions. Fig. 4.13 shows such a comparison for the rotation band of water vapour. Best agreement is expected with the universal curve (*a*) which represents an average for all computed ranges. The agreement is good, even bearing in mind that two parameters can be adjusted to give a best fit. The range 100–125 cm^{-1} was computed for four pressures and the agreement in (*b*) shows how well the effect of pressure broadening is taken into account by the model. The range 175 cm^{-1} to 250 cm^{-1} is of interest because the 25 cm^{-1} ranges computed failed to fit the model well, but when three are averaged good agreement results. In curve (*d*) discrepancies occur as the transmission approaches zero. The range 300–350 cm^{-1} contains only fourteen lines, so that deviations from a random model are perhaps not surprising.

The evidence shown in Fig. 4.13, taken together with the inherent plausibility of the random model, leaves little doubt as to its essential correctness. Nevertheless, on evidence similar to that shown, Yamamoto and Sasamori (1957) have questioned the applicability of the model to water vapour. It is not clear why their conclusion differs from that of other workers, unless they are using some technique for computing mean intensities and line spacings which is not based on a best fit.

Fig. 4.14 shows a comparison between the random model using an exponential distribution of line intensities, and laboratory measurements on water-vapour bands. The abscissa in this diagram is a/a_0, where a_0 is the amount of water-vapour required to give an absorption of 0·5 at 740 mm Hg pressure. Since intensities and amounts always occur as a product, this method of plotting eliminates the differences

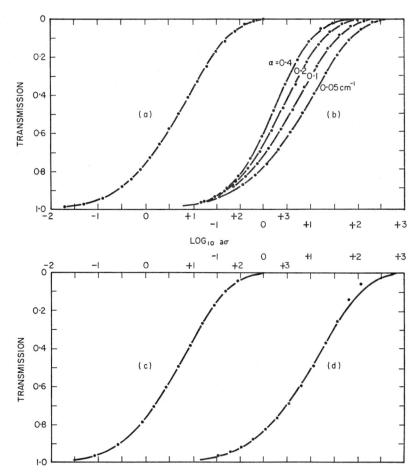

FIG. 4.13. Comparison between a random model based upon an exponential distribution of line intensities and Cowling's (1950) data for water vapour.

The points are Cowling's computations while the full lines are based on (4.94) and (4.28), x and y being chosen for best fit by a subjective method. (a) Cowling's *universal curve*, which is his estimate of the average behaviour of all frequency ranges and should therefore give the best fit to the random model. (b) 125 cm^{-1} to 100 cm^{-1}, at four different pressures, using the same values of σ and δ in each case. (c) 175 cm^{-1} to 250 cm^{-1}. This is the average transmission over three of Cowling's ranges. Each range separately did not fit the model well. (d) 300 cm^{-1} to 350 cm^{-1}. An average over two 25 cm^{-1} ranges.

from band to band which result from changes of line intensity. So, if α_L and δ are the same for all frequencies, all experimental points should fall on the same curve. With a scatter which does not greatly exceed

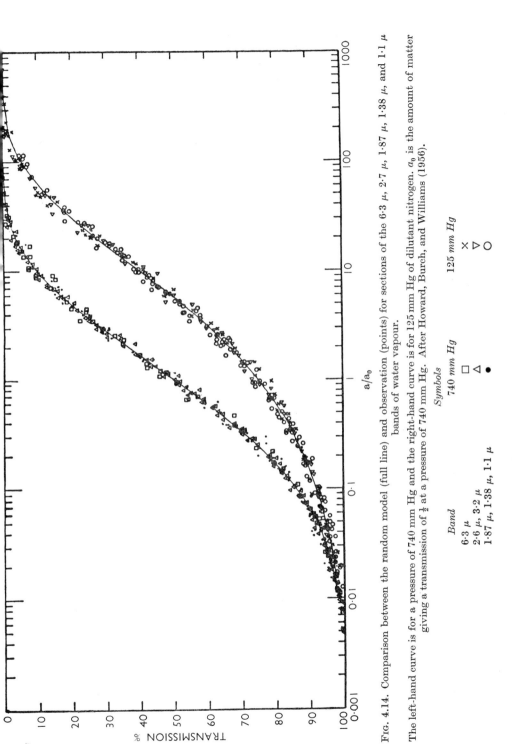

FIG. 4.14. Comparison between the random model (full line) and observation (points) for sections of the 6·3 μ, 2·7 μ, 1·87 μ, 1·38 μ, and 1·1 μ bands of water vapour.

The left-hand curve is for a pressure of 740 mm Hg and the right-hand curve is for 125 mm Hg of dilutant nitrogen. a_0 is the amount of matter giving a transmission of $\frac{1}{2}$ at a pressure of 740 mm Hg. After Howard, Burch, and Williams (1956).

the experimental error, this seems to be so although the use of $a_{1/2}$ as an adjustable parameter makes the test less discriminating than the figure may suggest.

The comparison of theory and experiment for pressure broadening can be facilitated by a method of plotting observed results which does not depend on the distribution of line intensities.

From (4.94) and (4.10)

$$\ln \overline{T} = -\overline{W}/\delta = \overline{-2\pi y_i \, L(u_i)}.$$

If, in accordance with the Lorentz theory, we write $\alpha_i = \alpha_i^0 p/p_0$, we have

$$\frac{1}{a}\ln \overline{T} = \overline{-2\pi \frac{\alpha_i^0}{\delta p_0} \frac{p}{a} L\!\left(\frac{S_i p_0}{2\pi \alpha_i^0} \frac{a}{p}\right)}$$

$$= -F(a/p), \tag{4.95}$$

regardless of the nature of the average.

If we plot $\log(-\ln \overline{T}/a)$ as a function of $\log a/p$ therefore we obtain a curve, similar to Fig. 4.2, with zero slope for small a/p and a slope of $-\frac{1}{2}$ for large values. Every distribution of line intensities gives the same two asymptotes but the transition region differs from one distribution to another.

Fig. 4.15 shows this method of presenting laboratory data for one frequency range in the $9\cdot6\,\mu$ band of ozone. The points for different pressures and amounts mingle, showing that a random model involves no systematic errors (see also § 5.6.3).

4.6. Generalized transmission functions

4.6.1. *Superposed regular and random bands*

In this and the next two sections we will discuss means of generalizing models to bands which are neither regular nor random. The means are devious, for the direct approach of attempting to calculate for an arbitrary distribution of line spacings leads to difficulties which, so far, have defied analysis.

One possibility is to superpose regular and random models using the multiplication principle. This gives rise to a wide variety of bands with different properties, but they should only be employed after careful examination of the circumstances. One circumstance in which it is obviously correct to combine models is when analysis of the fine structure of the band shows, by independent means, that such overlapping bands exist. For example, in the case of the band illustrated in Fig. 4.10, the usefulness of analysing the band into Elsasser-type sub-bands is clear, and in such a case information on line spacings and relative

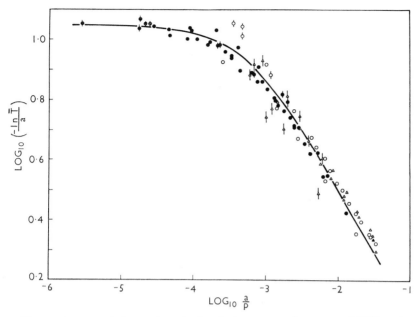

Fig. 4.15. Transmission of the 9·6 μ band of ozone at a frequency of 1057 cm⁻¹.

a is in cm s.t.p. and p in mm Hg.

● $a < 1\cdot70$,
○ $0\cdot190 < a < 0\cdot390$,
△ $0\cdot480 < a < 1\cdot50$.

The full line follows a statistical model with an exponential distribution of line
intensities. After Walshaw and Goody (1954).

intensities in the sub-bands is available from high-resolution data.
Using these data, adjustment of one intensity and one line-width parameter should produce a good fit to laboratory data, and confidence in
the applicability of the model to atmospheric conditions would be
justified.

This idea can be generalized into a simple three-parameter model,
which can include regular and random bands as asymptotic cases.
Consider the superposition of n bands, each with a line spacing $\delta' = n\delta$.
The transmission of one array is

$$\bar{T}(a,\delta') = 1 - \bar{A}(a, n\delta),$$

and that for n superposed arrays is

$$\bar{T}(n, a, \delta) = \bar{T}^n(a, \delta') = \{1 - \bar{A}(a, n\delta)\}^n. \qquad (4.96)$$

Regardless of the form of $\bar{A}(a, n\delta)$ we know from previous analysis
that a random model results in the limit $n \to \infty$. Thus we can create
a continuous series of models lying between any model we care to

choose and a random model. The transition from an Elsasser model
$\bar{A} = \bar{A}(64)$ and $n = 1$ to a random model with equal line intensities
($n = \infty$) is illustrated in Table 4.5.

TABLE 4.5

$-\log_{10} \bar{T}(n, u, y)$ from eq. (4.96) based on the Elsasser model

$\log_{10} u$		\multicolumn{4}{c}{$\log_{10} y$}
		0	-0.4	-0.8	-1.2
	$n = 1$	0·273	0·109	0·043	0·017
-1.0	$n = 10$	0·267	0·105	0·042	0·016
	$n = \infty$	0·260	0·103	0·041	0·016
	$n = 1$	2·729	1·068	0·376	0·131
0.0	$n = 10$	2·200	0·797	0·301	0·118
	$n = \infty$	1·836	0·731	0·291	0·116
	$n = 1$	—	—	2·550	0·660
$+1.0$	$n = 10$	—	—	1·215	0·450
	$n = \infty$	—	—	1·077	0·429

While this connexion between the two fundamental models is interesting, its usefulness is restricted. There are an unlimited number of three-
parameter models which can be made to include random and regular
bands as asymptotic cases. Equation (4.96) is perhaps the simplest,
but its use must be justified on theoretical grounds. Adjustment of the
three available parameters can usually give a reasonable fit to any
laboratory data for a restricted range of pressures and amounts, because
the data are rarely of high accuracy, but without independent justification extrapolation to unknown conditions is not permissible.

It is not necessary to choose arrays of identical bands as a starting
point; the sub-bands could, for example, have had different spacings.
It might perhaps be thought that any desired histogram of line spacings
could thus be synthesized. A few trials will soon convince the reader
that the histograms which can be generated from regular or random
models are limited.

4.6.2. Spectral 'gappiness'

From a qualitative view point the most important factors in determining transmission for long path lengths are occasional gaps in the
spectrum. For grey absorption the whole spectrum cuts off simultaneously; with an Elsasser band total extinction does not occur until
the valleys between the lines give strong absorption, but these all do
so together; with a random band, the occasional gap which is allowed

by the random line positions governs the behaviour for large amounts of absorbing material. However, this does not embrace the entire gamut of possibilities, for a band can have more large gaps than predicted by the random model. The following treatment pictures a gap of arbitrary extent between two random arrays as a first attempt to introduce a feature of real physical importance contained in none of the models discussed so far.

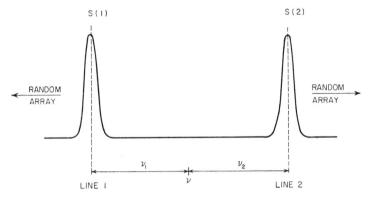

FIG. 4.16. An arbitrary gap in a random spectrum.

Consider two adjacent spectral lines (Fig. 4.16) surrounded by a random array. We will calculate the transmission at a point, between the lines, distance v_1 from the first line and v_2 from the second. Suppose that, to the left of line 1, there are N lines in a frequency range $N\delta$. Then, according to (4.89) by displacement of the zero frequency, we have

$$\bar{T} = \left[1 - \frac{1}{N} \int_{v_1}^{N\delta + v_1} \frac{dv}{\delta} \{1 - \exp(-ak_v)\}\right]^N. \tag{4.97}$$

Taking the limit $N \to \infty$ and following the steps used to derive (4.94)

$$\bar{T} = \exp\left[-\frac{1}{\delta} \int_0^\infty p(S)\,dS \int_{v_1}^\infty dv\{1 - \exp(-ak_v)\}\right]. \tag{4.98}$$

In addition we must allow for the effect of line 1. If the intensity of line 1 is subject to the same probability law, then the transmission at v_1 is

$$\int_0^\infty p(S)\,dS \exp(-ak_{v_1}),$$

and the transmission of all lines to the left of the point under consideration is

$$\bar{T}_1 = \int_0^\infty dS p(S) \exp(-ak_{\nu_1}) \exp\left[-\frac{1}{\delta}\int_0^\infty dS p(S) \int_{\nu_1}^\infty d\nu_1\{1-\exp(-ak_\nu)\}\right],$$

(4.99)

which can be conveniently simplified in terms of a generating function

$$Q(x) = \exp\left\{-\frac{1}{\delta}\int_0^\infty p(S)\,dS \int_0^x d\nu \exp(-ak_\nu)\right\}.$$

(4.100)

After a little manipulation we find

$$\bar{T}_1 = -\delta \exp(\nu_1/\delta).\bar{T}^{1/2}(94).Q'(\nu_1),$$

(4.101)

where $\bar{T}(94)$ is the random model transmission resulting from the
intensity distribution $p(S)$.

If the lines are symmetrical, the transmission is the product of two
terms like (4.101), one from the lines on the l.h.s. and one from those
on the r.h.s., and therefore

$$\bar{T} = \bar{T}(94)\delta^2 Q'(\nu_1)Q'(\nu_2)\exp\left(\frac{\nu_1+\nu_2}{\delta}\right).$$

(4.102)

Now suppose that the probability of a spacing between Δ and $\Delta+d\Delta$
is $\pi(\Delta)\,d\Delta$, then we must average (4.102) according to this weighting
function. There results

$$\bar{T} = \bar{T}(94)\frac{\int_0^\infty d\nu_2 \int_0^\infty d\nu_1 \delta^2 \pi(\nu_1+\nu_2)Q'(\nu_1)Q'(\nu_2)\exp(\{\nu_1+\nu_2\}/\delta)}{\int_0^\infty d\nu_2 \int_0^\infty d\nu_1 \pi(\nu_1+\nu_2)}.$$

(4.103)

The value of this expression is restricted to small deviations from
a random model, because the effect of non-random line spacings is only
taken into account for the two nearest lines; for strong overlapping
this is not sufficient.

Despite this restriction (4.103) is of more than academic interest, for
it permits any kind of deviation from a random distribution. The
generating functions (4.100) contain all the information about line
intensities and can be evaluated with the aid of the numerical data of
§ 4.3.1; since the integrals in (4.103) constitute a correcting factor only,
they do not have to be evaluated with the highest precision. There is,
of course, no difficulty in principle in using a regular band to the left

of line 1 and to the right of line 2, if this should appear to accord better
with the known line structure.

We must now show that (4.103) becomes a simple random model
under appropriate conditions. This is accomplished by substituting
(4.86) for the probability function π, with the same mean line spacing
as for the rest of the band

$$\overline{T} = \overline{T}(94) \frac{\int\limits_0^\infty dv_1 \int\limits_0^\infty dv_2 \, \delta^2 Q'(v_1) Q'(v_2)}{\int\limits_0^\infty dv_1 \int\limits_0^\infty dv_2 \exp\{-(v_1+v_2)/\delta\}}$$

$$= \overline{T}(94)\{Q(\infty)-Q(0)\}^2$$

$$= \overline{T}(94). \tag{4.104}$$

4.6.3. *An empirical approach*

Yamamoto and Sasamori (1957) have put forward a suggestion which
is not based upon deductive reasoning, but which may be useful in
presenting laboratory results in terms of a single variable. They suggested that the influences of line spacings and line intensities are
separable, in the sense that the line spacings determine the functional
form of the transmission while the line intensities determine the argument of the function. Following (4.94) we write for any distribution of
line spacings,

$$\overline{T} = F\left(\frac{1}{\delta}\overline{W}\right). \tag{4.105}$$

All random models satisfy this expression. To the extent to which
Godson's approximation to the Elsasser function (§ 4.4.4) is valid it is
also true for the regular model. However, with the value of $\overline{A}$ appropriate to an exponential distribution of line intensities in (4.83), we do
not obtain the Curtis model, even in rough approximation. The hypothesis of Yamamoto and Sasamori is therefore incorrect in this case.

This does not constitute a reason why (4.105) should not be employed
in attempts to correlate data. The possible usefulness of such a procedure has been demonstrated by Yamamoto and Sasamori in an
analysis of mean transmissions for regions of the water-vapour rotation
band which have too few lines for a good application of the random
model.

4.7. Restrictions on model theory

4.7.1. *Variations in band contour*

In practice an absorption band does not have the same statistical
properties at each frequency interval and an assumption of an infinite

homogeneous array is, in some degree, always at fault. The question arises as to the magnitude of errors involved, and whether simple first-order corrections can be made.

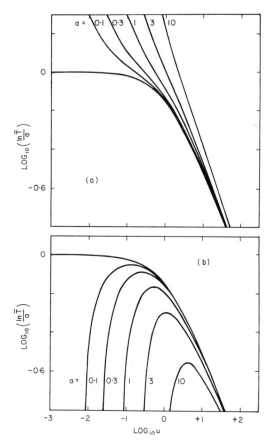

Fig. 4.17. Random model with a continuum.

(a) $\tau_c = +0.1 \ (2\pi y^2 u)$.
(b) $\tau_c = -0.07 \ (2\pi y^2 u)$.

τ_c is the optical thickness of the continuum. The lines are for constant values of $2\pi y u$, i.e. for constant amount but variable pressure, and consequently simulate a typical run of laboratory data. After Walshaw (1954).

One approach to this problem is to make a few exact computations on a real band and to compare with prediction from a band model. Such a computation made near $700.1 \ cm^{-1}$ in the $CO_2 \ \nu_2$ band indicated errors of the order of 20 per cent at this wave number. However, since band models consider averages this is not a strictly relevant comparison and compensating errors may well exist in other parts of the band.

Moreover, if band parameters are obtained from an empirical best fit between theory and laboratory data all such errors are automatically taken up to first order.

Near 1000 cm^{-1} in the spectrum of water vapour the absorption by distant strong lines exceeds that of the local, weaker lines (§ 5.4.7); in this circumstance a model based on statistical homogeneity is obviously unsatisfactory. A more suitable model consists of a grey continuum (to take account of the distant strong lines) combined with a random or regular model to describe the behaviour of the local lines. Since all theories of pressure broadening suggest wing coefficients proportional to the pressure [see § 3.6.5 (viii)] measurements at a single pressure suffice to establish the grey absorption coefficient for all pressures. The effect of temperature has however to be determined empirically, unless the detailed line profile is known.

A weak continuum, with positive or negative coefficients, can be introduced empirically into an examination of laboratory data in order to assess the value of a model based on spectral homogeneity. The curves in Fig. 4.17 show the appearance of a random model with weak 'positive' and 'negative' continua when plotted in a way that would give a single curve if there were no continuum. There is evidence for such behaviour in some laboratory data for the 9·6 μ ozone band.

4.7.2. *A modulated contour*

In addition to changes of band parameters outside the range of frequencies under consideration, there may be systematic changes within the interval. For example, we may suppose that the line strengths and widths vary continuously from one side to the other of a wide frequency interval of width Δ. Provided the variations across this interval are not too large, a number of different, plausible assumptions all lead to similar results. To make the treatment simple therefore we will assume that the logarithm of the transmission varies according to the relation

$$\ln \overline{T}_\nu = \left(1 + f\frac{\nu}{\Delta}\right)\ln \overline{T}_0, \tag{4.106}$$

where $\overline{T}_0$ is the transmission in the centre of the interval. The mean transmission is now

$$\overline{\overline{T}} = \frac{1}{\Delta}\int_{-\frac{1}{2}\Delta}^{+\frac{1}{2}\Delta} \overline{T}\, d\nu$$

$$= \overline{T}_0\frac{\sinh(\frac{1}{2}f\ln T_0)}{\frac{1}{2}f\ln T_0}. \tag{4.107}$$

From (4.106)
$$f^2 = \frac{\Delta^2}{\bar{T}} \frac{\partial^2 \bar{T}}{\partial \nu^2}.$$
(4.108)

Hence f^2 is proportional to the curvature of the band contour, and may be either positive or negative, leading to real or imaginary f. There is a qualitative resemblance between Fig. 4.17 (band plus continuum) and (4.107), so that the two effects are difficult to distinguish.

4.7.3. *Deviations from the Lorentz line shape*

The theory of band absorption is not restricted to any one line profile although in practice considerations have been restricted to Lorentz–Doppler profiles. Interpretation of data would be greatly complicated should such results not have wide validity. The Lorentz shape is never precise, however, and it is important to know the effect of deviations from this profile.

Let us consider the equivalent width (4.52) in the context of a random band with equal line intensities and compare it with a Lorentz line. From (4.94) we have

$$\frac{\Delta \bar{T}}{\bar{T}} = \frac{\Delta \bar{W}}{\delta},$$
(4.109)

and, taking the first term in (4.52), $n = 1 \cdot 73$ (as in § 3.6.5, (viii)) and $W \gg \alpha_L$ (which applies to all but trivial cases), we have

$$\Delta W = \frac{0 \cdot 27}{0 \cdot 73(2\pi)} \frac{W^2}{\nu_p},$$

and
$$\Delta \bar{T} = \bar{T}(\ln \bar{T})^2 \frac{0 \cdot 27}{0 \cdot 73(2\pi)} \frac{\delta}{\nu_p}.$$

But
$$\bar{T}(\ln \bar{T})^2 \leqslant 0 \cdot 54,$$

and
$$\Delta \bar{T} \leqslant 3 \cdot 2 \times 10^{-2} \frac{\delta}{\nu_p}.$$
(4.110)

Estimates of δ/ν_p are $0 \cdot 82$ for H_2O, $0 \cdot 050$ for O_3, and $0 \cdot 36$ for the $7 \cdot 6\mu$ CH_4 band giving maximum transmission errors of $2 \cdot 6$ per cent, $0 \cdot 16$ per cent, and $1 \cdot 1$ per cent, respectively. Computations for the Lindholm line shape using an exponential distribution of line intensities give almost identical results. These errors are probably not detectable because first-order corrections will automatically be performed if the line width and intensity data are derived from laboratory spectra. While the problem is not negligible, it is less important than others to be discussed in subsequent chapters.

BIBLIOGRAPHY

4.1. Introduction

A review of early work on band absorption in the infra-red spectrum is by

NIELSEN, J. R., THORNTON, V., and DALE, E. B., 1944, 'The absorption laws for gases in the infrared', *Rev. Mod. Phys.* **16**, p. 307.

More recent work is described by

PENNER, S. S., 1959, *Quantitative molecular spectroscopy and gas emissivities.* Reading, Mass.: Addison-Wesley.

Experimental evidence in favour of the multiplication property is given by

BURCH, D. E., HOWARD, J. N., and WILLIAMS, D., 1956, 'Infrared transmission of synthetic atmospheres. V. Absorption laws for overlapping bands', *J. Opt. Soc. Am.* **46**, p. 452.

Its correctness is also evidenced by the proper functioning of all commercial double-beam spectrometers. As a formal problem it seems first to have been mentioned by

KAPLAN, L. D., 1954, 'A quasi-statistical approach to the calculation of atmospheric transmission', *Proc. Toronto Met. Conf. (Roy. Met. Soc.)*, 1953, p. 43.

4.2. Single-line models

Definitions of the Bessel functions I_0 and I_1 and some numerical values are to be found in

WHITTAKER, E. T., and WATSON, G. N., 1915, *Modern Analysis.* Cambridge Univ. Press; and

WATSON, G. N., 1952, *A treatise on the theory of Bessel functions.* Cambridge Univ. Press.

The function $L(u)$ has been computed by

KAPLAN, L. D., and EGGERS, D. F., 1956, 'Intensity and line-width of the 15-micron CO_2 band, determined by a curve-of-growth method', *J. Chem. Phys.* **25**, p. 876.

The expression (4.10) was first derived by

LADENBERG, R., and REICHE, F., 1913, 'Über selektive Absorption', *Ann. Phys.* **42**, p. 181,

while (4.11) and (4.12) were given by

SCHNAIDT, F., 1939, 'Über die Absorption von Wasserdampf und Kohlensäure mit besonderer Berücksichtigung der Druck- und Temperaturabhängigkeit', *Beitr. Geophys.* **54**, p. 203.

The method of plotting data shown in Fig. 4.1 was introduced by

MATHESON, L. A., 1932, 'The intensity of infrared absorption bands', *Phys. Rev.* **40**, p. 813.

The data shown in Fig. 4.4 are from

SHAW, J. H., and FRANCE, W. L., 1956, *Intensities and widths of single lines of the 4·7 micron CO fundamental band*, AFCRC–TN–56–466. The Ohio State University Research Foundation.

The line intensities derived from this work agree well with independent measurements by

PENNER, S. S., and WEBER, D., 1951, 'Quantitative infrared intensity measurements, I. Carbon monoxide pressurized with infrared-inactive gases', *J. Chem. Phys.* **19**, p. 807.

Table 4.1 and Fig. 4.5 are taken from

BENEDICT, W. S., HERMAN, R., MOORE, G. E., and SILVERMAN, S., 1956, 'The strengths, widths, and shapes of infrared lines. Parts I and II', *Canad. J. Phys.* **34**, pp. 830 and 850.

Combinations of lines of unequal intensities are discussed by

GODSON, W. L., 1954, 'Spectral models and the properties of transmission functions', *Proc. Toronto Met. Conf. (Roy. Met. Soc.)*, 1953, p. 35;
—— 1955, 'The computation of infrared transmission by atmospheric water vapour. I and II', *J. Met.* **12**, pp. 272 and 533; and
GOODY, R. M., 1952, 'A statistical model for water-vapour absorption', *Quart. J. R. Met. Soc.* **78**, p. 165.

The equivalent width of a single line with Doppler profile has been discussed by many writers, e.g.

STRUVE, O., and ELVEY, C. T., 1934, 'The intensities of stellar absorption lines', *Astrophys. J.* **79**, p. 409; and
VAN DER HELD, E. F. M., 1931, 'Intensität und natürliche Breite von Spektrallinien', *Z. Phys.* **70**, p. 508.

The data in Table 4.2 are from

WALSHAW, C. D., 1954, *An experimental investigation of the 9·6 μ band of ozone*, Ph.D. Thesis, Cambridge University.

The expansion in (4.51) was developed by

PLASS, G. N., and FIVEL, D. I., 1953, 'Influence of Doppler effect and damping on line-absorption coefficient and atmospheric radiation transfer', *Astrophys. J.* **117**, p. 225.

4.3. Modified single-line models

The model discussed in § 4.3.1 was developed by SCHNAIDT (1939) (§ 4.2). A modified version has been given by

PLASS, G. N., 1958, 'Models for spectral band absorption', *J. Opt. Soc. Am.* **48**, p. 690.

The discussion in § 4.3.2 follows

REICHE, F., 1956, 'Total absorption and overlapping effect due to two spectral lines of equal width and strength', ibid. **46**, p. 590.

The general model of § 4.3.3 was developed by

MATOSSI, F., MAYER, R., and RAUSCHER, E., 1949, 'On the total absorption in spectra with overlapping lines', *Phys. Rev.* **76**, p. 760; and
—— and RAUSCHER, E., 1949, 'Zur Druckabhängigkeit der Gesamtabsorption in ultraroten Bandenspektren', *Z. Phys.* **125**, p. 418.

4.4. Regular models

The Elsasser model was the first physically satisfying band model to be developed, although it was given initially only in the strong-line approximation; see

ELSASSER, W. M., 1942, *Heat transfer by infrared radiation in the atmosphere*, Harvard Meteorological Studies No. 6. Harvard Univ. Press.

Details of the Mittag–Leffler theorem, required in this derivation, are given by WHITTAKER and WATSON (1915) (§ 4.2). Fig. 4.10 is taken from KAPLAN (1954) (§ 4.1). The three asymptotic limits for both regular and random models are discussed at length by PLASS (1958) (§ 4.3) and

PLASS, G. N., 1960, 'Useful representations for measurements of spectral band absorption', *J. Opt. Soc. Am.* **50**, p. 868.

A table of numerical values of $E(y, u)$ and references to computations published in the literature are given in Appendix 11.

The computations (§ 4.4.3) on an array of doublets were performed by

YAMAMOTO, G., and SASAMORI, T., 1958, 'Calculations of the absorption of the 15 μ carbon-dioxide band', *Science Rept., Tohoku Univ., Series 5, Geophysics*, **10**, No. 2, p. 37.

The experimental data discussed in § 4.4.5 are from KAPLAN and EGGERS (1956) (§ 4.1);

GOODY, R. M., and WORMELL, T. W., 1951, 'The quantitative determination of atmospheric gases by infra-red spectroscopic methods', *Proc. Roy. Soc.* A, **209**, p. 178; and

BURCH, D. E., and WILLIAMS, D., 1960, *Infrared absorption by minor atmospheric constituents*, Scientific Rept. No. 1 on R.F. Project 778. The Ohio State University Research Foundation.

4.5. Random models

The principle of the random model was first published by GOODY (1952) (§ 4.2). This particular investigation was concerned with the limit of a large number of lines. Ideas of a similar nature are said to have been in use a few years previously in the U.S. classified literature.

Several authors have shown how the random model results from the superposition of many sub-bands of various types. The superposition of Elsasser bands is discussed by KAPLAN (1954) (§ 4.1) and PLASS (1958) (§ 4.3). The superposition of random bands is discussed by GODSON (1955) (§ 4.2).

The distribution of line spacings in a random model follows

YAMAMOTO, G., and SASAMORI, T., 1957, 'Numerical study of water vapour transmission', *Science Rept., Tohoku Univ., Series 5, Geophysics*, **8**, No. 2, p. 146.

Theoretical computations, with which we may compare the random model are given by

COWLING, T. G., 1950, 'Atmospheric absorption of heat radiation by water vapour', *Phil. Mag.* **41**, p. 109.

The experimental data in Fig. 4.14 are from

HOWARD, J. N., BURCH, D. E., and WILLIAMS, D., 1956, 'Infrared transmission of synthetic atmospheres. IV. Application of theoretical band models', *J. Opt. Soc. Am.* **46,** p. 334,

while those in Fig. 4.15 are from

WALSHAW, C. D., and GOODY, R. M., 1954, 'Absorption by the 9·6 μ band of ozone', *Proc. Toronto Met. Conf. (Roy. Met. Soc.)*, p. 49.

4.6. Generalized transmission functions

The treatment of § 4.6.1 follows KAPLAN (1953) (§ 4.1), GODSON (1955) (§ 4.2), and PLASS (1958) (§ 4.3). That of § 4.6.3 follows YAMAMOTO and SASAMORI (1957) (§ 4.5).

4.7. Restrictions on model theory

Further details may be found in the following sources:

PLASS, G. N., and WARNER, D., 1952, 'Influence of line shift and asymmetry of spectral lines on atmospheric heat-transfer', *J. Met.* **9,** p. 333;
—— —— 1952, 'Pressure broadening of absorption lines', *Phys. Rev.* **86,** p. 138;
CURTIS, A. R., and GOODY, R. M., 1954, 'Spectral line shape and its effect on atmospheric transmission', *Quart. J. R. Met. Soc.* **80,** p. 58; and
WALSHAW (1954) (see § 4.2).

5

ABSORPTION BY ATMOSPHERIC GASES

5.1. Introduction

A DISCUSSION of the data for all bands of all atmospheric gases, regardless of line strengths or concentrations of the constituents, is impracticable; we have therefore selected data that are, or may be, important. An objective criterion of importance is difficult to find. The non-linearity of absorption processes can lead to the result that almost any absorption within reason can be important under some circumstances. For example, the heat balance of the stratosphere is importantly influenced by weak lines of the $15\,\mu$ carbon dioxide band, which laboratory spectroscopists might well overlook. In the ultraviolet spectrum any criterion of importance is confused by the many complex photochemical reactions which may result from the absorption of quanta. Only a detailed analysis of a complete closed group of reactions can justify neglect of any particular spectral feature. Faced with these problems we can only claim experience and common sense as a basis for selection; criteria which, unfortunately, may well fail as new ideas enter the subject.

For wavelengths longer than $0\cdot3\,\mu$ no data on continua will be presented apart from the Chappuis bands of ozone, and the water-vapour continuum between the $6\cdot3\,\mu$ and rotation bands. Measurements on the solar spectrum, and over long paths at ground level, indicate measurable extinction in all windows between major bands, but apart from these two regions the extinction is normally associated with dust and molecular scattering. Opinions on this point may well change, and attention may be directed more towards these spectral regions in the future; some information will probably emerge as military investigations become declassified.

The absorptive properties of atmospheric gases are being measured currently for a wide variety of purposes, mainly of a technical nature. They are published in many different journals, and there is a large unpublished literature, partly classified. The reader seeking the best available data is therefore cautioned that this chapter in particular may be out of date before it reaches his hands, and that he may have to invest some effort into the investigation of unpublished material.

5.2. Nitrogen

5.2.1. *Atomic absorptions*

The electron configuration of nitrogen in the lowest energy state is $(1s)^2(2s)^2(2p)^3$, with three electrons in the incomplete L_2 sub-shell. Permitted electric dipole transitions from the ground configuration to the next highest principal quantum number involve jumps of more than 10 eV, placing the absorptions in the far ultra-violet, where they can be neglected when compared to others to be discussed in this section. Transitions in the ground configuration between the ^4S ground state and ^2D or ^2P are forbidden by dipole selection rules, but are observed in the spectrum of the aurora together with ground state transitions of the ionized atom.

More important for the absorption of solar energy are the bound-free transitions at wavelengths shorter than 852 Å, corresponding to the first ionization potential of 14·54 eV. Theoretical computations for this continuum are shown in Fig. 5.1. Additional ionization potentials of 16·44 and 18·60 eV, corresponding to an excited ionized state, also give absorption edges, but these are not included in Fig. 5.1. Ionizations from complete L_1 or K shells give X-ray absorption edges at 608 Å, 367 Å, and 30 Å.

The important continua below 852 Å originate from the ^4S ground state. Since the next level above the ground state has an excitation of 2·37 eV (cf. $\mathbf{k}\theta \simeq 0\cdot017$ eV), the ground-state population will not vary noticeably with the ambient temperature. Absorption coefficients should not therefore depend markedly upon temperature. The same applies to other atomic spectra and particularly to X-ray spectra which, originating from a closed shell, are uninfluenced by external conditions.

5.2.2. *Molecular absorptions*

The Lyman–Birge–Hopfield bands are the lowest frequency bands of molecular nitrogen (see Fig. 5.2), with a (0—0) transition at 1450 Å, the threshold of nitrogen absorption. Lyman–Birge–Hopfield bands up to the (14—0) band at 1114·2 Å have been positively identified, but below 1123 Å they are overlapped by the Tanaka bands.

Absorption coefficients in the region of the Tanaka and Lyman–Birge–Hopfield bands are not known with accuracy. There appears to be no dissociation continuum (one might be expected for wavelengths less than 1270 Å, i.e. for energies greater than the dissociation energy, 9·76 eV) and measurements have not been made with a resolution sufficient to delineate individual lines. It is not known whether the absorption depends upon temperature and pressure. The existing information is summarized in Table 5.1.

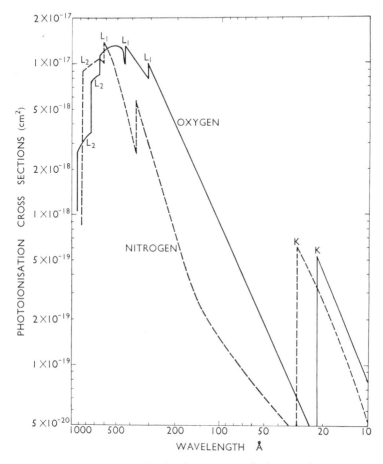

Fig. 5.1. Photo ionization by oxygen and **nitrogen** atoms.

The full line is for oxygen; the broken line for nitrogen. The shell from which the photo-electron is ejected is indicated at the absorption edges. The cross-sections have mainly been derived theoretically. After Dalgarno and Parkinson (1960).

Between the Tanaka bands and the first ionization edge at 796 Å ($\equiv 15 \cdot 58$ eV) a group of bands known as the Worley bands is partially identified. The absorption coefficients are uncertain, varying between 10^{-17} and 10^{-19} cm^2. A continuum has been claimed to exist but is not identified with any particular dissociation.

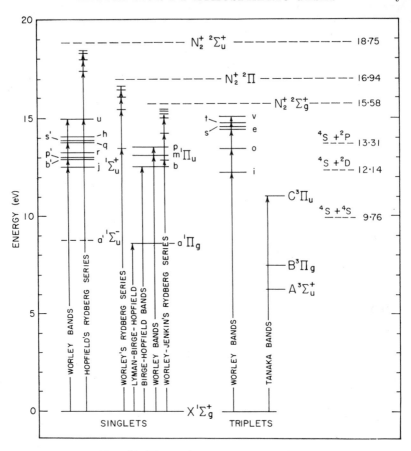

FIG. 5.2. Electronic energy levels of $N^{14}N^{14}$.

Only those transitions relevant to studies of solar absorption are shown.
Additional transitions within the triplet states and ionized states are observed
in the auroral spectrum. A more complete scheme of transitions is given by
Herzberg (1950). After Watanabe (1958).

TABLE 5.1

Molecular absorption coefficients in the Tanaka and Lyman–Birge–
Hopfield bands of N_2

After Watanabe (1958)

Position in spectrum	Coefficient
	(cm^2)
Maxima in L–B–H bands	10^{-20}
Maxima in Tanaka bands	10^{-21}
1215·6 Å ($L\alpha$)	$< 6 \times 10^{-23}$
1025·7 Å ($L\beta$)	$< 4 - 11 \times 10^{-21}$

At wavelengths below 800 Å the spectrum simplifies, and is dominated by an ionization continuum, with a peak cross-section of $2\text{--}3\times10^{-17}$ cm^2 near 600 Å. Some of the bands in this spectral region have been identified as belonging to Hopfield's and Worley's Rydberg series.

At wavelengths less than 30·3 Å (the K-shell photo ionization, see Fig. 5.1) absorption by a nitrogen molecule corresponds to that by two nitrogen atoms.

5.3. Oxygen

5.3.1. *Atomic absorptions*

Oxygen, with eight electrons, has a ground configuration $(1s)^2(2s)^2(2p)^4$, the ground state being ^3P. Dipole transitions within the ground configuration are forbidden, as for nitrogen, and intensities are too small to affect solar absorption; the 5577 ($^1S \rightarrow {}^1D$) line and the 6300–6364 ($^1D \rightarrow {}^3P$) multiplet are, however, two of the most prominent features of the airglow spectrum.

Bates (1951) has drawn attention to the important possibility of magnetic dipole transitions between the three ^3P levels in the ground configuration. The Einstein coefficients are small, $1\cdot7\times10^{-5}$ s^{-1} for $^3P_0 \rightarrow {}^3P_2$ and $8\cdot9\times10^{-5}$ s^{-1} for $^3P_1 \rightarrow {}^3P_2$, but the energy gaps are only 226 and 161 cm^{-1} respectively, so that all three states can be readily populated by thermal excitation and can emit thermal radiation. While the transition probabilities are small compared with vibration bands of polyatomic molecules the ease of excitation together with the high concentration of atomic oxygen in the ionosphere gives importance to these transitions.

Permitted transitions involving a change in the principal quantum number have not been observed because they are overlaid by strong ultra-violet continua. However, the $3d(^3D) \rightarrow 2p(^3P)$ transition of the O atom coincides with the Lyman β line of hydrogen (1026 Å) which is one of the important sources of solar ultra-violet radiation. There may therefore be circumstances under which this particular permitted line of atomic oxygen becomes important.

The X-ray spectrum of atomic oxygen is shown in Fig. 5.1.

5.3.2. *The 'atmospheric bands' of molecular oxygen*

The potential curves of O_2 and O_2^+ are shown in Fig. 3.15. Magnetic dipole transitions between the three lowest electronic levels lead to the *red* ($^3\Sigma_g^- \rightarrow {}^1\Sigma_g^+$) and *infra-red* ($^3\Sigma_g^- \rightarrow {}^1\Delta_g$) bands of molecular oxygen, sometimes called the 'atmospheric bands'. The $O^{16}O^{18}$ and $O^{16}O^{17}$ bands can also be detected in the solar spectrum.

The main infra-red bands of oxygen are the (0–0) band at $1 \cdot 2683 \, \mu$ and the (0–1) band at $1 \cdot 0674 \, \mu$. The former is the stronger and is shown in Fig. 5.3 as it appears in the solar spectrum. The only intensity

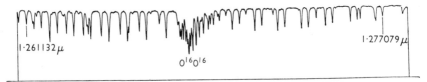

FIG. 5.3. The solar spectrum in the region of the $1 \cdot 2683 \, \mu \, O^{16}O^{16}$ band. This spectrum, recorded with a photo-conductive detector, has a lower resolving power than photographic spectra in the same region. After Goldberg (1954).

measurement available for the infra-red bands gives $A(0, 0) = 1 \cdot 9 \times 10^{-4}$ s^{-1}. Intensities and line widths for the *red* oxygen bands have been derived from measurements on P-branch lines in the solar spectrum (Table 5.2). The (0–0), (0–1), and (0–2) bands of $O^{16}O^{18}$ lie at $0 \cdot 7620 \, \mu$, $0 \cdot 6901 \, \mu$, and $0 \cdot 6317 \, \mu$. Owing to the decreased symmetry, alternate J lines are not missing for the isotopic molecule as they are for $O^{16}O^{16}$. No significant differences in line width, either inside or between the red bands has been detected, and an average figure of $0 \cdot 047 \, cm^{-1}$ at s.t.p. corresponds to a collision diameter for O_2–air collisions of $4 \cdot 1 \, Å$.

TABLE 5.2

Molecular band intensities for the red atmospheric bands of $O^{16}O^{16}$

After van de Hulst (1945)

Transition	Band centre	Band intensity†	Strength in solar spectrum‡
	(μ)	(cm)	
0–0	0·7621	$4 \cdot 3 \times 10^{-22}$	s
0–1	0·6884	$2 \cdot 0 \times 10^{-23}$	s
0–2	0·6288	$5 \cdot 1 \times 10^{-25}$	m
0–3	0·5796	$2 \cdot 0 \times 10^{-26}$	m
0–4	0·5384	$8 \cdot 1 \times 10^{-28}$	—
1–1	0·7710	$2 \cdot 7 \times 10^{-27}$	w
1–2	0·6970	$2 \cdot 3 \times 10^{-23}$	w
1–3	0·6379	—	m

† Intensities refer to the number of molecules in the *initial* state. Other tables in this chapter refer intensities to the total number of molecules and therefore include a Boltzmann factor.

‡ The classification according to strength in the solar spectrum follows Goldberg (1954): vs (very strong) complete black-out at band centre; s (strong) some very narrow gaps between lines; m (medium) structure clear for 1 air mass; w (weak) structure just detectable for 1 air mass; vw (very weak) structure only detectable with low sun.

The atmospheric bands of O_2 have been observed in emission in the auroral and airglow spectrum. A downward transition which terminates in an unpopulated, excited, vibrational state will not be subject to reabsorption, since upper state bands are of negligible intensity. However, transitions to the lowest vibrational level of the ground electronic state will suffer partial or complete reabsorption.

5.3.3. *Ultra-violet molecular absorptions*

The ultra-violet absorption spectrum of $O^{16}O^{16}$ commences with the weak Herzberg bands at 2600 Å (transition (i) of Fig. 3.15). Below 2420 Å the transition becomes dissociative, with end-products $O^{16}(^3P)$ $+O^{16}(^3P)$, and the weak Herzberg continuum sets in. The absorption coefficient per molecule is very small, being between 10^{-23} and 10^{-24} cm² at the threshold, and of little importance as regards energy absorption. It may, nevertheless, be important as regards the production of atomic oxygen, and hence have an influence on a number of chemical reactions. The Herzberg bands appear in the blue and ultra-violet regions of the airglow spectrum.

The Schumann–Runge bands occupy the spectral region 1950 Å to 1750 Å. They result from the transition (ii) ($^3\Sigma_g^- \rightarrow {}^3\Sigma_u^-$) in Fig. 3.15. Approximate absorption coefficients are given in Fig. 5.4. At 1750 Å the bands merge into a stronger dissociation continuum,

$$O_2\,{}^3\Sigma_g^- \rightarrow O^{16}(^3P)+O^{16}(^1D),$$

which extends to 1300 Å and is the most important single feature of the absorption spectrum of molecular oxygen (the Schumann–Runge continuum). The three bands at 1293 Å, 1332 Å, and 1352 Å may indicate dissociation products with higher energy than $^3P+{}^1D$.

The bands between 1060 Å and 1280 Å have not yet been identified. Particular attention has been paid to measuring the absorption at the hydrogen $L\alpha$ line (1215·7 Å), which happens to lie in a deep absorption minimum. Most recent measurements of absorption at very low pressures give $1·00 \times 10^{-20}$ cm² and a self-broadening pressure effect of $1·47 \times 10^{-23}$ cm² mb^{-1}. The mechanism involved in this pressure effect is unclear, but for atmospheric conditions it is unimportant.

Between 850 and 1100 Å are a series of distinct Rydberg bands, known as the Hopfield bands, with peak cross-sections near 950 Å as great as 5×10^{-17} cm². Below 1026·5 Å (12·08 eV) the absorption is partly caused by bound-free ionizing transitions. At the hydrogen $L\beta$ line (1025·7 Å) ionization is responsible for 58 per cent of the observed coefficient of $1·55 \times 10^{-18}$ cm².

Below 850 Å ionization dominates the absorption. A maximum cross-section close to 3×10^{-17} cm² is observed between 400 and 600 Å, and hereafter the absorption falls to small values below 100 Å. Below 300 Å absorption by an oxygen molecule is probably close to that of two oxygen atoms (see Fig. 5.1).

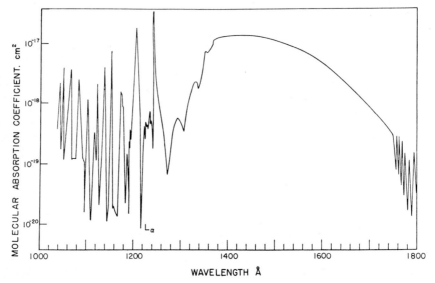

Fig. 5.4. Absorption coefficients of $O^{16}O^{16}$ in the far ultra-violet spectrum. After Watanabe (1958).

5.3.4. *Microwave absorption*

The ground state of molecular oxygen is a $^3\Sigma$ state, for which $\Lambda = 0$. The important coupling between angular momenta is between the spin vector **S** and the vector **K** (see § 3.5; in this particular case the nuclear angular momentum) to form the total angular momentum

$$\mathbf{J} = \mathbf{S} + \mathbf{K}. \qquad (5.1)$$

Since **J**, **S**, and **K** are quantized, they must be added according to the quantal addition rule, which permits $(2S+1)$ possibilities, provided $K \geqslant S$. For oxygen, $S = 1$, and there are three states with $J = (K+1)$, K, or $(K-1)$. The nuclear rotation generates a very small magnetic moment, whose interaction with the spin magnetic moment is small and differs in the three cases. The magnetic moment differs from state to state so that a magnetic dipole transition is possible between these close sub-levels. The transition probabilities are small, but this is offset (e.g. with respect to water vapour) by the large amount of oxygen in the atmosphere. Interaction energies vary slowly with K provided $K > 1$ and all transitions from $(K-1) \to K$ or $K \to (K+1)$ ($K(-)$ and $K(+)$ lines respectively) give lines in the neighbourhood of 2 cm^{-1}.

Outside the 2 cm^{-1} band lies the $1(-)$ transition, a single line close to 4 cm^{-1}. Also a small non-resonant absorption, corresponding to the classical Debye effect, gives a line centred on zero frequency.

Matrix elements of the magnetic moment for these transitions can be computed from standard formulae and require only a knowledge of the magnetic moment of the molecule (two Bohr magnetons in this case). No satisfactory, direct measurements of line intensity have yet been made, but widths have been determined in both bands.

5.3.5. *Oxygen polymers*

Dufay (1942) claims that a weak band at 4774 Å in the solar spectrum, visible only for zenith angles greater than 85°, is caused by the oxygen polymer $(O_2)_2$. The spectrum of the polymer has been recorded in the laboratory. Three bands lie at 6290 Å, 5770 Å, and 4774 Å. They are diffuse (as is to be expected since free rotation will be hindered) and the optical thickness varies as the square of the oxygen pressure. Optical thicknesses for 1 km of pure oxygen at s.t.p. are 0·018, 0·035, and 0·021 for the maxima of the above three bands respectively. Two weaker bands have been claimed at 5325 Å and 4470 Å.

Since the density of $(O_2)_2$ depends upon the square of the pressure, it must form a very low-lying layer. The absorption is normally small compared with that of ozone, but for zenith angles close to 90° the geometric path through the low-lying $(O_2)_2$ is much longer than that through the ozone lying near 25 km. The cosine of the zenith angle at height z and at ground level for the same ray are related by

$$\xi_z = \sqrt{\left/ \left(1 - \frac{1-\xi_0^2}{(1+z/r)^2}\right) \right.}, \tag{5.2}$$

where r is the radius of the earth. For example, if $(\xi_0)^{-1}$ is infinite, $(\xi_z)^{-1}$ at 25 km is only about 10. According to Dufay this effect should cause the 4774 Å polymer band to emerge from the stronger ozone Chappuis bands (§ 5.6.1) for very large zenith angles. The 5770 Å and 6290 Å bands are overlaid by water vapour, which also forms a low-lying layer, and this means of differentiation is not available.

5.4. Water vapour

5.4.1. *Electronic bands*

The electronic energy levels of water vapour have not yet been classified, although Rydberg series have been identified among the many observed bands, and the first ionization potential (12·59 eV) has been inferred from the series limits. The long-wave limit of detectable electronic absorption by water vapour is about 1860 Å. From here a continuum reaches to 1450 Å, with a maximum absorption of 5×10^{-18} cm² near 1650 Å. Between 1450 Å and the first ionization at 986 Å, there are a number of well-marked bands. Below 936 Å is again a continuum, with an ionization yield of 34 per cent at 931 Å and 75 per cent at 850 Å. There is some disagreement as to whether these absorptions are pressure-dependent or not.

5.4.2. *The vibration-rotation spectrum*

The vibration-rotation spectrum of water vapour and its isotopes is rich and complex, with irregularly spaced lines detectable in the solar spectrum from the visible to the microwave region. The large dipole moments (H_2O, $M_B = 1·94$ debye; D_2O, $M_B = 1·87$ debye; HDO, $M_A = 0·64$ debye, $M_B = 1·70$ debye) give rise to intense rotation bands. The two hydrogen atoms mean small moments of inertia and hence widely spaced rotation lines. The three moments of inertia differ greatly from one another and the rotational structure appears complicated and disorganized. The comparatively large amount of water vapour in the earth's atmosphere gives observable absorption in the solar spectrum for some very weak transitions. These four features combine to account for the ubiquitous water-vapour lines, in every part of the solar spectrum, and their unique importance to atmospheric problems.

Four isotopic forms of water have identifiable lines in the solar spectrum: HHO^{16}, HHO^{18}, HHO^{17}, and HDO^{16}, which, according to the average abundances shown in Table 1.2, are present in the proportions 99·73:0·2039:0·0373:0·0298 respectively. Each of these molecules has a different vapour pressure and the abundances depend to some extent on the evaporation-condensation cycle; abundances in the atmosphere may well differ by 10 per cent from the values quoted. The HHO^{18} and HHO^{17} molecules have vibrational and rotational constants differing very little from those of HHO^{16}. Relative to the latter molecule, HHO^{18} lines are shifted by 1 to -11 cm^{-1} while shifts for HHO^{17} are half as great. HDO^{16} bands stand out separately, however, since the molecule has completely different vibrational frequencies from the other isotopic forms.

Water vapour has three fundamentals (see Fig. 3.5 (a) for vibrations of the XY_2 molecule), the bending vibration ν_2 having the lowest frequency while ν_1 and ν_3 both have approximately twice this frequency. The close coincidence between ν_1, ν_3, and $2\nu_2$ implies complex interactions between states. Table 5.3 shows observed positions of the fundamental bands of HHO¹⁶ and HDO¹⁶. HHO bands are type B if ν_3 changes by an even number, otherwise they are type A. For HDO, the lower symmetry allows hybrid bands of mixed type A and B.

TABLE 5.3

Observed positions of H_2O *fundamentals*

Band	Transition	Band centre (cm^{-1}) HHO¹⁶	HDO¹⁶
ν_1	000–100	3657·05	2723·66
ν_2	000–010	1594·78	1403·3
ν_3	000–001	3755·92	3707·47

The characteristically random appearance of the water vapour spectrum (see Figs. 3.4 and 3.13) is caused by the great differences between the three moments of inertia. Rotational constants for the ground electronic and vibrational state are shown in Table 5.4; they differ slightly in excited vibrational states.

TABLE 5.4

Rotational constants of water vapour in the ground state (cm^{-1})

Axis	HHO¹⁶	HDO¹⁶
A	27·79	23·38
B	14·51	9·06
C	9·29	6·38

5.4.3. *The rotation band*

A complete account of a molecular band requires a knowledge of line positions, intensities, shapes, and widths, for ground state and upper state bands, and for all isotopic species. Largely through the work of Benedict and his collaborators, most of this information is now available for the rotation band of water vapour.

Line positions can be computed when energy levels and selection rules are known. The selection rules are type B, and the energy levels have been tabulated. The dipole moments for water vapour are known (§ 5.4.2), and line intensity computations based on a rigid rotator approximation are straightforward, if tedious. An early result of such a computation is given in Fig. 3.4. More recent but unpublished work by Benedict applies a small first-order correction for centrifugal stretching, varying with both J and τ. This correction has been included in the data in Table 5.5, which also include isotopic lines as if they were additional, weak lines of HHO[16].

The discussion in § 3.6.4.3 suggests that the Anderson theory should apply to water-vapour vibration-rotation lines, and that the Lorentz shape will hold out to frequency displacements of the order of a line spacing. In the main part of an absorption band therefore transmission functions based on the Lorentz profile should be reasonably accurate; in band wings, where line-wing contributions are of great importance, a semi-empirical treatment is necessary. The Anderson theory has been used to compute line widths of water vapour broadened by nitrogen. Since the results are of unique importance to our problem they will be briefly described.

The postulated molecular interaction is between a water-vapour dipole and a nitrogen quadrupole, with an additional catastrophic effect at a separation of 3·20 Å. This latter figure is the kinetic-theory collision diameter derived from an analysis of the equation of state. The quadrupole moment was fixed at $q = 1·87 \times 10^{-18}$ e.s.u. by a trial-and-error procedure which gave the correct value for the measured width of the $(6_{-5}$—$5_{-1})$ line in the microwave spectrum (see below). These data are then sufficient to compute all other line widths.

The computed line widths depend more strongly on the quantum numbers of the lower state than the upper state, and in Fig. 5.5 are shown the results of averaging over all upper states with weights proportional to the line intensities. Smooth curves have been fitted to the points, which are omitted to avoid confusion. The smallest individual line width computed is 0·03200 cm^{-1} at s.t.p. for 13_{-13}–14_{-14}. This is close to the value obtained if the kinetic-theory collision diameter alone is considered. The largest line width is 0·11115 cm^{-1} at s.t.p. for 1_{-1}–1_{+1}. For line widths greater than 0·08 cm^{-1} at s.t.p. the assumption of a catastrophic encounter at 3·20 Å no longer influences the computation. The weighted average line width over the whole band is 0·087 cm^{-1} at 300° K and 1 atmosphere pressure.

The temperature-dependence of line width was obtained by means of repeated computations at $\theta = 220°$ K, $260°$ K, and $300°$ K. In this temperature range a relation of the type

$$\frac{\alpha(\theta)}{\alpha(300)} = \left(\frac{\theta}{300}\right)^{-n}, \tag{5.3}$$

holds at constant pressure, implying σ proportional to $\theta^{(0.5-n)/2}$. For the narrow lines, when the collision diameters approach kinetic-theory values, σ does not vary with θ, and $n = 0.5$. Under this condition the modified Lorentz profile of § 3.6.4.2 will apply. For the broadest lines, however, there is close to complete resonance and, according to these

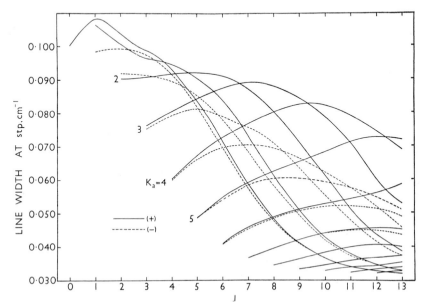

FIG. 5.5. Weighted mean H_2O line widths for collisions with N_2 molecules in terms of quantum numbers for the lower state (at 300° K).
After Benedict and Kaplan (1959).

computations $\sigma \sim \theta^{-1/6}$ with $n = 0.833$. Under this circumstance the line profile should be closer to the Lorentz profile.

In some sample computations line widths were estimated to be from 0.00585 cm^{-1} lower to 0.00160 cm^{-1} higher in the ν_2 band than for the same quantum numbers in the rotation band. For the $2\nu_2$ band the differences were about twice as great.

The use of these extensive data depends to some degree upon having a suitable band model; in this case the statistical model is the obvious choice. For all distributions of line intensities discussed in § 4.2.2, we saw that the parameters y and u could be expressed in terms of the sums $\sum_i S(i)$ and $\sum_i \sqrt{\{S(i)\alpha(i)\}}$. In Table 5.5 these quantities are given for 20 cm^{-1} intervals throughout the rotation band, at three representative temperatures. The final row in the table gives the total band intensity.

TABLE 5.5

Smoothed data for the H_2O rotation band

The first-order correction for centrifugal stretching has been included. Line widths are at 1 atmosphere pressure. The final row is the band intensity. After Benedict and Kaplan (1959) and unpublished data by Benedict.

ν (cm^{-1})	$\sum_i S_m(i)$ $(g^{-1}\,cm)$			$\sum_i \sqrt{\{\alpha(i)S_m(i)\}}$ $(g^{-1/2})$		
	220° K	260° K	300° K	220° K	260° K	300° K
0–20	$3\cdot639\times10^3$	$2\cdot500\times10^3$	$1\cdot805\times10^3$	$3\cdot068\times10$	$2\cdot506\times10$	$2\cdot106\times10$
20–40	$2\cdot716\times10^4$	$2\cdot069\times10^4$	$1\cdot614\times10^4$	$1\cdot463\times10^2$	$1\cdot218\times10^2$	$1\cdot036\times10^2$
40–60	$8\cdot005\times10^4$	$6\cdot147\times10^4$	$4\cdot900\times10^4$	$2\cdot886\times10^2$	$2\cdot524\times10^2$	$2\cdot244\times10^2$
60–80	$8\cdot600\times10^4$	$7\cdot347\times10^4$	$6\cdot309\times10^4$	$3\cdot454\times10^2$	$3\cdot133\times10^2$	$2\cdot860\times10^2$
80–100	$1\cdot854\times10^5$	$1\cdot497\times10^5$	$1\cdot237\times10^5$	$4\cdot214\times10^2$	$3\cdot831\times10^2$	$3\cdot517\times10^2$
100–120	$1\cdot296\times10^5$	$1\cdot187\times10^5$	$1\cdot078\times10^5$	$3\cdot368\times10^2$	$3\cdot281\times10^2$	$3\cdot172\times10^2$
120–140	$2\cdot459\times10^5$	$2\cdot307\times10^5$	$2\cdot141\times10^5$	$4\cdot804\times10^2$	$4\cdot798\times10^2$	$4\cdot743\times10^2$
140–160	$2\cdot492\times10^5$	$2\cdot312\times10^5$	$2\cdot142\times10^5$	$4\cdot223\times10^2$	$4\cdot158\times10^2$	$4\cdot082\times10^2$
160–180	$1\cdot345\times10^5$	$1\cdot300\times10^5$	$1\cdot249\times10^5$	$2\cdot893\times10^2$	$2\cdot912\times10^2$	$2\cdot918\times10^2$
180–200	$3\cdot754\times10^4$	$4\cdot953\times10^4$	$6\cdot054\times10^4$	$1\cdot431\times10^2$	$1\cdot657\times10^2$	$1\cdot838\times10^2$
200–220	$2\cdot267\times10^5$	$2\cdot274\times10^5$	$2\cdot204\times10^5$	$3\cdot579\times10^2$	$3\cdot534\times10^2$	$3\cdot480\times10^2$
220–240	$1\cdot091\times10^5$	$1\cdot176\times10^5$	$1\cdot214\times10^5$	$2\cdot189\times10^2$	$2\cdot313\times10^2$	$2\cdot398\times10^2$
240–260	$1\cdot216\times10^5$	$1\cdot444\times10^5$	$1\cdot580\times10^5$	$2\cdot307\times10^2$	$2\cdot485\times10^2$	$2\cdot601\times10^2$
260–280	$3\cdot927\times10^4$	$5\cdot404\times10^4$	$6\cdot452\times10^4$	$1\cdot172\times10^2$	$1\cdot350\times10^2$	$1\cdot485\times10^2$
280–300	$2\cdot806\times10^4$	$3\cdot910\times10^4$	$4\cdot864\times10^4$	$1\cdot219\times10^2$	$1\cdot424\times10^2$	$1\cdot584\times10^2$
300–320	$3\cdot591\times10^4$	$5\cdot594\times10^4$	$7\cdot415\times10^4$	$1\cdot021\times10^2$	$1\cdot250\times10^2$	$1\cdot436\times10^2$
320–340	$2\cdot280\times10^4$	$3\cdot445\times10^4$	$4\cdot636\times10^4$	$1\cdot123\times10^2$	$1\cdot335\times10^2$	$1\cdot512\times10^2$
340–360	$1\cdot274\times10^4$	$2\cdot467\times10^4$	$3\cdot950\times10^4$	$9\cdot127\times10$	$1\cdot210\times10^2$	$1\cdot472\times10^2$
360–380	$3\cdot380\times10^3$	$7\cdot247\times10^3$	$1\cdot287\times10^4$	$3\cdot500\times10$	$4\cdot998\times10$	$6\cdot479\times10$
380–400	$3\cdot589\times10^3$	$7\cdot065\times10^3$	$1\cdot240\times10^4$	$4\cdot343\times10$	$6\cdot287\times10$	$7\cdot795\times10$
400–420	$2\cdot879\times10^3$	$4\cdot801\times10^3$	$7\cdot310\times10^3$	$3\cdot010\times10$	$4\cdot150\times10$	$5\cdot291\times10$
420–440	$4\cdot067\times10^2$	$1\cdot058\times10^3$	$2\cdot522\times10^3$	$1\cdot458\times10$	$2\cdot218\times10$	$3\cdot152\times10$
440–460	$1\cdot338\times10^3$	$2\cdot305\times10^3$	$3\cdot737\times10^3$	$2\cdot971\times10$	$4\cdot027\times10$	$5\cdot179\times10$
460–480	$5\cdot353\times10^2$	$9\cdot816\times10^2$	$1\cdot748\times10^3$	$1\cdot473\times10$	$2\cdot106\times10$	$2\cdot846\times10$
480–500	$3\cdot266\times10^2$	$4\cdot725\times10^2$	$7\cdot165\times10^2$	$8\cdot949$	$1\cdot247\times10$	$1\cdot704\times10$
500–520	$5\cdot320\times10^2$	$1\cdot059\times10^3$	$2\cdot023\times10^3$	$1\cdot501\times10$	$2\cdot260\times10$	$3\cdot137\times10$
520–540	$2\cdot056\times10^2$	$3\cdot762\times10^2$	$5\cdot861\times10^2$	$6\cdot055$	$8\cdot368$	$1\cdot124\times10$
540–560	$1\cdot230\times10^2$	$2\cdot126\times10^2$	$3\cdot318\times10^2$	$7\cdot803$	$1\cdot044\times10$	$1\cdot337\times10$
560–580	$1\cdot532\times10^2$	$2\cdot761\times10^2$	$4\cdot350\times10^2$	$7\cdot016$	$9\cdot253$	$1\cdot160\times10$
580–600	$1\cdot090\times10^2$	$2\cdot405\times10^2$	$4\cdot137\times10^2$	$6\cdot770$	$1\cdot017\times10$	$1\cdot368\times10$
600–620	$4\cdot473\times10$	$9\cdot160\times10$	$1\cdot660\times10^2$	$3\cdot594$	$5\cdot056$	$6\cdot557$
620–640	$4\cdot738\times10$	$1\cdot155\times10^2$	$2\cdot241\times10^2$	$3\cdot804$	$5\cdot954$	$8\cdot291$
640–660	$1\cdot811\times10$	$4\cdot632\times10$	$1\cdot149\times10^2$	$2\cdot611$	$4\cdot046$	$5\cdot646$
660–680	$1\cdot752\times10$	$2\cdot861\times10$	$4\cdot432\times10$	$2\cdot187$	$2\cdot988$	$3\cdot866$
680–700	$1\cdot510\times10$	$4\cdot446\times10$	$9\cdot799\times10$	$2\cdot371$	$3\cdot992$	$5\cdot825$
700–720	$9\cdot878$	$2\cdot375\times10$	$4\cdot898\times10$	$1\cdot857$	$3\cdot026$	$4\cdot421$
720–740	$2\cdot759$	$5\cdot297$	$8\cdot913$	$6\cdot89\times10^{-1}$	$1\cdot034$	$1\cdot434$
740–760	$3\cdot936$	$1\cdot282\times10$	$3\cdot219\times10$	$1\cdot223$	$2\cdot243$	$3\cdot553$
760–780	$1\cdot395$	$3\cdot682$	$8\cdot347$	$6\cdot80\times10^{-1}$	$1\cdot138$	$1\cdot685$
780–800	$3\cdot779$	$9\cdot047$	$1\cdot816\times10$	$9\cdot77\times10^{-1}$	$1\cdot601$	$2\cdot327$
800–820	$7\cdot679\times10^{-1}$	$2\cdot382$	$6\cdot481$	$5\cdot20\times10^{-1}$	$9\cdot65\times10^{-1}$	$1\cdot565$
820–840	$3\cdot820\times10^{-1}$	$9\cdot012\times10^{-1}$	$2\cdot015$	$3\cdot31\times10^{-1}$	$5\cdot57\times10^{-1}$	$8\cdot56\times10^{-1}$
840–860	$6\cdot959\times10^{-1}$	$2\cdot179$	$5\cdot539$	$5\cdot37\times10^{-1}$	$1\cdot010$	$1\cdot676$
860–880	$2\cdot528\times10^{-1}$	$7\cdot373\times10^{-1}$	$1\cdot677$	$2\cdot22\times10^{-1}$	$4\cdot02\times10^{-1}$	$6\cdot20\times10^{-1}$
880–900	$4\cdot597\times10^{-1}$	$9\cdot512\times10^{-1}$	$1\cdot855$	$3\cdot20\times10^{-1}$	$5\cdot24\times10^{-1}$	$7\cdot86\times10^{-1}$
900–920	$1\cdot411\times10^{-1}$	$5\cdot639\times10^{-1}$	$1\cdot630$	$1\cdot34\times10^{-1}$	$3\cdot03\times10^{-1}$	$5\cdot68\times10^{-1}$
920–940	$1\cdot956\times10^{-1}$	$5\cdot298\times10^{-1}$	$1\cdot239$	$1\cdot64\times10^{-1}$	$2\cdot94\times10^{-1}$	$4\cdot75\times10^{-1}$
940–960	$1\cdot890\times10^{-1}$	$5\cdot588\times10^{-1}$	$1\cdot327$	$2\cdot60\times10^{-1}$	$4\cdot70\times10^{-1}$	$7\cdot50\times10^{-1}$
960–980	$8\cdot159\times10^{-1}$	$3\cdot323\times10^{-1}$	$1\cdot029\times10^{-1}$	$2\cdot30\times10^{-1}$	$4\cdot20\times10^{-1}$	$6\cdot63\times10^{-1}$
980–1000	$2\cdot814\times10^{-1}$	$1\cdot044\times10^{-1}$	$2\cdot615\times10^{-1}$	$6\cdot1\times10^{-2}$	$1\cdot22\times10^{-1}$	$1\cdot98\times10^{-1}$
0–1000	$1\cdot785\times10^6$	$1\cdot792\times10^6$	$1\cdot795\times10^6$	—	—	—

While experimental confirmation of these data is fragmentary it all points to the conclusion that they are reliable. Measurements on individual high-J lines have been given in Table 3.3 which show agreement with computations to within ± 10 per cent. In the same spectral region averaged absorption data have been shown to agree reasonably well with the theory. A number of low-resolution measurements in the

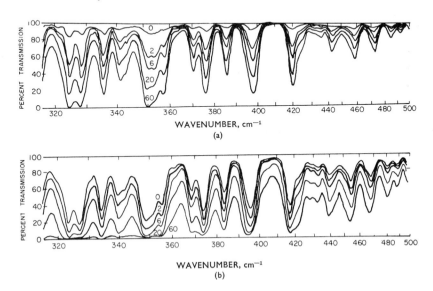

FIG. 5.6. The absorption of water vapour mixed with nitrogen from $20\ \mu$ to $31\cdot7\ \mu$.

Temperature approx. 293° K; resolution $4\ cm^{-1}$. After Palmer (1957). (a) $7\cdot7 \times 10^{-3}$ g cm^{-2}; H_2O partial pressure $0\cdot04$ cm Hg; N_2 partial pressure in cm Hg shown on each curve. (b) $7\cdot5 \times 10^{-2}$ g cm^{-2}; H_2O partial pressure $0\cdot4$ cm Hg; N_2 partial pressure in cm Hg shown on each curve.

laboratory are available (see Bibliography) and some results from one are shown in Fig. 5.6.

The data in Fig. 5.6 have been compared with theoretical computations using the detailed spectrum. The computations were in satisfactory agreement with experiment after a slight increase of the theoretical intensities. The computed transmissions over wide frequency intervals were found to 'approach statistically calculated forms'. The results were only in the strong line region and hence both band parameters were not determined.

The microwave spectrum is a more promising region for an exacting test of the theory, but unfortunately not very much quantitative work has been done. Strong lines occur at $10\cdot80$ cm^{-1} $(5_{-2}-4_0)$, $10\cdot78$ cm^{-1}

$(10_{-7}-9_{-3})$, $6\cdot11$ cm^{-1} $(3_{-2}-2_{-2})$, and $0\cdot742$ cm^{-1} $(6_{-5}-5_{-1})$, and there are many other weak lines as well as HDO and D_2O lines in this spectral region, some of whose positions are known to seven significant figures. The only critical test of the theory is for the $(6_{-5}-5_{-1})$ line. The single unknown parameter in the comparison between theory and observation was the line width, for which a value of $0\cdot087$ cm^{-1} at s.t.p. has been found for best fit. For resonant broadening in pure water vapour a width $4\cdot7$ times greater was found. With these lines excellent agreement was found with the Van Vleck and Weisskopf line shape, except for a small and constant difference of absorption coefficient, probably attributable to the far wings of other strong lines.

5.4.4. *The ν_1, ν_2, ν_3, and $2\nu_2$ bands*

The ν_2 band, at $1594\cdot78$ cm^{-1} $(6\cdot25\,\mu)$, is the most important vibration-rotation band of water vapour. The ν_1 and ν_3 bands at $3657\cdot05$ cm^{-1} and $3755\cdot92$ cm^{-1} $(2\cdot74\,\mu$ and $2\cdot66\,\mu$ respectively) are at wavelengths which are not very important for either solar or planetary radiation. The ν_1 band is considerably weaker than ν_3. The $2\nu_2$ band, centred at $3161\cdot60$ cm^{-1} $(3\cdot17\,\mu)$ is considered together with the ν_1 and ν_3 bands, since with some weak CO_2 bands they together make up the strong χ-band in the solar spectrum. Since all four bands involve transitions from the ground state, they all have the same temperature dependence.

The selection rules for ν_1, ν_2, and $2\nu_2$ are type B, while those for ν_3 are type A. Line positions can be calculated from the published energy levels. Line intensities and widths are not so well established for these bands as for the rotation band. We may tentatively accept the line widths computed for the rotation spectrum as applicable to any type B band, and in the absence of evidence to the contrary they may also be applied to type A bands.

Owing to the anharmonicity of the vibrations we have no theoretical knowledge of vibrational transition probabilities. Fortunately, measurements of band intensities are available (Table 5.6). In the absence of interactions between vibration and rotation the band intensity can be partitioned between lines according to the probability of the rotational transition. Agreement between theory and observation for the ν_2 and $2\nu_2$ bands suggests that this is valid for strong lines, and it is reasonable to suppose that it is also valid for the strong lines of ν_1 and ν_3. However, the close frequency coincidence between ν_1, ν_3, and $2\nu_2$ gives large interactions, which seriously modify the strength of weak lines of the ν_2 band. Measurements on the $(9_{-5}-10_{-3})$ and $(6_{-5}-7_1)$ lines of ν_2, for example, yield intensities only 10 per cent of expectation on a non-interacting model.

In addition to these details there exist two extensive empirical investigations of the four bands at low resolution by Howard, Burch, and Williams (1956) and Burch, Singleton, France, and Williams (1960).

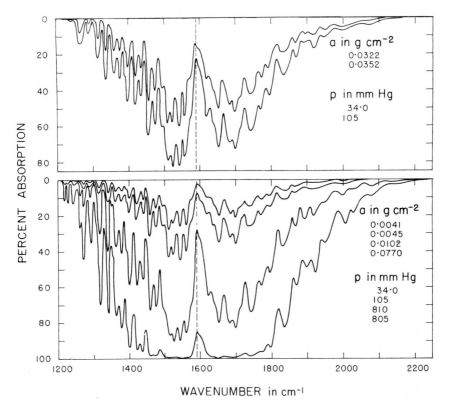

FIG. 5.7. Absorption by the ν_2 band of H_2O at low resolution.
p refers to the pressure of nitrogen. After Burch, Singleton,
France, and Williams (1960).

Specimen data are shown in Figs. 5.7 and 5.8. Observations were made with path lengths up to 1320 m in a multiple reflection cell, and the water vapour was mixed with nitrogen in a dilute solution. Howard *et al.* interpreted their data in terms of a statistical model with an exponential distribution of line intensities (4.28) and (4.94), and the precision of the fit between theory and observation has already been demonstrated in Fig. 4.14. Results are not given directly in terms of the parameters u and y, but, assuming that y $(= \alpha/\delta)$ varies little from band to band, the authors use the relation

$$\bar{T} = \exp\left(\frac{-1\cdot97a/a_0}{(1+6\cdot56a/a_0)^{1/2}}\right),\tag{5.4}$$

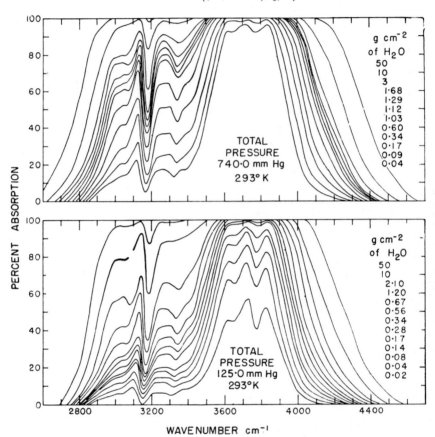

FIG. 5.8. Absorption by the ν_1, ν_3, and $2\nu_2$ bands of H_2O at low resolution.
The total pressure refers to the combination of water vapour and nitrogen.
After Howard, Burch, and Williams (1956).

where a_0 is the amount of water required to give a transmission of $\frac{1}{2}$.†
From (4.28) and (5.4)

$$\frac{1\cdot97}{a_0} = \frac{\sigma}{\delta},\tag{5.5}$$

$$\frac{6\cdot56}{a_0} = \frac{\sigma}{\pi\alpha}.\tag{5.6}$$

† From (5.4) with $a = a_0$, $\bar{T} = 0\cdot491$. This discrepancy is presumably an oversight
on the part of the authors.

There is a difficulty with these equations. It is claimed that a_0 is proportional to p^{-1} and yet σ and δ are independent of pressure. It appears that the numbers in (5.4) must be pressure dependent and should be $6{\cdot}56\times(p_0/p)^2$ and $1{\cdot}97\times(p_0/p)$ respectively, where p_0 is a pressure of one atmosphere.

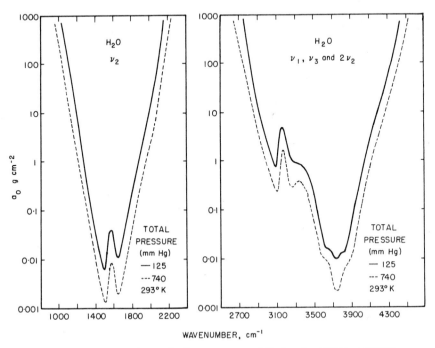

Fig. 5.9. The quantity a_0 for the ν_1, ν_2, ν_3, and $2\nu_2$ bands of H_2O at 293° K. The total pressure refers to the mixture of nitrogen and water vapour. After Howard, Burch, and Williams (1956).

From (5.5) and (5.6)

$$\alpha/\delta = 0{\cdot}0955 \text{ at s.t.p.} \tag{5.7}$$

Using the mean line width $0{\cdot}087$ cm^{-1} at s.t.p. computed for the rotation band (§ 5.4.3) the mean line spacing is $0{\cdot}91$ cm^{-1}, a not unreasonable value for the centre of a water-vapour band. Although we will make use of these data, the reader is warned that the relations (5.4), (5.5), (5.6), and the behaviour of a_0 as shown in Fig. 5.9 are only consistent because all of the data happen to fall in the *strong-line* region. Consequently we do not have enough data to determine σ, α, and δ independently, but must bring in some extra information. We will use the value $0{\cdot}087$ cm^{-1} for the mean line width at s.t.p. for this purpose.

From (5.5) the band intensity is

$$S = \int \frac{\sigma}{\delta}\, dv = 1 \cdot 97 \int \frac{dv}{a_0}. \qquad (5.8)$$

Band intensities derived on this basis are given in Table 5.6. Independent data are shown in the last column. These employ trial and error to fit measured relative intensities and emissivities (§ 5.4.8).

TABLE 5.6

Molecular band intensities of H_2O *from the data of Howard et al.* (1956)

The ratio $S(v_1)/S(v_2)$ is taken from Benedict (1956) and is used to separate the combined data on v_1 and v_3. The final column is from Penner (1959)

Band	Band centre	Type	$\int dv/a_0$	S_n (5.8)	S_n (Penner)
	(cm^{-1})		(g^{-1} cm)	(cm)	(cm)
v_2	1594·78	B	$1 \cdot 4 \times 10^5$	$8 \cdot 3 \times 10^{-18}$	$6 \cdot 5 \times 10^{-18}$
$2v_2$	3151·60	B	$1 \cdot 1 \times 10^3$	$6 \cdot 5 \times 10^{-20}$	
v_1	3657·05	B	$\Big\}\ 1 \cdot 1 \times 10^5$	$3 \cdot 0 \times 10^{-19}$	$\Big\}\ 4 \cdot 7 \times 10^{-18}$
v_3	3755·92	A		$6 \cdot 2 \times 10^{-18}$	

Table 5.7 gives empirical data for the total band area. For *strong absorption* (defined by $\int A_v\, dv$ greater than column 5) Howard *et al.* (1956) give

$$\int A_v\, dv = ca^d p^k, \qquad (5.9)$$

where c, d, and k are constants. For weak absorption ($\int A_v\, dv$ less than column 5) they give

$$\int A_v\, dv = C + D \log a + K \log p, \qquad (5.10)$$

where C, D, and K are constants, a is in g cm^{-2} and p in mm Hg.

TABLE 5.7

Empirical constants for the $(v_1 + v_3)$, v_2, *and* $2v_2$ *bands of* H_2O

Use with (5.9) and (5.10) where a is in g cm^{-2} and p in mm Hg and $d = 0 \cdot 5$.
After Howard *et al.* (1956)

Band	Limits	c	k	$A_v\, dv$ (transition)	C	D	K
	(cm^{-1})			(cm^{-1})			
v_2	1150–2050	356	0·30	160	302	218	157
$2v_2$	2800–3340	40·2	0·30	500	—	—	—
$v_1 + v_3$	3340–4400	316	0·32	200	337	246	150

5.4.5. *The overtone and combination bands*

The solar spectrum contains identifiable lines from a large number of water-vapour bands besides the fundamentals. Those in the visible spectrum are all relatively weak and arise from ground state transitions (see Table 5.8).

TABLE 5.8

The visible bands of H_2O

The band intensities follow Benedict (1948) and Goldberg (1954)

Transition	Band centre	Type	S_n
	(cm^{-1})		(cm)
000–411	18 394	A	2×10^{-23}
000–203	17 495	A	1×10^{-22}
000–401	16 899	A	3×10^{-22}
000–302	16 898	B	3×10^{-23}
000–321	16 822	A	2×10^{-22}
000–113	15 832	A	2×10^{-23}
000–311	15 348	A	2×10^{-22}
000–103	14 319	A	1×10^{-21}
000–400	14 221	B	1×10^{-22}
000–301	13 831	A	3×10^{-21}
000–202	13 828	B	$< 2 \times 10^{-23}$
000–221	13 653	A	6×10^{-21}
000–013	12 565	A	1×10^{-22}
000–112	12 408	B	6×10^{-21}
000–211	12 151	A	6×10^{-23}
000–210	12 140	B	1×10^{-22}
000–131	11 813	A	2×10^{-21}

At longer wavelengths appears a stronger group of water-vapour bands, some of which show regions of complete black-out in the solar spectrum. They absorb an important amount of solar radiation in the lower atmosphere and are commonly identified in groups by the Greek letters ρ, σ, τ, ϕ, ψ, and Ω. A number of investigations at low resolution have been made and constants for the empirical expressions (5.9) and (5.10) are known (Table 5.9). Curves of a_0 for the ϕ, ψ, and Ω bands are shown in Fig. 5.10. Band intensities have been derived from these data on the lines of equation (5.8) and are shown in Table 5.10.

These bands are all blends. Benedict (see Goldberg, 1954) has assigned band intensities, and with his relative values and the band intensities of Table 5.9 the data in Table 5.10 have been prepared.

Complete lists of rotational energy levels for all the above-mentioned bands are not available, although Benedict (1956) has identified many

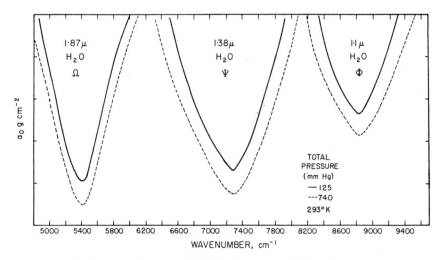

FIG. 5.10. The quantity a_0 at 293° K for the ϕ, ψ, and Ω bands of water vapour.
After Howard *et al.* (1956).

TABLE 5.9

Empirical constants for ρ, σ, τ, ϕ, ψ, and Ω bands of H_2O at 293° K

After Howard *et al.* (1956). $d = 0.5$ for all bands. The final column is from
Penner (1959)

Band	Limits	c	k	$\int A_\nu \, d\nu$ (transition)	C	D	K	S_n Howard	S_n Penner
	(cm^{-1})			(cm^{-1})				(cm)	(cm)
ρ, σ, τ	10 100–11 500	38	0.27	200	—	—	—	9×10^{-21}	1.2×10^{-20}
ϕ	8 300–9 300	31	0.26	200	—	—	—	1.6×10^{-19}	3.1×10^{-19}
ψ	6 500–8 000	163	0.30	350	202	460	198	1.6×10^{-19}	3.1×10^{-19}
Ω	4 800–5 900	152	0.30	275	127	232	144	2.3×10^{-19}	3.7×10^{-19}

lines in the solar spectrum. With these identifications the energy levels
could be calculated, if required.

Upper-state bands of water vapour with (010) as the initial state
have been observed in the solar spectrum. For practical purposes the
ground-state populations of atmospheric molecules can be taken to be
independent of temperature. It is very different with the upper-state
bands. For water vapour, intensities of the strongest upper-state bands
change by an order of magnitude between 200° K and 250° K. The
situation can arise where variations in the optical depth are affected
more by temperature than density changes, even for a gas whose density
varies as greatly as water vapour. Intensities for four observed upper-
state bands are given in Table 5.11.

TABLE 5.10

The near infra-red bands of H_2O

Name	Transition	Band centre	Type	S_n
		(cm^{-1})		(cm)
ρ	000–003	11 032	A	2×10^{-21}
	000–102	10 869	B	4×10^{-22}
σ	000–201	10 613	A	1×10^{-20}
	000–300	10 600	B	6×10^{-22}
τ	000–121	10 329	A	2×10^{-21}
	000–220	10 284	B	$< 4 \times 10^{-23}$
	000–041	9 834	A	6×10^{-23}
ϕ	000–012	9 000	B	3×10^{-22}
	000–121	8 807	A	8×10^{-21}
	000–210	8 762	B	1×10^{-23}
	000–130	8 274	B	7×10^{-24}
	000–031	8 374	A	3×10^{-23}
ψ	000–002	7 445	B	1×10^{-21}
	000–101	7 250	A	$1\cdot5 \times 11^{-19}$
	000–200	7 201	B	$1\cdot5 \times 10^{-20}$
	000–021	6 871	A	1×10^{-20}
	000–120	6 775	B	2×10^{-22}
Ω	000–011	5 331	A	$2\cdot2 \times 10^{-19}$
	000–110	5 235	B	7×10^{-21}
	000–030	4 667	B	3×10^{-22}

TABLE 5.11

Upper-state bands of H_2O

Intensities refer to the total number of molecules and contain a Boltzmann factor. After Benedict (1956)

Transition	Frequency	Type	S_n at 259° K
	(cm^{-1})		(cm)
010–001	2161·14	A	2×10^{-22}
010–100	2062·27	B	8×10^{-23}
010–020	1556·82	B	9×10^{-22}
010–010	0–500	B	$4\cdot6 \times 10^{-20}$

5.4.6. *Isotopic bands*

Many lines of HHO[18] and HHO[17] have been identified in both the rotation and ν_2 bands, and the intensities relative to HHO[16] lines shown to agree with the isotopic abundances. The selection rules are the same as for HHO¹⁶ and line positions are known throughout the spectrum whether they have been observed or not. Line widths will differ slightly from those of HHO¹⁶, but the difference is not likely to be an important factor in atmospheric studies. Two sections of the solar spectrum showing HHO¹⁸ and HHO¹⁷ lines are shown in Fig. 5.11.

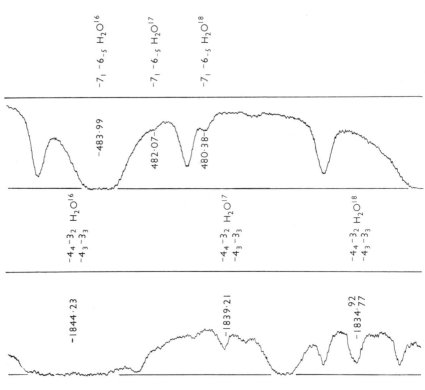

Fig. 5.11. H_2O^{16}, H_2O^{17}, and H_2O^{18} lines in the solar spectrum.

(a) The rotation band. (b) R-branch of the ν_2 band. After Benedict (1957b).

The bands of HDO, on the other hand, are well separated from the corresponding HHO bands and form distinct features in the solar spectrum. The molecule has a lower symmetry than HHO and, as a result, type A and type B transitions take place simultaneously (Table 5.12). The combined ν_1 and $2\nu_2$ bands have been studied by Howard et al. (1956) who find that c, d, and k in (5.9) are 0·325, 0·5, and 0·37 respectively. Data on the rotation band are available from theoretical computations and from observations in the microwave region of the spectrum (see Bibliography).

TABLE 5.12

Vibration-rotation bands of HDO

After Benedict, Gailard, and Plyler (1953)

Transition	Frequency	S_n, Type A component	S_n, Type B component
	(cm^{-1})	(cm)	(cm)
000–010	1403·3	5×10^{-23}	$2·3 \times 10^{-22}$
000–100	2723·66	5×10^{-22}	$< 5 \times 10^{-24}$
000–020	2782·16	$1·4 \times 10^{-23}$	9×10^{-24}
000–001	3707·47	3×10^{-22}	$1·4 \times 10^{-22}$
000–110	4100·05	2×10^{-23}	$< 1 \times 10^{-24}$
000–030	4145·59	2×10^{-23}	$< 1 \times 10^{-24}$
000–011	5089·59	3×10^{-23}	$1·6 \times 10^{-23}$
000–200	5363·59	3×10^{-23}	$< 2 \times 10^{-24}$
000–101	6415·64	1×10^{-24}	9×10^{-24}
000–021	6452·05	2×10^{-24}	$< 5 \times 10^{-25}$
000–012	8611·22	5×10^{-24}	$< 5 \times 10^{-25}$

5.4.7. *The* 1000 cm^{-1} *continuum*

The telluric spectrum from 800 to 1200 cm^{-1} contains many weak lines and the moderately strong 9·6 μ band of ozone. Under high resolution gaps between lines can be distinguished, some being as wide as 9 cm^{-1}. Measurements indicate an absorption coefficient in these windows of the order of 0·1 per gramme of water vapour which is too great to be accounted for in terms of unresolved lines or of wings nearby strong lines.

The continuum follows Lambert's law, and its spectrum is shown in Fig. 5.12. Recent work indicates that in clean air it is almost entirely caused by water vapour, although some aerosol extinction is probably present at all frequencies. After eliminating other factors Saiedy (1960) finds mass absorption coefficients for water vapour of 0·1144, 0·0925, 0·0545 g^{-1} cm^{-2} at 832 cm^{-1}, 901 cm^{-1}, and 1159 cm^{-1} respectively. A less direct determination by Vigroux (1959) gives 0·098 g^{-1} cm^{-2} near to 1040 cm^{-1}. A tendency for the different sets of points in Fig. 5.12 to separate is attributable to temperature variation of the absorption; e.g. k_{850}/k_{901} increases by about 1 per cent per °K.

The only satisfactory explanation of this continuum is that of Elsasser (1942), who believed it to be caused by the far wings of very strong lines near the absorption maxima of the ν_2 and rotation bands (see § 3.6.5). We may reasonably infer from the theory that whatever interactions may be involved the continuum has a (p^{-1}) pressure dependence, and therefore the relevant physical data for all pressures are known, even if not well understood.

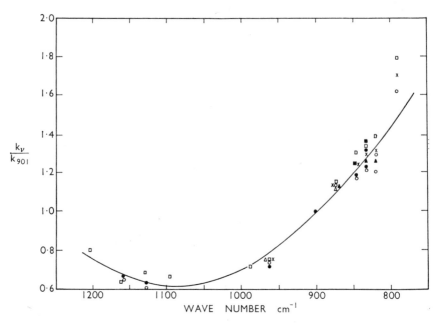

Fig. 5.12. The 1000 cm⁻¹ continuum in the solar spectrum.

Observations made on the same day are represented by the same symbol. It is believed that some contribution near 800 cm⁻¹ comes from wings of lines in the ν_2 band of CO_2. On each day a measurement was made at 901 cm⁻¹, and the results are normalized to unity at this frequency ($k_{901} = 0.0925$ g⁻¹ cm²). After Saiedy (1960).

5.4.8. *The emissivity of water vapour*

If $\overline{T}_i$ is the mean absorption of the ith spectral range (presumed narrow, see § 4.1), and B_i is the Planck function at its centre, the *column emissivity* is defined by

$$\epsilon_c(p, \theta, a) = \frac{\sum_i \{1 - \overline{T}_i(a, \theta, p)\} B_i(\theta)}{\sum_i B_i(\theta)}, \qquad (5.11)$$

and relates the emission of a column of gas to that of a black body at the same temperature. We will find in Chapter 6 that, under some circumstances, this quantity alone is involved in heat-transfer problems. It is also the parameter most generally used in engineering applications. It can be measured directly by simple (but often not too precise) experiments, or it can be computed from the transmission functions, where known.

Emissivities can be measured in the laboratory with black, or non-selective detectors, but a number of difficulties of interpretation have not yet been satisfactorily settled. Measurements both in the laboratory and in the atmosphere involve the emission of other gases besides water vapour (notably CO_2). Since the measurements are non-selective, only a theoretical correction is possible, and this correction can be large. Further, for measurements over long atmospheric paths, the possible influence of haze cannot be overlooked, especially since the two most widely used sets of measurements took place near large industrial cities. The use of atmospheric measurements, regardless of possible aerosol contributions, is rationalized by Brooks (1950) on the grounds that, even if incorrect, it nevertheless should give the correct results in computations of radiative fluxes for similar atmospheric conditions. This is not a very satisfying conclusion, but perhaps the best which can be achieved with the information at present available. Brooks' estimates are given in Table 5.13. These data are so-called *slab emissivities* (ϵ_s) obtained by integrating (5.11) over all solid angles in one hemisphere, for an infinite sheet containing amount a per cm^2 of the sheet. The problems associated with this angular integration will be discussed in Chapter 6. It will suffice here to note that Brooks assumes

$$\epsilon_s(a) = \epsilon_c(1\cdot66a). \tag{5.12}$$

TABLE 5.13

Slab emissivities of H_2O at about $300°$ K and 1 atmosphere pressure

After Brooks (1950)

a	ϵ_s	a	ϵ_s
(g cm^{-1})		(g cm^{-1})	
0·0001	0·0244	0·1	0·3959
0·001	0·1155	0·2	0·4479
0·002	0·1532	0·5	0·5192
0·005	0·2050	0·7	0·5465
0·007	0·2249	1·0	0·5752
0·01	0·2456	2·0	0·6308
0·02	0·2881	5·0	0·7020
0·05	0·3470	7·0	0·7272
0·07	0·3700	10·0	0·7540

The effects of temperature and pressure upon emissivity have been inferred from theoretical considerations. Fig. 5.13 shows computations of the column emissivity at three pressures. The data for one atmosphere pressure do not agree precisely with Table 5.13, but the relative pressure effect should not be seriously in error. In Table 5.14 are shown figures for the variation of ϵ_s between $300°$ K and $220°$ K at $0·25$ atmosphere pressure. The three rows show the effect of changing θ from $300°$ K to $220°$ K separately in the two factors $\bar{T}_i$ and B_i in (5.11). These two changes cancel to a certain extent and the net temperature effect is not large.

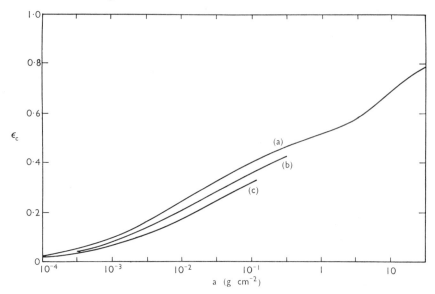

FIG. 5.13. Column emissivity of water vapour at about $300°$ K for (a) 1, (b) $0·5$, (c) $0·25$ atmospheres pressure.
After Cowling (1950).

TABLE 5.14

Column emissivity of H_2O *at* $0·25$ *atmosphere (per cent)*

(a) $\theta = 220°$ K in both terms of (5.11)
(b) $\theta = 220°$ K in B_i, but $\theta = 300°$ K in $\bar{T}_i$
(c) $\theta = 300°$ K in both terms of (5.11)

a (g cm^{-2})	10^{-4}	$10^{-3·5}$	10^{-3}	$10^{-2·5}$	10^{-2}	$10^{-1·5}$	10^{-1}
(a)	2·3	4·5	8·2	13·6	20·65	28·9	37·4
(b)	2·6	5·3	9·7	16·0	24·2	33·7	43·0
(c)	1·7	3·4	6·4	11·1	17·2	24·8	32·5

5.5. Carbon dioxide

5.5.1. *Electronic bands*

The electronic bands in the ultra-violet spectrum of carbon dioxide have not yet been satisfactorily analysed, but some absorption coefficients are known. Absorption is first detected at about 1750 Å and there is a maximum close to 1475 Å, with a molecular absorption coefficient of about 6×10^{-19} cm². Near 1325 Å there is another weak maximum with a coefficient of 8×10^{-19} cm². Below 1175 Å the absorption coefficient increases rapidly until it is about 10^{-16} at 1125 Å.

5.5.2. *The vibration-rotation spectrum*

The $C^{12}O_2^{16}$ molecule is linear and symmetric (OCO), with a C—O bond length of 1·1632 Å in the ground vibrational state, and a corresponding rotational constant of 0·3895 cm⁻¹.

The fundamental modes of vibration for linear molecules have been discussed in § 3.3. The ν_2 bending frequency is degenerate and we must take account of the selection rules relating to the quantized vibrational angular momentum. The ν_2 fundamental involves the transition (00^00–01^10) and is perpendicular. The ν_3 fundamental is a (00^00–00^01) transition, which is parallel and lacks a Q-branch.

Owing to the symmetry of the molecule, the ν_1 vibration involves no change of dipole moment and is therefore inactive in the infra-red spectrum. It has a frequency nearly twice that of the ν_2 fundamental, with the result that groups of levels such as (02^00, 10^00), (03^10, 11^10), (04^01, 12^01, 20^01), etc., interact. Also owing to its symmetry, $C^{12}O_2^{16}$ has no pure rotation spectrum.

Since oxygen has zero nuclear spin, the statistical weights of levels with odd J are zero, and alternate lines are missing from the vibration-rotation fine structure.

The same remarks apply to the $C^{13}O_2$ molecule, which forms 1·108 per cent of the total carbon dioxide. $CO^{16}O^{17}$ and $CO^{16}O^{18}$ are present in concentrations of 0·0646 and 0·4078 per cent and, being of lower symmetry, have a rotational structure which differs considerably from that of CO_2^{16}. The concentrations are too small to justify a detailed discussion, but both molecules will resemble more closely the nitrous oxide molecule (§ 5.7.1). Molecules with two rare isotopes can be neglected. Isotopic shifts in carbon dioxide are shown in Table 5.15.

TABLE 5.15

The ν_2 and ν_3 fundamentals of CO_2

Species	Percentage abundance	Band centre (cm⁻¹)	
		ν_2	ν_3
$C^{12}O^{16}O^{16}$	98·420	667·40	2349·16
$C^{13}O^{16}O^{16}$	1·108	648·52	2283·48
$C^{12}O^{16}O^{18}$	0·408	662·39	2333

5.5.3. *The ν_2 bands*

The ν_2 bands near $15\,\mu$ are probably the most intensively studied bands in the solar spectrum. By 1941 a considerable body of data existed and was summarized in a paper by Callendar (1941), who introduced an empirical transmission function of the form

$$\bar{T} = (1+na^c p^{c/2})^{-1}, \tag{5.13}$$

where n and c are constants, with different values for each $1\,\mu$ region of the spectrum. Although it is without any theoretical foundation, (5.13) is remarkably similar in form to some of the band transmission functions discussed in Chapter 4. Since this work has been superseded by more reliable and elaborate data, it will not further be discussed.

From 1947 a series of papers by Kaplan and later by Yamamoto explored the possibility of deriving absorption data from consideration of the detailed band structure. This requires a full knowledge of line positions, intensities, and shapes, which in the P- and Q-branches can then be put together by multiplying Elsasser bands, one for each of the many transitions which are involved. In the Q-branches (this being a perpendicular band) the lines are a few hundredths of a wave number apart, and therefore at tropospheric pressures they can be treated as single, pressure-independent 'lines'. This 'line profile' is temperature-dependent because different points correspond to groups of lines with different J-values. Above 30 km it is possible to make the opposite assumption, namely that the lines do not overlap and that the equivalent widths can be summed; in the stratosphere neither assumption is valid.

Besides the ν_2 fundamental, fourteen overtone and combination bands have been detected in the $15\,\mu$ region, with a total intensity about 10 per cent of the fundamental, providing many weak lines of potential importance to atmospheric studies. Of these fourteen, half show l-type doubling. All are perpendicular, and the rotational constants are known for all relevant levels; the spectrum can therefore be reconstructed in detail. Some of the relevant data are shown in Table 5.16. The energy of the lower state is given so that intensities may be computed at temperatures other than 300° K. The theoretical fine structure of this band was illustrated in Fig. 4.10. Fig. 5.14 shows a section of the band under high resolution, which does not expose all the fine structure. Fig. 5.15 shows the general appearance of the band under lower resolutions.

Yamamoto and Sasamori (1958, 1961) have published lists of line positions and line intensities derived from the data in Table 5.16.

TABLE 5.16

Band intensities at 300° K *and other molecular constants for the* 15 μ CO_2 *bands*

Isotopic species	Transition	Band centre		Molecular band intensity†	E''	References
		(cm⁻¹)	(μ)	(cm)	(cm⁻¹)	
$C^{12}O^{16}O^{16}$	00^00-01^10	667·40	15·0	$7·89 \times 10^{-18}$	0·0	1
$C^{13}O^{16}O^{16}$	00^00-01^10	648·52	15·4	$7·9 \times 10^{-20}$	0·0	2
$C^{12}O^{18}O^{16}$	00^00-01^10	662·39	15·1	$3·7 \times 10^{-20}$	0·0	2
$C^{12}O^{16}O^{16}$	01^10-02^00	618·03	16·2	$1·75 \times 10^{-19}$	667·4	2
—	01^10-10^00	720·83	13·9	$2·3 \times 10^{-19}$	667·4	3
—	01^10-02^20	667·76	15·0	$6·2 \times 10^{-19}$	667·4	1
—	02^00-03^10	647·02	15·5	$4·2 \times 10^{-20}$	1285·43	2
—	02^00-11^10	791·48	12·6	$8·2 \times 10^{-22}$	1285·43	4
—	02^20-03^10	597·29	16·7	$5·83 \times 10^{-21}$	1335·16	2
—	02^20-11^10	741·75	13·5	$5·2 \times 10^{-21}$	1335·16	3
—	02^20-03^30	668·3	15·0	$3·2 \times 10^{-20}$	1335·16	1
—	10^00-03^10	544·26	18·4	$1·64 \times 10^{-22}$	1388·19	2
—	03^30-04^20	581·2	17·2	$1·56 \times 10^{-22}$	2003·28	4
—	03^30-12^20	756·75	13·2	$2·2 \times 10^{-22}$	2003·28	4
—	03^10-12^20	828·18	12·1	$1·8 \times 10^{-23}$	1932·45	4
—	03^10-12^00	740·5	13·5	$5·2 \times 10^{-22}$	1932·45	3
—	04^40-13^30	769·5	13·0	$1·5 \times 10^{-23}$	2674·76	4

† The band intensity is reckoned with respect to the total number of molecules of *all* CO_2 isotopes, not only those of the initial state of the particular species involved. Each intensity therefore involves the Boltzmann factor for the initial state, and the relative intensities can vary greatly with temperature.

References:

    1. Kaplan and Eggers (1956).   3. Kostkowsky and Kaplan (1957).
    2. Madden (1957).   4. Yamamoto and Sasamori (1958).

These lists are deficient in two respects: they omit corrections for Coriolis interaction; and they take no account of isotopic bands. According to Madden (1957) the Coriolis interaction requires that the intensities be multiplied by a factor $(1+\zeta m)^2$ where m is the ordinal number of the line and ζ is a constant differing from band to band. For the $000-01^10$ and 01^10-02^00 bands Madden gives $\zeta = 0·0016$ and 0·0035 respectively. For $m = 50$ the corresponding Coriolis factors are 1·17 and 1·38, and this may be one reason why Yamamoto and Sasa-mori find some disagreement between experiment and predictions based on their data.

Line widths in the 15 μ band have been the subject of a number of investigations, the most relevant being that of Kaplan and Eggers (1956), who fitted a curve of growth to the entire band using the band intensity and mean line width as adjustable parameters. The best fit

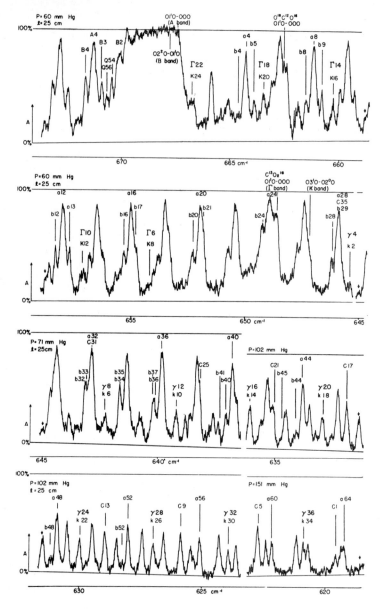

FIG. 5.14. The observed fine structure of part of the 15 μ CO$_2$ bands.
After Madden (1957). (a) $675-620$ cm^{-1}. (b) $620-580$ cm^{-1}. $p =$ pressure of
pure CO$_2$; $l =$ absorption path length.

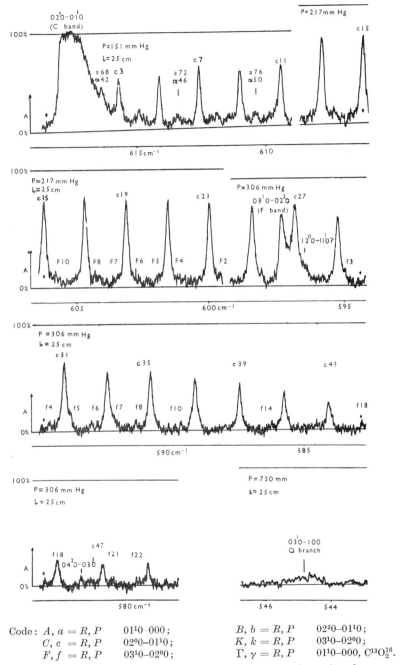

Code: $A, a = R, P$ $01^1 0 - 000$;
$C, c = R, P$ $02^0 0 - 01^1 0$;
$F, f = R, P$ $03^1 0 - 02^0 0$;

$B, b = R, P$ $02^2 0 - 01^1 0$;
$K, k = R, P$ $03^1 0 - 02^0 0$;
$\Gamma, \gamma = R, P$ $01^1 0 - 000, C^{13}O_2^{16}$.

A more complete spectrum, but at lower resolution, has been given by Rossman *et al.* (1956).

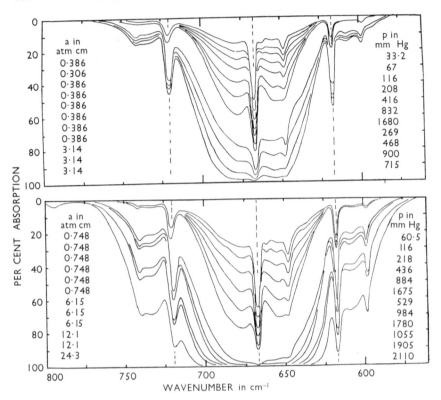

FIG. 5.15. The 15 μ CO_2 bands under low resolution.
After Burch *et al.* (1960).

gave the band intensity for the ν_2 fundamental which is listed in Table
5.16 and a mean line width for broadening by air at s.t.p. of 0·064 cm^{-1}.
Kaplan and Eggers made use of short-path data only, and this width
is therefore representative of the stronger lines; it is probable that the
weak lines have rather different widths. For example Madden (1957),
working with pure CO_2 on the P- and R-branches of 01^10–02^00 and
on the ν_2 fundamental found significant variations of line width both
inside and between bands.

With the above line width and intensity data, Yamamoto and Sasamori have computed low-resolution absorption spectra using the modified Elsasser model described in § 4.4.3 and single lines of complex
profile to represent the Q-branches. The spectra extend from 550 to
830 cm^{-1}, and the range of averaging is slightly less than 2 cm^{-1}. The
amount of CO_2 varies from 0·03 to 1000 cm s.t.p., the temperature from
218° K to 300° K, and the pressure from 0·008 to 1 atmosphere. It is

impracticable to reproduce all of these data, and the reader is referred
to the original papers for details.

Laboratory data at comparable resolution now exist (Burch, Gryvnak, and Williams, 1960), while data at lower resolution are also available (Howard *et al.*, 1956). The former cover unevenly the following
ranges: amounts 0·0054 to 11 200 cm s.t.p.; pressures 0·26 to 3800 mm

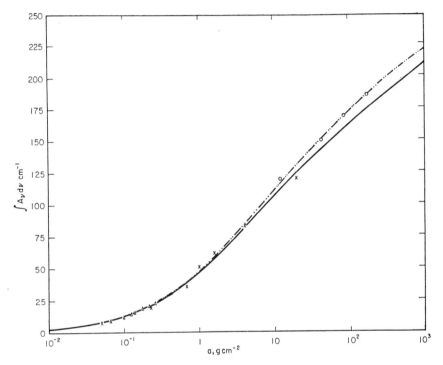

FIG. 5.16. Band area of the 15 μ CO_2 band at 1 atmosphere pressure.

The full line is theoretical and is based upon a modified Elsasser model (§ 4.4.3);
the broken line is the best fit to empirical data. O Howard *et al.* (1956);
× Callendar (1941); △ Kaplan and Eggers (1956). Computations have also been
made at 0·2 atmosphere pressure. The discrepancies between theory and
observation are attributed to insufficiently accurate line widths, but may also
be caused by neglect of isotopic bands and Coriolis interactions. After
Yamamoto and Sasamori (1958).

Hg; temperatures 23·5 to 65° C. Two sets of data are shown in Fig.
5.15. No attempt has yet been made to compare these detailed results
with the theory.

Constants for an empirical fit to the total absorption area are given
in Table 5.17. The original data upon which these constants are based
are compared with theory in Fig. 5.16. For large amounts of absorbing

matter the agreement is unsatisfactory, but since the theory does not include Coriolis interactions or weak isotopic bands, and assumes a single line width, this is perhaps not surprising. Comparison of narrow ranges through the spectrum should serve to clarify these problems.

TABLE 5.17

Empirical constants for carbon dioxide bands at 293° K (*see* (5.9) *and* (5.10))

After Howard *et al.* (1956) and Burch, Gryvnak, and Williams (1960)

Band	*Limits*	*c*	*d*	*k*	$\int A_\nu \, d\nu$ (*transition*)	*C*	*D*	*K*
(μ)	(cm^{-1})				(cm^{-1})			
15	550–800	3·16	0·50	0·44	50	−58	55	47
10·4	850–1000	0·016	0·78	0·20	35	—	—	—
9·4	1000–1110	0·023	0·75	0·23	40	—	—	—
5·2	1870–1980	0·024	0·50	0·40	30	—	—	—
4·8	1980–2160	0·12	0·50	0·37	60	—	—	—
4·3	2160–2500	15·0	0·54	0·41	50	27·5	34	31·5
2·7	3480–3800	3·5	0·58	0·38	50	−137	77	68
2·0	4750–5200	0·492	0·50	0·39	80	−536	138	114
1·6	6000–6550	0·063	0·50	0·38	80	—	—	—
1·4	6650–7250	0·058	0·50	0·41	80	—	—	—

For computations at great altitudes, when the overlapping of lines in Q-branches can be neglected, line positions become of small importance and a histogram of line intensities is sufficient to represent the data. At 273° K the intensities can be fitted by the expression

$$N(S) = \frac{n_\mathrm{s}}{\sigma_\mathrm{s}} e^{-S/\sigma_\mathrm{s}} + \frac{n_\mathrm{w}}{\sigma_\mathrm{w}} e^{-S/\sigma_\mathrm{w}}, \qquad (5.14)$$

where $N(S)\,dS$ is the number of lines with intensities between S and $S+dS$, $n_\mathrm{s} = 108$, $n_\mathrm{w} = 372$, $\sigma_\mathrm{s} = 7\cdot45\times10^{-20}$ cm, $\sigma_\mathrm{w} = 2\cdot63\times10^{-21}$ cm. σ_w refers mainly to the weak lines whose lower state is (01^{1}0) and will have the temperature dependence of the number density in that state. σ_s, on the other hand, refers mainly to transitions from the (00^{0}0) state, and will hardly vary within the range of atmospheric temperatures.

5.5.4. *The ν_3 bands*

These strong bands are responsible for the great opacity of the atmosphere near 4·3 μ. There are three bands superposed: the ν_3 band of $C^{12}O^{16}O^{16}$ at 2349·16 cm^{-1}, the ν_3 band of $C^{13}O^{16}O^{16}$ at 2283·48, and the 02^{0}0–10^{1}1 (or $\nu_1+\nu_3-2\nu_2$) combination band of $C^{12}O^{16}O^{16}$ at

2429·37 cm^{-1}. All are parallel bands and are therefore without Q-branches. The rotational constants are listed by Migeotte *et al.* (1956) and laboratory spectra have been obtained by many investigators (see Bibliography). In addition to these bands there are upper-state and isotopic bands, but none has been identified in the solar spectrum. According to Eggers and Arends (1957) the intensity per species molecule of ν_3 $C^{12}O^{18}O^{18}$ is one-half the intensity of ν_3 $C^{12}O^{16}O^{16}$.

Empirical data upon this band (see Table 5.17) have been given by Howard *et al.* (1956) and Burch *et al.* (1960). The latter authors give spectra over the ranges: amount 0·0108 to 22·8 cm s.t.p.; pressure 7·9 to 2115 mm Hg. The band intensity of all three bands together was found to be $(0·93\pm0·15)\times10^{-16}$ cm, consistent with a value of $1·01\times10^{-16}$ cm given by Benedict and Plyler (1954).

The detailed data could be compared with computations by Plass (1959), but this has yet to be done. Plass's computations contain a number of approximations (e.g. he neglects combination bands and Coriolis interactions) and the predicted emissivities disagree considerably with measurements by Tourin (1961).

5.5.5. *Weaker bands of carbon dioxide*

Overtone and combination bands of carbon dioxide are comparatively strong, and a large number have been identified, both in the laboratory and in the solar spectrum. They fall into a number of distinct groups, for which empirical data are given in Table 5.17.

The 10 μ bands

Two medium strong bands appear in the solar spectrum at 1063·8 and 961·0 cm^{-1}; the former is $(02^00\text{--}001)$ and the latter $(100\text{--}001)$. Both are parallel; rotational constants are given by Migeotte *et al.* (1956). A line width of 0·084 cm^{-1} at 1 atmosphere and 298° K in pure CO_2 has been measured at the R-branch maximum by Kostkowsky (1955) and the width has been shown to vary as $\theta^{-0·56}$. Burch, Gryvnak, and Williams (1960) give molecular band intensities at 26° C of $8·6\times10^{-22}$ cm and $2·7\times10^{-21}$ cm respectively.

The 5 μ bands

Perpendicular bands near $5\,\mu$ can be separated into two groups. Near $5·2\,\mu$ is the $3\nu_2$ band and near $4·8\,\mu$ are the $(\nu_1+\nu_2)$ bands of $C^{12}O^{16}O^{16}$ and $C^{13}O^{16}O^{16}$, the $(\nu_1+2\nu_2^2-\nu_2^0)$ and the $(2\nu_1-\nu_2)$ bands. Table 5.18 gives some details. No measurements of line widths are at present available.

TABLE 5.18

The 5 μ bands of CO_2

After Benedict (see Migeotte *et al.* (1956))

Band	Isotopic species	Transition	Band centre	E''	Molecular band intensity at 259° K†
(μ)			(cm^{-1})	(cm^{-1})	(cm)
5·2	$C^{12}O_2^{16}$	000–03^{1}0	1932·45	0	$2\cdot4\times10^{-21}$
	$C^{13}O_2^{16}$	000–11^{1}0	2037·08	0	$4\cdot0\times10^{-22}$
4·8	$C^{12}O_2^{16}$	000–11^{1}0	2076·86	0	$3\cdot2\times10^{-20}$
	$C^{12}O_2^{16}$	01^{1}0–12^{2}0	2093·35	667·40	$2\cdot0\times10^{-21}$
	$C^{12}O_2^{16}$	01^{1}0–200	2129·79	667·40	$8\cdot0\times10^{-22}$

† Intensity per molecule of CO_2 of all species (see Table 5.16).

The 2·7 μ bands

Even though they are not fundamentals these bands appear very strongly in the solar spectrum. There are at least four combination bands of $C^{12}O_2^{16}$ and one of $C^{13}O_2^{16}$ in the region 3613–3723 cm^{-1}. Two intensities are quoted by Goldberg (1954), while that of the isotopic band can be inferred from the isotopic abundance. These data are shown in Table 5.19.

TABLE 5.19

The 2·7 μ bands of CO_2

After Goldberg (1954), France and Dickey (1955), and Burch, Gryvnak, and Williams (1960, figures in parentheses)

Isotopic species	Transition	Band centre	E''	Molecular band intensity at 259° K†
		(cm^{-1})	(cm^{-1})	(cm)
$C^{12}O_2^{16}$	000–02^{0}1	3613·03	0	$1\cdot0\times10^{-18}$ $(1\cdot4\times10^{-18})$
—	000–101	3714·56	0	$1\cdot3\times10^{-18}$ (2×10^{-18})
—	01^{1}0–03^{1}0	3580·81	667·40	—
—	01^{1}0–11^{1}1	3723·05	667·40	—
$C^{13}O_2^{16}$	000–02^{0}1	3527·70	0	$1\cdot1\times10^{-20}$
—	000–101	3632·92	0	$1\cdot4\times10^{-20}$
—	01^{1}0–03^{1}1	3498·72	667·40	—

† Intensity per molecule of CO_2 of all species (see Table 5.16).

The 2·0 μ, 1·6 μ, and 1·4 μ bands

The remaining CO_2 bands in the solar spectrum are weaker than most of the bands discussed above. Table 5.20 gives some available data, which are incomplete. Bands are parallel or perpendicular depending upon whether ν_3 changes by unity or not.

Goldberg (1954) has measured the curve of growth of a number of lines in the $1 \cdot 6 \mu$ band; the best fit for a Lorentz profile is for $\alpha_L = 0 \cdot 07$ cm^{-1} at s.t.p.

TABLE 5.20

The $2 \cdot 0 \mu$, $1 \cdot 6 \mu$ and $1 \cdot 4 \mu$ bands of CO_2.

Intensities follow Goldberg (1954)

Band	Isotopic species	Transition	Band centre	E''	Molecular band intensity at $259° K$†
(μ)			(cm^{-1})	(cm^{-1})	(cm)
	$C^{12}O_2^{16}$	00^00-04^01	5100	0	$1 \cdot 6 \times 10^{-20}$
	$C^{12}O_2^{16}$	00^00-12^01	4978	0	$3 \cdot 7 \times 10^{-20}$
	$C^{12}O_2^{16}$	00^00-20^01	4853	0	$1 \cdot 0 \times 10^{-20}$
	$C^{12}O_2^{16}$	01^10-05^10	5132	$667 \cdot 40$	—
	$C^{12}O_2^{16}$	01^10-13^11	4965	$667 \cdot 40$	—
	$C^{12}O_2^{16}$	01^10-21^11	4808	$667 \cdot 40$	—
$2 \cdot 0$	$C^{13}O_2^{16}$	00^00-04^01	5046	0	$1 \cdot 8 \times 10^{-22}$
	$C^{13}O_2^{16}$	00^00-12^01	4887	0	$4 \cdot 1 \times 10^{-22}$
	$C^{13}O_2^{16}$	00^00-20^01	4748	0	$1 \cdot 1 \times 10^{-22}$
	$C^{12}O^{16}O^{18}$	00^00-04^01	5042	0	7×10^{-23}
	$C^{12}O^{16}O^{18}$	00^00-12^01	4905	0	$1 \cdot 5 \times 10^{-22}$
	$C^{12}O^{16}O^{18}$	00^00-20^01	4791	0	4×10^{-23}
	$C^{12}O_2^{16}$	00^00-06^01	6503	0	—
$1 \cdot 6$	$C^{12}O_2^{16}$	00^00-14^01	6350	0	—
	$C^{12}O_2^{16}$	00^00-22^01	6228	0	$2 \cdot 9 \times 10^{-22}$
	$C^{12}O_2^{16}$	00^00-20^01	6076	0	—
$1 \cdot 4$	$C^{12}O_2^{16}$	00^00-00^03	6973	0	$8 \cdot 6 \times 10^{-22}$

† Intensity per molecule of CO_2 of all species (see Table 5.16).

Visible and photographic infra-red bands

In his survey of the solar spectrum Goldberg (1954) lists no bands with wavelengths shorter than $1 \cdot 4 \mu$. Many weak bands are, however, detectable. Mohler (1955) lists lines from the (00^00-10^03) and (00^00-02^03) bands between 8151 cm^{-1} and 9342 cm^{-1}. Herzberg (1945) records five very weak bands between 8195 cm^{-1} and 12 774$\cdot$4 cm^{-1}. Herzberg and Herzberg (1953) have found thirteen bands between 8089 cm^{-1} $(00^00-02^03$ of $C^{13}O_2)$ and 12 774 cm^{-1} $(00^00-10^05$ of $C^{12}O_2)$. These probably play no significant role in atmospheric processes.

5.6. Ozone

5.6.1. *Electronic bands*

The spectrum of ozone is dominated by the Hartley bands, centred at 2553 Å, with a peak cross-section of $1 \cdot 08 \times 10^{-17}$ cm^2. A typical solar beam reaching the ground traverses about $1 \cdot 4 \times 10^{19}$ mol. cm^{-2} and the transmission at 2553 Å in the solar spectrum is therefore about 10^{-66}.

The Hartley bands consist of a large number of weak bands, about 10 Å apart, on a very strong continuum. Measurements by three investigators are shown in Fig. 5.17. Above 2700 Å Vigroux and Inn and Tanaka agree well. At shorter wavelengths an investigation by Hearn (1961),

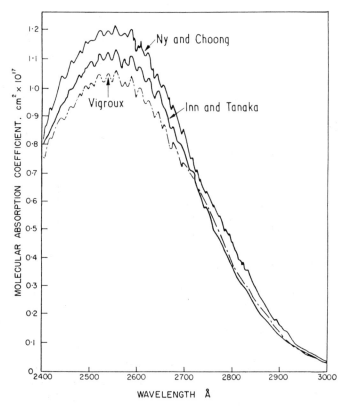

FIG. 5.17. Molecular absorption coefficients in the Hartley bands of ozone at 18° C.

After Ny Tsi-Ze and Choong Shin-Piaw (1933), Vigroux (1953), and Inn and Tanaka (1953). The results of Ny and Choong extend to about 2100 Å. Tanaka, Inn, and Watanabe (1953) have made measurements to 1050 Å.

based on a technique differing greatly from the previous investigators, and probably of higher accuracy, tends to confirm the results of Inn and Tanaka.

Absorption in the Hartley bands is slightly dependent upon temperature. The ratio $k(\theta)/k(18° C)$ for θ between $-72°$ C and $-46°$ C varies from about 0·88 at 3100 Å to 0·97 near 2500 Å. At $\theta = -30°$ C the ratio is greater, being about 0·92 at 3100 Å and 0·98 near 2500 Å.

On the short-wave wing of the Hartley bands the absorption falls initially to a minimum of 3×10^{-19} cm^2 at 2000 Å. Thereafter it increases again to a series of maxima below 1400 Å, the highest recorded being 2×10^{-17} cm^2 at 1220 Å.

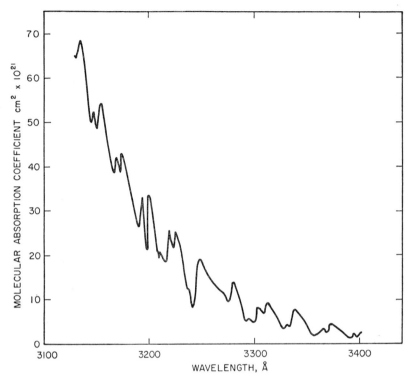

Fig. 5.18. Molecular absorption coefficients in the Huggins bands of ozone at 18° C.

After Vigroux (1953).

The spectral region from 3100 Å to 3400 Å, on the long-wave wing of the Hartley bands, has a more marked band structure than most other wavelengths (Fig. 5.18). These weak bands (the Huggins bands) can be found in the spectrum of the low sun and were responsible for the first positive identification of ozone as an atmospheric constituent. Absorption in the Huggins bands is sensitive to temperature and the magnitude of the temperature dependence differs from absorption maximum to absorption minimum. Further details are given by Vigroux (1953).

Careful tests have been made to discover whether absorption in the Huggins bands depends upon pressure. Vigroux (1953) has made measurements over a pressure range of 200:1; Strong (1941) over a range of 750:1; Vassy (1937) over a range of 40:1. All investigators agree that there is no detectable pressure effect. Until measurements are available from other spectral regions it seems reasonable to assume that the result is valid for the ozone electronic spectrum as a whole.

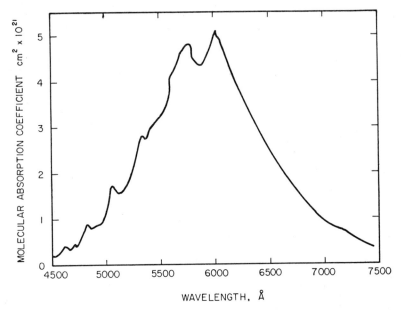

FIG. 5.19. Molecular absorption coefficients at 18° C in the Chappuis bands of ozone.

After Vigroux (1953).

Between 3400 Å and 4500 Å lies a relatively transparent region in the ozone spectrum and very long paths are required to produce measurable absorption.

Between 4500 Å and 7400 Å lie the Chappuis bands of ozone (Fig. 5.19). The maximum molecular absorption coefficient in this band is 5×10^{-21} cm^2 and the peak absorption for solar radiation traversing two air masses is about 7 per cent. Such small absorptions are, however, significant, both as regards direct solar heating of the atmosphere and for twilight optical effects. Temperature coefficients for these bands appear to be negligible.

5.6.2. *The vibration-rotation spectrum*

The infra-red absorption spectrum of ozone shows strong bands at 710 cm^{-1}, 1043 cm^{-1}, and 2105 cm^{-1}, with weaker absorptions at

1740 cm^{-1}, 2800 cm^{-1}, and 3050 cm^{-1}. Until 1948 it proved impossible to fit a satisfactory molecular model to these bands. In that year, however, Wilson and Badger pointed out that a weak band at 1110 cm^{-1} had escaped detection because it was merged with the much stronger 1043 cm^{-1} band, and that both must be fundamentals. On this basis the derived molecular structure can be reconciled with evidence from other sources. It is to be expected that two fundamentals which overlap will result in very strong Coriolis interactions, and the 1110 cm^{-1} fundamental is difficult to analyse on this account.

Table 5.21 shows the assignment of fundamentals for O_3^{16}, $O^{16}O^{18}O^{16}$, and $O^{16}O^{16}O^{18}$. The first two are symmetric molecules and have alternate levels missing. The third gives rise to a quite different spectrum from the other two.

TABLE 5.21

Fundamentals of the O_3 *molecules*

After Hughes (1956)

	O_3^{16}	$O^{16}O^{18}O^{16}$	$O^{16}O^{16}O^{18}$
Percentage abundance	99·4	0·21	0·41
ν_1	1110 (?)	1080	1095
ν_2	701·42	697	688
ν_3	1045·16	1008	1029

The O_3^{16} molecule has a dipole moment of 0·58 Debye and therefore a fairly strong rotation spectrum. This has been detected and measured with great precision in the microwave spectrum. On the basis of these measurements the molecule has been shown to have an apical angle of 116° 45′ and a bond length of 1·26 Å in the ground state.

We have quantitative information only on the ν_3 fundamental of ozone, which will be described in the next section. Line positions in the ν_2 and rotation bands have been analysed, and line positions in the 2800 cm^{-1}, 3050 cm^{-1}, 2105 cm^{-1}, and 1110 cm^{-1} bands have been recorded (see Bibliography).

5.6.3. *The* ν_3 *fundamental*

The fine structure of this band has been recorded with a resolution of about 0·14 cm^{-1}. Although this is insufficient to resolve the individual rotation lines it has nevertheless proved possible to derive the molecular constants for the excited state by a trial-and-error procedure (those for the ground state being known from analysis of the microwave spectrum). The result of this analysis is shown in Fig. 5.20. Agreement

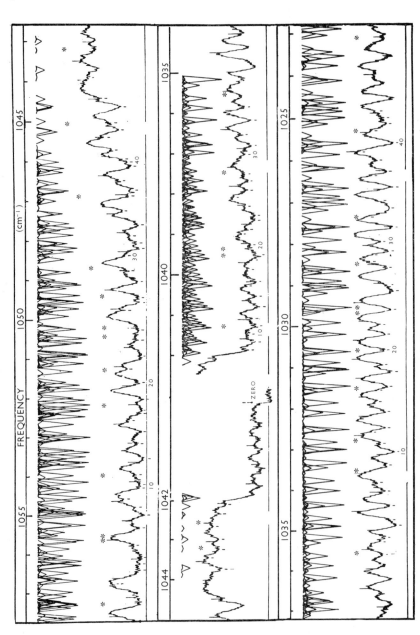

Fig. 5.20. Calculated and observed absorption spectra of ozone near 1043 cm⁻¹. The asterisks indicate water-vapour absorption lines. After Kaplan, Migeotte, and Neven (1956).

between observation and prediction is excellent up to $J = 17$, but higher J-values are affected by Coriolis interaction with the ν_1-band.

An extensive quantitative study of this band has been made by Walshaw (1954 and 1957), who finds the following empirical relation for the integrated band area at 20° C:

$$\int A_\nu \, d\nu = 138(1-10^{-a\zeta f(\phi)}) \text{ cm}^{-1}, \qquad (5.15)$$

$$\phi = a\zeta(a)^{2\cdot11}/p,$$

$$f(\phi) = 1\cdot185(1+734\phi)^{-1/2}\eta(\phi),$$

where η is a tabulated quantity running from $\eta = 1$ for $\phi < 10^{-4}$ to $0\cdot977$ for $\phi = 10^{-3}$ to $1\cdot079$ for $\phi = 10^{-2}$ and

$$\zeta(a) = \begin{cases} \dfrac{1+0\cdot1025a}{1+1\cdot61a} & (a \leqslant 0\cdot1), \\ 0\cdot984 \times 10^{-0\cdot53a} & (0\cdot1 \leqslant a \leqslant 0\cdot4), \\ 0\cdot317 \times a^{-0\cdot74} & (a \geqslant 0\cdot4). \end{cases}$$

In these formulae a is in cm of ozone at s.t.p. and p is the dilutant air pressure in mm Hg. The pressure range of the experiment was from $p = 0\cdot5$ to $p = 760$ mm Hg and from $a = 3 \times 10^{-3}$ to $1\cdot5$ cm s.t.p. The above formulae do not, however, apply below 10 mm Hg, where Doppler broadening becomes important. Measurements at different temperatures indicated a very slight temperature dependence of absorption.

Extrapolating $(1/a) \int A_\nu \, d\nu$ to $a = 0$ from (5.15) we find for the band intensity $1\cdot40 \times 10^{-17}$ cm^2. Some of these observations for small a are in the linear law region, and these results alone lead to a value of $1\cdot34 \times 10^{-17}$ cm^2.

Walshaw's observations included point by point measurements over the band contour, and attempts have been made to fit these to a statistical model with an exponential distribution of line intensities. At some frequencies this proved possible (see Fig. 4.15) but at others large deviations occurred and an empirical fit was therefore made using the correcting factor (4.107). Fig. 5.21 shows values of α/δ at s.t.p., σ/δ and $(\frac{1}{2}f)^2$ which give a good fit to all points of the band.

Averaging between $1053\cdot3$ cm^{-1} and $1029\cdot6$ cm^{-1}, Kaplan (1959) obtained an excellent fit between a statistical model and observation, and hence derived a mean line width of $0\cdot089$ cm^{-1} at s.t.p. which agrees reasonably well with an interesting indirect determination by Walshaw (1955). This author derived the parameters of Fig. 5.21 from pressures above 10 cm Hg and then sought a value of the parameter (d) for a mixed Doppler–Lorentz profile which gave a good fit between theory ((4.94) and Table 4.2) and observation for low pressures. A single value was found to be sufficient for all frequencies and its best mean value is 138 at s.t.p. From (3.52) for a temperature of 20° C there results $\alpha_L = 0 \cdot 076$ cm^{-1}.

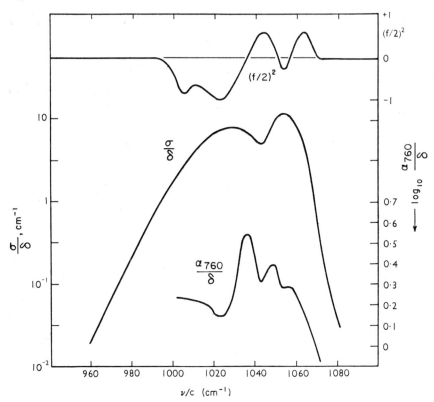

FIG. 5.21. Values of $(\tfrac{1}{2}f)^2$, α/δ at s.t.p. and σ/δ for the ν_3 band of ozone. After Walshaw and Goody (1954).

5.7. Other atmospheric gases

Among the many minor atmospheric constituents those discussed below are the most likely to have significant thermal effects.

5.7.1. *Nitrous oxide*

Nitrous oxide has a rich ultra-violet spectrum. Six Rydberg series have been reported, converging to limits at 12·72, 16·39, 16·55, and 20·10 eV. A number of continua have been observed, with maxima at 1820 Å, 1450 Å, 1275 Å, and 1080 Å. Below 1000 Å three more continua with maximum cross-sections of 3×10^{-17} cm^2 have been detected.

The molecule is linear and asymmetric, with the configuration NNO. $N^{14}N^{14}O^{16}$ has a dipole moment of 0·166 Debye and a detectable rotation spectrum (rotational constant 0·4182 cm^{-1}).

There are three fundamentals: ν_1 at 1285·6 cm^{-1}; ν_2 at 588·8 cm^{-1}; ν_3 at 2223·5 cm^{-1}. The distinction between parallel and perpendicular bands is the same as for carbon dioxide, and only ν_2 has a Q-branch. A high resolution spectrum of ν_1 has been given in Fig. 3.10. A spectrum of ν_3 is shown in Fig. 5.22.

Besides the fundamentals many overtone, combination, and upper-state bands have been reported by various authors, e.g.: 23 with frequencies greater than 7998·52 cm^{-1} ($2\nu_1+2\nu_2+2\nu_3$); 18 between 4417 cm^{-1} and 1880 cm^{-1}; 14 between 4062 cm^{-1} and 7431 cm^{-1}; 7 between 2209·53 cm^{-1} and 2798·30 cm^{-1}; and a large number of upper state bands between 2395 cm^{-1} and 3570 cm^{-1}. Most are too weak to be of any importance, and Table 5.22 includes only those bands which have been observed in the solar spectrum. Extensive empirical data are available on the ν_1, $2\nu_2$, and ν_3 bands at low resolution (see Table 5.22 for references).

TABLE 5.22

Nitrous oxide bands in the solar spectrum

Band centre (cm^{-1})	(μ)	Transition	Type	Molecular band intensity at s.t.p. (cm)	References	Line width at s.t.p. (cm^{-1})	References
4730·86	2·11	00⁰0–20⁰1	para.	—	—	—	—
4630·31	2·16	00⁰0–12⁰1	para.	—	—	—	—
4417·51	2·27	00⁰0–00⁰2	para.	—	—	—	—
4389·06	2·28	01¹0–01¹2	para.	—	—	—	—
3481·2	2·87	00⁰0–10⁰1	para.	$1·29 \times 10^{-18}$	1	—	—
3365·6	2·97	00⁰0–02⁰1	para.	—	—	—	—
2798·6	3·57	00⁰0–01¹1	perp.	$9·0 \times 10^{-20}$	2	—	—
2577	3·88	01¹0–21¹0	para.	—	—	—	—
2563·5	3·90	00⁰0–20⁰0	para.	$1·6 \times 10^{-18}$	5	—	—
2461·5	4·06	00⁰0–12⁰0	para.	$4·3 \times 10^{-19}$	5	—	—
2223·5	4·50	00⁰0–00⁰1	para.	$6·88 \times 10^{-17}$	2	—	—
2210	4·52	01¹0–01¹1	para.	—	—	—	—
1285·0	7·78	00⁰0–10⁰0	para.	$9·78 \times 10^{-18}$	3, 5	0·15	3
1167·0	9·56	00⁰0–02⁰0	para.	$4·08 \times 10^{-19}$	3, 5	0·16	3
588·8	17·0	00⁰0–01¹0	perp.	$7·75 \times 10^{-19}$	1, 5	0·16(?)	4

References:

1. Eggers and Crawford (1951).
2. Burch and Williams (1960).
3. Goody and Wormell (1951).
4. Adel and Barker (1944).
5. Burch, Singleton, France, and Williams (1961).

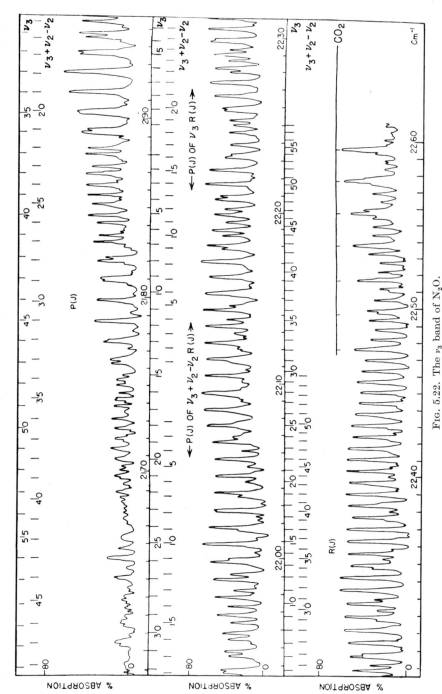

Fig. 5.22. The ν_3 band of N_2O.

Spectrum of a 15 cm cell containing 10 mm Hg of N_2O. After Lakshmi and Shaw (1955).

5.7.2. *Methane*

The ultra-violet spectrum of methane starts at 1450 Å, and has been measured by a number of workers down to 400 Å. The first ionization potential is 12·99 eV.

The infra-red spectrum of methane confirms the classical tetrahedral model developed by organic chemists. The molecule is a spherical top, and consequently highly degenerate; it has no rotation spectrum. Of the fundamentals, only ν_3 and ν_4 are active in the infra-red spectrum and are centred at 3020·3 cm^{-1} and 1306·2 cm^{-1} respectively. The inactive ν_1 and ν_2 fundamentals lie at 2914·2 cm^{-1} and 1526 cm^{-1} respectively. Since $2\nu_2 \simeq \nu_1 \simeq \nu_3$ there are numerous possibilities for interaction both by Fermi resonance and through Coriolis effects, and the degeneracies in ν_3 and ν_4 are resolved, giving line structures of exceptional complexity. Some P-branch lines of ν_3 at high resolution are shown in Fig. 5.23.

In addition to the two fundamentals, methane possesses an exceptionally rich spectrum of overtone and combination bands which are, however, difficult to analyse. Nine bands have been identified in the solar spectrum, and many more have been found in the reflection spectra of Uranus and Neptune, which have atmospheres rich in methane. Table 5.23 gives the observed bands and such information on band intensities as is available. Measurements on individual lines in the 1·67 μ

TABLE 5.23

Methane bands observed in the solar spectrum

Band centre		Transition	Molecular band intensity at s.t.p.	References
(cm^{-1})	(μ)		(cm)	
6005	1·67	0000–0020	$3\cdot6 \times 10^{-20}$	1
5861	1·71	0000–1101	—	—
5775	1·73	0000–0111	—	—
4420	2·20	0000–0110	—	—
4313	2·32	0000–0011	—	—
4216	2·37	0000–1001	—	—
4123	2·43	0000–0102	—	—
3019	3·31	0000–0010	$1\cdot26 \times 10^{-17}$	2
3823	3·55	0000–0101	—	—
2600	3·85	0000–0002	$1\cdot04 \times 10^{-17}$	3
1306	7·66	0000–0001	$6\cdot9 \times 10^{-18}$	2

References:
  1. Goldberg, Mohler, and Donovan (1952).   3. Thorndike (1947).
  2. Burch and Williams (1960).

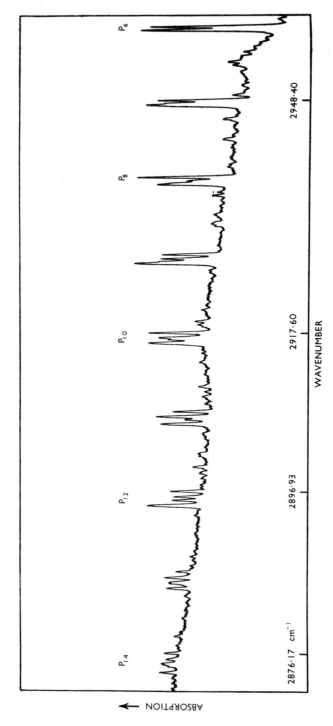

FIG. 5.23. *P*-branch lines in the ν_3 band of methane. After Allen and Plyler (1957).

band suggest a line width of $0 \cdot 08$ cm^{-1} at s.t.p. Detailed quantitative data on ν_3 and ν_4 at low resolution have been obtained by Burch and Williams (1960).

5.7.3. *Carbon monoxide*

The carbon monoxide molecule is, on account of the simplicity of its structure, one of the most carefully studied of molecules. Details of the

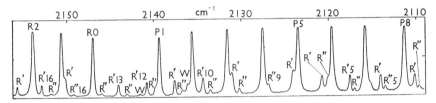

FIG. 5.24. The $C^{12}O^{16}$ fundamental.

Path length 10 cm; gas pressure 8 cm Hg. P, $R = C^{12}O^{16}$, P', $R' = C^{13}O^{16}$; P'', $R'' = C^{12}O^{18}$, $W =$ water. After Mills and Thompson (1953).

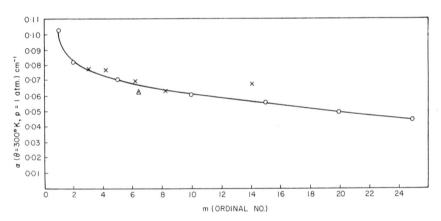

FIG. 5.25. Line widths at 300° K and 1 atmosphere pressure in the CO fundamental.

$\times$, Shaw and France (1956). $\bigcirc$, Benedict (1956). $\triangle$, from Fig. 4.12.

electronic spectrum are given by Herzberg (1950) and absorption coefficients in the ultra-violet spectrum are discussed by Watanabe (1959). These are of little significance to atmospheric studies.

The CO molecule has a dipole moment of $0 \cdot 1$ Debye, and pure rotational lines have been detected in the spectral range 100 to $600\,\mu$, while the $J = 0 \to 1$ lines of both $C^{12}O^{16}$ and $C^{13}O^{16}$ have been found in the microwave spectrum.

The only fundamental band of $C^{12}O^{16}$ is at 2143·2 cm^{-1} and the first overtone lies at 4260 cm^{-1}. Both bands have simple P- and R-branches. Details of the 2143·2 cm^{-1} band are shown in Fig. 5.24. $C^{13}O^{16}$ lines can be detected in the P-branch, but are much weaker in the R-branch, for which reason the R-branch maximum was used to test the Elsasser model in Chapter 4 (see Fig. 4.12).

Molecular band intensities in the fundamental and first overtone have been measured to be $8·82 \times 10^{-18}$ and $6·1 \times 10^{-20}$ cm, respectively. Line widths for carbon monoxide broadened by nitrogen are shown in Fig. 5.25. A typical curve of growth was given in Fig. 4.4.

5.7.4. *Nitric oxide*

No bands of this gas have been detected in the solar spectrum but it is, nevertheless, widely held responsible for the creation of the D-layer of the ionosphere, because of its low first ionization potential of 9·25 eV. Watanabe (1959) gives details of the ultra-violet spectrum and Herzberg (1950) gives data and references to the visible spectrum. The vibration-rotation spectrum is of no interest to atmospheric studies.

BIBLIOGRAPHY

5.1. Introduction

A number of books and reviews contain material for more than one of the following sections. The three books of G. HERZBERG, already mentioned in the bibliography of Chapter 3, are important sources of further references.

A valuable review of ultra-violet spectra of atmospheric gases, containing 264 references, is by

WATANABE, K., 1958, 'Ultraviolet absorption processes in the upper atmo-sphere', *Advances in Geophysics*, **5**, p. 153. New York: Academic Press.

The infra-red spectrum is given prominence in the exhaustive review by

GOLDBERG, L., 1954, 'The absorption spectrum of the atmosphere', p. 434 in *The earth as a planet*, ed. G. P. Kuiper. Chicago University Press.

Many valuable data, particularly on emissivities, are to be found in PENNER (1959) (§ 3.1). A recent survey of data on atmospheric gases is contained in

ELSASSER, W. M., and CULBERTSON, M. F., 1960, *Atmospheric radiation tables*, Meteor. Monographs **4**, No. 23. Am. Met. Soc.

Maps of the solar spectrum, with tables of identifications and other associated material prepared for astrophysical purposes can be instructive for quick estimates of the relative importance of atmospheric absorptions. These will be discussed at greater length in a later chapter, but will be listed here:

MINNAERT, M., MULDERS, G. F. W., and HOUTGAST, J., 1940, *Photometric altas of the solar spectrum, from λ3612 to λ8771*. Sterrewacht 'Sonnenborgh', Utrecht;

St. John, C. E., Moore, C. E., Ware, L. M., Adams, E. F., and Babcock, H. D., 1928, *Revision of Rowland's preliminary table of solar spectrum wavelengths*. Carnegie Institution of Washington, Publication No. 396;

Babcock, H. D., Moore, C. E., and Coffeen, M. F., 1948, 'The ultra-violet solar spectrum, λλ2935–3060', *Astrophys. J.* **107**, p. 287;

——— 1947, *The solar spectrum λ6600 to λ13495*. Carnegie Institution of Washington, Publication No. 579;

Mohler, O. C., Pierce, A. K., McMath, R. R., and Goldberg, L., 1950, *Photometric atlas of the near infrared solar spectrum λ8465 to λ25242*. University of Michigan Press;

——— 1955, *A table of solar spectrum wavelengths 11984 Å to 25578 Å*. University of Michigan Press;

Shaw, J. H., Oxholm, M. L., and Claassen, H. H., 1952, 'The solar spectrum from 7 μ to 13 μ', *Astrophys. J.* **116**, p. 554;

———, Chapman, R. M., Howard, J. N., and Oxholm, M. L., 1951, 'A grating map of the solar spectrum from 3·0 to 5·2 microns', ibid. **113**, p. 268;

Migeotte, M., Neven, L., and Swensson, J., 1956, *The solar spectrum from 2·8 to 23·7 microns*: Part I. *Photometric Atlas*; Part II. *Measures and Identifications*. Institut d'Astrophysique de l'Université de Liège; and

Vigroux, E., Migeotte, M., Neven, L., and Swensson, J., 1957, *An atlas of nitrous oxide, methane and ozone infrared absorption bands*: Part I. *The photometric records*; Part II. *Measures and Identifications*. Institute d'Astrophysique de l'Université de Liège.

The effect of the non-linearity of absorption in emphasizing particular lines in different circumstances is demonstrated theoretically by

Kaplan, L. D., 1954, 'Energy transfer by infrared bands', *J. Met.* **11**, p. 16;

and in an actual example by

Murgatroyd, R. J., and Goody, R. M., 1958, 'Sources and sinks of radiative energy from 30 to 90 km', *Quart. J. R. Met. Soc.* **84**, p. 225.

5.2. Nitrogen

Most of the data in this section were collected together by Watanabe (1958) (§ 5.1). Additional details of N_2 and N_2^+ levels can be found in Herzberg (1950) (§ 3.1).

The status of knowledge of auroral and airglow spectra is reviewed by

Chamberlain, J. W., 1961, *Physics of the airglow and the aurora*. New York: Academic Press; and

Bates, D. R., 1960, 'The airglow' and 'The auroral spectrum and its interpretation', chapters 5 and 7 in *Physics of the upper atmosphere*, ed. J. A. Ratcliffe. New York: Academic Press.

The ionization continua of nitrogen and oxygen atoms are treated theoretically by

Bates, D. R., and Seaton, M. J., 1949, 'The quantal theory of continuum absorption of radiation by various atoms in their ground states. II. Further calculations on oxygen, nitrogen and carbon', *M.N.R.A.S.*, **109**, p. 698;

Nicolet, M., 1952, Chapter xii, *Physics and medicine of the upper atmosphere*. University of New Mexico Press; and

DALGARNO, A., and PARKINSON, D., 1960, 'Photoionisation of atomic oxygen and atomic nitrogen', *J. At. Terr. Phys.* **18**, p. 335.

5.3. Oxygen

The importance of the transitions in the ground state of atomic oxygen was first pointed out by

BATES, D. R., 1951, 'The temperature of the upper atmosphere', *Proc. Phys. Soc. B*, **64**, p. 805.

The spectrum in Fig. 5.3 is after GOLDBERG (1954) (§ 5.1). High resolution spectra of the infra-red bands have been recorded by

HERZBERG, L., and HERZBERG, G., 1947, 'Fine structure of the infrared atmospheric oxygen bands', *Astrophys. J.* **105**, p. 353.

The intensity of the (0, 0) band was measured by

JONES, A. V., and HARRISON, A. W., 1958, '$^1\Delta_g$—$^3\Sigma_g$—O_2 infrared emission band in the twilight airglow spectrum', *J. At. Terr. Phys.* **13**, p. 45.

The data on the red atmospheric bands were collected by

VAN DE HULST, H. C., 1945, 'The atmospheric oxygen bands', *Annales d'astrophys.* **8**, p. 1;

and detailed analysis of the fine structure is given by

BABCOCK, H. D., and HERZBERG, L., 1948, 'Fine structure of the red system of atmospheric oxygen bands', *Astrophys. J.* **108**, p. 167.

The discussion of ultra-violet absorptions in § 5.3.3 follows WATANABE (1958) and GOLDBERG (1954) (both § 5.1).

The first quantitative measurements on the microwave absorption of oxygen were published by

BERINGER, R., 1946, 'The absorption of one-half centimeter electromagnetic waves in oxygen', *Phys. Rev.* **70**, p. 53.

The theory of oxygen absorption in the microwave spectrum was given first by

VAN VLECK, J. H., 1947, 'The absorption of microwaves by oxygen', ibid. **71**, p. 413.

The 2 cm^{-1} band has been resolved by

BURKHALTER, J. H., ANDERSON, R. S., SMITH, W. V., and GORDY, W., 1950, 'The fine structure of the microwave absorption spectrum of oxygen', ibid. **79**, p. 651;

MIZUSHIMA, M., and HILL, R. M., 1954, 'Microwave spectrum of O_2', ibid. **93**, p. 745; and

STRANDBERG, M. W. P., MENG, C. Y., and INGERSOLL, J. G., 1949, 'The microwave absorption spectrum of oxygen', ibid. **75**, p. 1524.

The 4 cm^{-1} line is treated by

ANDERSON, R. S., JOHNSON, C. M., and GORDY, W., 1951, 'Resonant absorption of oxygen at 2·5-millimeter wavelength', ibid. **83**, p. 1061.

Line widths in the microwave spectrum have been measured by BURKHALTER *et al.* (1950) and by

GOKHALE, B. V., and STRANDBERG, M. W. P., 1951, 'Line breadths in the 5-mm microwave absorption of oxygen', ibid. **84**, p. 844;

ANDERSON, R. S., SMITH, W. V., and GORDY, W., 1951, 'Line breadths of the fine structure of the microwave spectrum of oxygen', *Phys. Rev.* **82**, p. 264;

HILL, R. M., and GORDY, W., 1954, 'Zeeman effect and line breadth studies of the microwave lines of oxygen', ibid. **93**, p. 1019; and

ARTMAN, J. O., and GORDON, J. P., 1954, 'Absorption of microwaves by oxygen in the millimeter wavelength region', ibid. **96**, p. 1237.

The oxygen polymer band in the solar spectrum has been described by

DUFAY, J., 1942, 'Notes sur l'absorption selective dans l'atmosphère terrestre I—Description du spectra du soleil couchant, de 4600 à 6900 Å', *Annales d'astrophys.* **5**.

5.4. Water vapour

Details of the data on the electronic bands of water vapour are given by WATANABE (1958) (§ 5.1).

For general information about the vibration-rotation spectrum the reader is referred to HERZBERG (1945) (§ 3.1), GOLDBERG (1954) (§ 5.1), and PENNER (1959) (§ 3.1). In addition W. S. BENEDICT (1956) has written extensive introductions, amounting to reviews, for MOHLER (1955) (§ 5.1) and MIGEOTTE, NEVEN, and SWENSSON (1956) (§ 5.1). In the latter publication he gives the energies of all relevant rotational levels for the vibrational ground state, as well as for upper states ν_1, ν_2, $2\nu_2$, and ν_3. Energy levels and matrix elements for the rigid asymmetric rotator are given by CROSS, HAINER, and KING (1954) and SCHWENDEMANN and LAURIE (1958) (§ 3.2).

Line widths according to Anderson's theory are computed by

BENEDICT, W. S., and KAPLAN, L. D., 1959, 'Calculation of line widths in H_2O—N_2 collisions', *J. Chem. Phys.* **30**, p. 388.

Measurements of average transmission for the water-vapour rotation band have been made by

ROACH, W. T., and GOODY, R. M., 1958, 'Absorption and emission in the atmospheric window from 770 to 1,250 cm^{-1}', *Quart. J. R. Met. Soc.* **84**, p. 319;

DAW, H., 1954, 'Generalized absorption coefficients of water vapor', *Phys. Rev.* **94**, p. 1424;

CLOUD, W. H., 1952, *Pressure broadening of pure rotation lines of the water vapor spectrum from 490 to 590 cm*$^{-1}$, unpublished report from Johns Hopkins Univ. on Contract N–onr 248–01;

PALMER, C. H., 1957, 'Long path water vapor spectra with pressure broadening. I. 20 μ to 31·7 μ. II. 29 μ to 40 μ', *J. Opt. Soc. Am.* **47**, pp. 1024 and 1028; and

—— 1960, 'Experimental transmission functions for the pure rotation band of water vapor', ibid. **50**, p. 1232.

Work on the rotation band before 1950 is reviewed by

COWLING, T. G., 1950, 'Atmospheric absorption of heat radiation by water vapour', *Phil. Mag.* **41**, p. 109; and

—— 1943, 'The absorption of water vapour in the far infra-red', *Rep. on Progress in Physics*, **9**, p. 29.

Comments on Palmer's measurements have been made by

BENEDICT, W. S., 1957, 'Transmission functions for atmospheric water vapor', *J. Opt. Soc. Am.* **47**, p. 1056.

Work on the microwave spectrum is by

TOWNES, C. H., and MERRITT, F. R., 1946, 'Water spectrum near one-centimeter wave-length', *Phys. Rev.* **70**, p. 558;

KING, W. C., and GORDY, W., 1954, 'One-to-two millimeter wave spectroscopy', ibid. **93**, p. 407;

BECKER, G. E., and AUTLER, S. H., 1946, 'Water vapor absorption of electromagnetic radiation in the centimeter wave-length range', ibid. **70**, p. 300;

KING, G. W., HAINER, R. M., and CROSS, P. C., 1947, 'Expected microwave absorption coefficients of water and related molecules', ibid. **71**, p. 433; and

ROGERS, T. F., 1954, 'Absolute intensity of water-vapor absorption at microwave frequencies', ibid. **93**, p. 248.

High-resolution analysis of the ν_1, ν_2, ν_3, and $2\nu_2$ bands of water vapour has been performed by

PLYLER, E. K., and TIDWELL, E. D., 1957, 'The precise measurement of the infra-red spectra of molecules of the atmosphere', in *Les Molécules dans les Astres*, Mém. de la Soc. R. des Sciences de Liège, **18**, p. 426;

DALBY, F. W., and NIELSEN, H. H., 1956, 'Infrared spectrum of water vapor. Part I. The 6·26 μ region', *J. Chem. Phys.* **25**, p. 934; and

NIELSEN, H. H., 1941, 'The near infra-red spectrum of water vapor. Part I. The perpendicular bands ν_2 and $2\nu_2$', *Phys. Rev.* **59**, p. 565.

For measurements and analysis of high-J ν_2 lines in the laboratory, see

BENEDICT, W. S., CLAASSEN, H. H., and SHAW, J. H., 1952, 'Absorption spectrum of water vapor between 4·5 and 13 microns', *J. Res. Nat. Bur. Stand.* **49**, p. 91.

Low-resolution empirical data on the vibration-rotation bands prior to 1950 are discussed by COWLING (1950 and 1943). This is almost entirely superseded by the work of

HOWARD, J. N., BURCH, D. E., and WILLIAMS, D., 1956, 'Infrared transmission of synthetic atmospheres', *J. Opt. Soc. Am.* **46**, pp. 186, 237, 334, 452; and

BURCH, D. E., SINGLETON, E. B., FRANCE, W. L., and WILLIAMS, D., 1960, *Infrared absorption by minor atmospheric constituents.* Report on RF project 778. The Ohio State University Research Foundation.

Other data are listed by PENNER (1959) (§ 3.1).

The following additional papers deal with the overtone and combination bands of water vapour:

BENEDICT, W. S., 1948, 'New bands in the vibration-rotation spectrum of water vapor', *Phys. Rev.* **74**, p. 1246;

MOHLER, O. C., and BENEDICT, W. S., 1948, 'Atmospheric absorption of water vapor between 1·42 μ and 2·50 μ', ibid., p. 702;

NELSON, R. C., and BENEDICT, W. S., 1948, 'Absorption of water vapor between 1·34 μ and 1·97 μ', ibid., p. 703; and

WHITE, J. U., ALPERT, N. L., DeBELL, A. G., and CHAPMAN, R. M., 1957, 'Infrared grating spectrophotometer', *J. Opt. Soc. Am.* **47**, p. 358.

Details of isotopic bands are treated by

BENEDICT, W. S., GAILAR, N., and PLYLER, E. K., 1956, 'Rotation-vibration spectra of deuterated water vapor', *J. Chem. Phys.* **24**, p. 1139;
—— —— —— 1953, 'The vibration-rotation spectrum of HDO', ibid. **21**, p. 1302;
CHAPMAN, R. M., and SHAW, J. H., 1950, 'Fine structure of HDO near $3\cdot7\ \mu$ in the solar spectrum', *Phys. Rev.* **78**, p. 71; and
BARKER, E. F., and SLEATOR, W. W., 1935, 'The infrared spectrum of heavy water', *J. Chem. Phys.* **3**, p. 660.

HDO bands were first identified in the telluric spectrum by

ADEL, A., 1941, 'The grating infrared solar spectrum. I. Rotational structure of the heavy water (HDO) band ν_2 at $7\cdot12\ \mu$', *Astrophys. J.* **93**, p. 506; and
GEBBIE, H. A., HARDING, W. R., HILSUM, C., PRYCE, A. W., and ROBERTS, V., 1951, 'Atmospheric transmission in the 1 to $14\ \mu$ region', *Proc. Roy. Soc. A*, **206**, p. 87.

Work on the rotation spectrum of heavy water in the microwave region is by KING, HAINER, and CROSS (1947), and ROGERS (1954).

The first quantitative measurements on the 1000 cm⁻¹ continuum were made at low resolution by

ADEL, A., and LAMPLAND, C. O., 1940, 'Atmospheric absorption of infrared solar radiation at the Lowell Observatory', *Astrophys. J.* **91**, pp. 1 and 481.

Attempts to separate off line absorption have been made by

ROACH, W. T., and GOODY, R. M., 1958, 'Absorption and emission in the atmospheric window from 770 to 1,250 cm⁻¹', *Quart. J. R. Met. Soc.* **84**, p. 319; and
VIGROUX, E., 1959, 'Émission continue de l'atmosphère terrestre a $9\cdot6\ \mu$', *Annales de Géophysique*, **15**, p. 453.

The separation was unsatisfactory until high-resolution measurements were made by

SAIEDY, F., and GOODY, R. M., 1959, 'The solar emission intensity at $11\ \mu$', *M.N.R.A.S.* **119**, p. 313; and
SAIEDY, F., 1960, *Absolute measurements on infra-red radiation in the atmosphere.* Ph.D. Thesis, London University.

The original theoretical work on the continuum is by ELSASSER (1941) (§ 1.1).

The work on water-vapour emissivities divides into three groups: theoretical, by ELSASSER (1942) (§ 1.1), COWLING (1950), and

YAMAMOTO, G., 1952, 'On a radiation chart', *Science Rept. Tohoku Univ., Series 5, Geophysics*, **4**, No. 1, p. 9;

experimental, over short paths in the laboratory by

ELSASSER, W. M., 1941, 'A heat radiation telescope and the measurement of the infrared emission of the atmosphere', *Mon. Weath. Rev.* **69**, p. 1; and
FALKENBERG, G., 1939, 'Experimentellen zur Eigenstrahlung dünner wasserdampf-haltiger Luftschichten', *Meteor. Z.* **56**, p. 72;

observational on the entire atmosphere by

BROOKS, F. A., 1941, 'Observations of atmospheric radiation', *Pap. Phys. Ocean. Met.* (M.I.T., W.H.O.I.), **8**, No. 2; and

ROBINSON, G. D., 1947 and 1950, 'Notes on the measurement and estimation of atmospheric radiation', *Quart. J. R. Met. Soc.* **73**, p. 127 and **76**, p. 37.

Numerous attempts to summarize these data have been made; in Table 5.13 we have followed

BROOKS, D. L., 1950, 'A tabular method for the computation of temperature change by infrared radiation in the free atmosphere', *J. Met.* **7**, p. 313.

The data required by engineers are discussed at length by PENNER (1959) (§ 3.1) and by

HOTTEL, H. C., Chapter iii in *Heat transmission*, by W. H. McAdams. New York: McGraw-Hill.

5.5. Carbon dioxide

References to data on the electronic bands of CO_2 are given by WATANABE (1958) (§ 5.1).

The following refer to the ν_2 band:

CALLENDAR, G. S., 1941, 'Infra-red absorption by carbon dioxide with special reference to atmospheric radiation', *Quart. J. R. Met. Soc.* **67**, p. 263;

KAPLAN, L. D., 1950, 'Line intensities and absorption for the 15-micron carbon dioxide band', *J. Chem. Phys.* **18**, p. 186;

—— and EGGERS, D. F., 1956, 'Intensity and line-width of the 15-micron band, determined by a curve-of-growth method', ibid. **25**, p. 876;

KOSTKOWSKI, J. H., and KAPLAN, L. D., 1957, 'Absolute intensities of the 721 and 742 cm^{-1} bands of CO_2', ibid. **26**, p. 1252;

YAMAMOTO, G., and SASAMORI, T., 1958, 'Calculation of the absorption of the 15 μ carbon dioxide band', *Sci. Rept., Tohoku Univ., Series 5, Geophysics*, **10**, No. 2, p. 37;

SASAMORI, T., 1959, The temperature effect on the absorption of 15 μ carbon dioxide band, ibid. **11**, p. 149;

YAMAMOTO, G., and SASAMORI, T., 1961, 'Further studies on the absorption by the 15-micron carbon dioxide bands', ibid. **13**, p. 1;

ROSSMAN, K., RAO, K. N., and NIELSEN, H. H., 1956, 'Infrared spectrum and molecular constants of carbon dioxide. Part I. ν_2 of $C^{12}O_2^{16}$ at 15 μ', *J. Chem. Phys.* **24**, p. 103;

—— FRANCE, W. L., RAO, K. N., and NIELSEN, H. H., 1956, 'Infrared spectrum and molecular constants of carbon dioxide. Part II. Levels 10^00 and 02^00, 10^01 and 02^01 coupled by Fermi resonance', ibid., p. 1007;

MADDEN, R. F., 1957, *Study of CO_2 absorption spectra between 15 and 18 microns*, unpublished report by Johns Hopkins University on Contract Nonr 248(01);

HOWARD, et al. (1956), (§ 5.4); and

BURCH, D. E., GRYVNAK, D., and WILLIAMS, D., 1960, *Infrared absorption by carbon dioxide*. Report on Project 778: Ohio State Univ. Research Foundation.

Equation (5.14) follows MURGATROYD and GOODY (1952) (§ 5.1).

Rotational constants for all important vibrational levels of carbon dioxide are given by MIGEOTTE et al. (1956) (§ 5.1). The following papers refer especially ν_3-bands:

PLYLER, E. K., BLAINE, L. R., and TIDWELL, E. D., 1955, 'Infrared absorption and emission spectra of carbon monoxide in the region from 4 to 6 microns', J. Res. Nat. Bur. Stand. **55**, p. 183;

BENEDICT, W. S., HERMAN, R. C., and SILVERMAN, S., 1951, 'R-branch heads of some CO_2 infrared bands in the $CO + O_2$ flame spectrum', J. Chem. Phys. **19**, p. 1325;

NIELSEN, A. H., and YAO, Y. T., 1945, 'The analysis of the vibration-rotation band ω_3 for $C^{12}O_2^{16}$ and $C^{13}O_2^{16}$', Phys. Rev. **68**, p. 173;

EGGERS, D. F., and ARENDS, C. B., 1957, 'Infrared intensities and bond moments in $CO^{16}O^{18}$', J. Chem. Phys. **27**, p. 1405;

PLASS, G. N., 1959, 'Spectral emissivity of carbon dioxide from 1800–2500 cm^{-1}', J. Opt. Soc. Am. **49**, p. 821;

BENEDICT, W. S., and PLYLER, E. K., 1954, in Energy transfer in hot gases, Washington, D.C., Govt. Printing Office, p. 57; and

TOURIN, R. H., 1961, 'Measurements of infrared spectral emissivities of hot carbon dioxide in the $4\cdot3\text{-}\mu$ region', J. Opt. Soc. Am. **51**, p. 175.

The following refer to the weaker bands:

KOSTKOWSKI, H. J., 1955, 'Line half-width and its variation with temperature in the 10·4-micron band of CO_2', ibid. **45**, p. 406;

GAILAR, N. M., and PLYLER, E. K., 1952, 'The $3\nu_3$ bands of carbon disulphide and carbon dioxide', J. Res. Nat. Bur. Stand. **48**, p. 392;

FRANCE, W. L., and DICKEY, F. P., 1955, 'Fine structure of the 2·7 micron carbon dioxide rotation-vibration band', J. Chem. Phys. **23**, p. 471; and

HERZBERG, G., and HERZBERG, L., 1953, 'Rotation-vibration spectra of di-atomic and simple polyatomic molecules with long absorbing paths. XI. The spectrum of carbon dioxide (CO_2) below 1·25 μ', J. Opt. Soc. Am. **43**, p. 1037.

Also

GOLDBERG (1954) (§ 5.1), HERZBERG (1945) (§ 3.1), MIGEOTTE et al. (1956) (§ 5.1) and MOHLER (1955) (§ 5.1).

5.6. Ozone

The following are some of the papers besides WATANABE (1958) (§ 5.1) which refer to the electronic bands of ozone:

NY TSI-ZE and CHOONG SHIN-PIAW, 1933, 'Sur l'absorption ultra-violette de l'ozone', Chinese J. Phys. **1**, p. 38;

VIGROUX, E., 1953, 'Contribution à l'étude expérimentale de l'absorption de l'ozone', Annales de Phys. **8**, p. 709;

INN, E. C. Y., and TANAKA, Y., 1953, 'Absorption coefficient of ozone in the ultraviolet and visible regions', J. Opt. Soc. Am. **43**, p. 870;

HEARN, A. G., WALSHAW, C. D., and WORMELL, T. W., 1957, 'The absorption coefficient of ozone at 3021 Å', Quart. J. R. Met. Soc. **83**, p. 364;

—— 1961, 'The absorption of ozone in the ultra-violet and visible regions of the spectrum', Proc. Phys. Soc. A, **78**, p. 932;

TANAKA, Y., INN, E. C. Y., and WATANABE, K., 1953, 'Absorption coefficients of gases in the vacuum ultraviolet. Part IV, Ozone', *J. Chem. Phys.* **21**, p. 1651;

STRONG, J., 1941, 'On a new method of measuring the mean height of the ozone in the atmosphere', *J. Franklin Inst.* **231**, p. 121;

VASSY, E., 1937, 'Sur quelques propriétés de l'ozone et leurs conséquences géophysiques', *Annales de Phys.* **8**, p. 679; and

HERRON, J. T., and SCHIFF, H. I., 1956, 'Mass spectroscopy of ozone', *J. Chem. Phys.* **24**, p. 1266.

The vibration-rotation spectrum of ozone is discussed by

WILSON, M. K., and BADGER, R. M., 1948, 'A reinvestigation of the vibration spectrum of ozone', ibid. **16**, p. 741;

HUGHES, R. H., 1956, 'Structure of ozone from the microwave spectrum between 9000 and 45 000 Mc', ibid. **24**, p. 131;

NEXSEN, W. E., 1956, *Measurement and analysis of the rotational fine structure of the ν_2 fundamental of ozone.* Scientific Rept. No. 1, Ohio State Univ. Research Foundation Project 587;

PIERCE, L., 1956, 'Determination of the potential constants of ozone from centrifugal distortion effects', *J. Chem. Phys.* **24**, p. 139;

DANTI, A., and LORD, R. C., 1959, 'Pure rotational absorption of ozone and sulphur dioxide from 100 to 200 microns', ibid. **30**. p. 1310;

VIGROUX, E., MIGEOTTE, M., and NEVEN, L., 1953, 'Étude, à grande dispersion, des bandes d'absorption de l'ozone a 9 μ, 4·75 μ, 3·59 μ et 3·27 μ', *Physica* **19**, p. 140; and

GEBBIE, H. A., STONE, N. W. B., and WALSHAW, C. D., 1960, *Pure rotational absorption spectrum of ozone from 125 microns to 500 microns,* unpublished report from National Physical Laboratory, Basic Physics Div. Rept. No. 2.

Theoretical analysis of the rotation band has been carried out by

GORA, E. K., 1959, 'The rotational spectrum of ozone', *J. Mol. Spec.* **3**, p. 78, following the work of KING, HAINER, and CROSS (1943) (§ 3.2).

For work on the ν_3 band see BENEDICT, CLAASSEN, and SHAW (1952) (§ 5.4);

KAPLAN, L. D., MIGEOTTE, M. V., and NEVEN, L., 1956, '9·6-micron band of telluric ozone and its rotational analysis', *J. Chem. Phys.* **24**, p. 1183;

WALSHAW, C. D., 1954, *An experimental investigation of the 9·6 μ band of ozone.* Thesis, Cambridge University;

—— 1955, 'Line widths in the 9·6 μ band of ozone', *Proc. Phys. Soc.* A, **68**, p. 530;

—— 1957, 'Integrated absorption by the 9·6 μ band of ozone', *Quart. J. R. Met. Soc.* **83**, p. 315;

—— and GOODY, R. M., 1954, 'Absorption by the 9·6 μ band of ozone', *Proc. Toronto Met. Conf. 1953* (Roy. Met. Soc.), p. 49;

KAPLAN, L. D., 1959, 'A method for calculation of infra-red flux for use in numerical models of atmospheric motion', in *The atmosphere and the sea in motion,* Rockefeller Inst. Press, p. 170.

5.7. Other atmospheric gases

The literature on the electronic spectrum of nitrous oxide and all other relevant atmospheric constituents is detailed by Watanabe (1958) (§ 5.1).

The rotation spectrum of nitrous oxide has been investigated by

PALIK, E. D., and RAO, K. N., 1956, 'Pure rotational spectra of CO, NO, and N_2O between 100 and 600 microns', *J. Chem. Phys.* **25**, p. 1174; and

JOHNSON, C. M., TRAMBARULO, R., and GORDY, W., 1951, 'Microwave spectroscopy in the region from two to three millimeters. Part II', *Phys. Rev.* **84**, p. 1178.

Fig. 5.22 is after

LAKSHMI, K., and SHAW, J. H., 1955, 'Absorption bands of N_2O near $4.5\,\mu$', *J. Chem. Phys.* **23**, p. 1887.

High resolution spectra of ν_2 and ν_1 have been published by WHITE *et al.* (1957) (§ 5.4); and

LAKSHMI, K., RAO, K. N., and NIELSEN, H. H., 1956, 'Molecular constants of nitrous oxide from measurements of ν_2 at $17\,\mu$', *J. Chem. Phys.* **24**, p. 811.

Investigations of ν_1, $2\nu_2$, and ν_3 at low resolution have been made by

GOODY, R. M., and WORMELL, T. W., 1951, 'The quantitative determination of atmospheric gases by infra-red spectroscopic methods', *Proc. Roy. Soc. A*, **209**, p. 178;

BURCH, D. E., and WILLIAMS, D., 1960, *Infrared absorption by minor atmospheric constituents*. Scientific Rept. No. 1 on R.F. project 778. The Ohio State University; and

—— SINGLETON, E. B., FRANCE, W. L., and WILLIAMS, D., 1961, *Infrared absorption by minor atmospheric constituents*. Final Rept. on R.F. project 778. The Ohio State University.

Overtone, combination, and upper-state bands of N_2O have been investigated by

HERZBERG, G., and HERZBERG, L., 1950, 'Rotation-vibration spectra of diatomic and simple polyatomic molecules with long absorbing paths. VI. The spectrum of nitrous oxide (N_2O) below $1.2\,\mu$', *J. Chem. Phys.* **18**, p. 1551;

PLYLER, E. K., TIDWELL, E. D., and ALLEN, H. C., 1956, 'Near infrared spectrum of nitrous oxide', ibid. **24**, p. 95;

DOUGLAS, A. E., and MØLLER, C. K., 1954, 'The near infrared spectrum and the internuclear distances of nitrous oxide', ibid. **22**, p. 275;

THOMPSON, H. W., and WILLIAMS, R. L., 1953, 'Vibration bands and molecular rotational constants of nitrous oxide', *Proc. Roy. Soc. A*, **220**, p. 435; and

TIDWELL, E. D., PLYLER, E. K., and BENEDICT, W. S., 1960, 'Vibration-rotation bands of N_2O', *J. Opt. Soc. Am.* **50**, p. 1243.

The following additional sources were used in the construction of Table 5.21:

EGGERS, D. F., and CRAWFORD, B. L., 1951, 'Vibrational intensities. III. Carbon dioxide and nitrous oxide', *J. Chem. Phys.* **19**, p. 1554; and

ADEL, A., and BARKER, E. F., 1944, 'Grating infra-red measurements at oblique incidence. Line width in the spectrum of N_2O', *Rev. Mod. Phys.* **16**, p. 236.

The following refer to the vibration-rotation spectrum of methane:

ALLEN, H. C., and PLYLER, E. K., 1957, 'ν_3 band of methane', *J. Chem. Phys.* **26**, p. 972;

GOLDBERG, L., MOHLER, O. C., and DONOVAN, R. E., 1952, 'Experimental determination of absolute f-values for methane', *J. Opt. Soc. Am.* **42**, p. 1;

THORNDIKE, A. M., 1947, 'The experimental determination of the intensities of infra-red absorption bands. III. Carbon dioxide, methane and ethane', *J. Chem. Phys.* **15**, p. 868;

GOLDBERG, L., 1951, 'The abundance and vertical distribution of methane in the earth's atmosphere', *Astrophys. J.* **113**, p. 567;

NIELSEN, A. H., and MIGEOTTE, M. V., 1952, 'Abundance and vertical distribution of telluric methane from measurements at 3580 meters elevation', *Annales d'astrophys.* **15**, p. 134; and

BURCH and WILLIAMS (1960).

The following refer to the spectrum of carbon monoxide:

HERZBERG (1950) (§ 3.1); WATANABE (1958);

GILLIAM, O. R., JOHNSON, C. M., and GORDY, W., 1950, 'Microwave spectroscopy in the region from two to three millimeters', *Phys. Rev.* **78**, p. 140;

MILLS, I. M., and THOMPSON, H. W., 1953, 'The fundamental vibration-rotation bands of $C^{13}O^{16}$ and $C^{12}O^{18}$', *Trans. Faraday Soc.* **49**, p. 224;

PENNER, S. S., and WEBER, D., 1951, 'Quantitative infrared intensity measurements. I. Carbon monoxide pressurized with infrared-inactive gases. II. Studies on the first overtone of unpressurized CO', *J. Chem. Phys.* **19**, pp. 807 and 817;

SHAW, J. H., and FRANCE, W. L., 1956, *Intensities and widths of single lines of the 4·7 micron CO fundamental*. Scientific Rept. 4 on Project 587. Ohio State University R.F.;

LOCKE, J. L., and HERZBERG, L., 1953, 'The absorption due to carbon monoxide in the infrared solar spectrum', *Canad. J. Phys.* **31**, p. 504; and

BENEDICT, W. S., 1956, *Theoretical studies of infra-red spectra of atmospheric gases*. Unpublished report on contract No. AF 19(604)–1001.

6

COMPUTATION OF FLUXES AND HEATING RATES

6.1. Transmission along non-homogeneous paths

6.1.1. *An exact solution*

THE first step towards a solution of realistic atmospheric problems is to extend the considerations of Chapter 4 to non-homogeneous atmospheric paths. Only approximate solutions will be sought, but in one ideal case an exact solution is available. Let us assume the product $cS = S\rho/\rho_a$ to be independent of height, where c is the mixing ratio, ρ_a the air density, ρ that for the absorbing gas, and S the line intensity. This model is attractive at first sight because the mixing ratio is approximately constant for carbon dioxide and some other, less-important gases (§ 1.3). However, the dependence of line intensity on temperature and the variation of temperature with height rob the model of direct practical importance.

Consider the case of a single line of Lorentz shape. The optical path between points (1) and (2) is

$$\tau_\nu(1,2) = \int_1^2 \frac{S\,da}{\pi} \frac{\alpha_L}{(\nu^2 + \alpha_L^2)}. \tag{6.1}$$

According to the hydrostatic approximation, the mass of air between two close levels is equal to $-dp/g$ where dp is the pressure difference between them. For $Sc = \text{const}$, we can therefore write

$$-S\,da = S_m \frac{c\,dp}{g\xi} = \text{const}\,dp = \text{const}\,d\alpha_L, \tag{6.2}$$

where ξ is the cosine of the zenith angle. By analogy with (4.8) we define

$$\tilde{u} = \frac{S}{2\pi} \frac{da}{d\alpha_L}, \tag{6.3}$$

which according to (6.2) is constant. Hence

$$\tau_\nu(1,2) = \int_2^1 \frac{2\tilde{u}\alpha_L\,d\alpha_L}{\xi(\nu^2 + \alpha_L^2)}$$

$$= \frac{\tilde{u}}{\xi} \ln\left(\frac{\nu^2 + \alpha_L^2(1)}{\nu^2 + \alpha_L^2(2)}\right). \tag{6.4}$$

Substituting in (4.6) we find for the mean absorption

$$\bar{A}(1,2) = \frac{1}{\delta} \int\limits_{-\infty}^{+\infty} \left\{ 1 - \left(\frac{\nu^2 + \alpha_L^2(2)}{\nu^2 + \alpha_L^2(1)} \right)^{\tilde{u}/\xi} \right\} d\nu. \tag{6.5}$$

For $\alpha(2) = 0$, i.e. a ray passing from outer space to the level (1) (the problem of solar radiation), this integral can be expressed in terms of Γ-functions,

$$\bar{A} = 2y(1)\pi^{1/2} \frac{\Gamma(\tilde{u}/\xi + \frac{1}{2})}{\Gamma(\tilde{u}/\xi)}, \tag{6.6}$$

where $y(1)$ is the value of y at level (1).

Equations (6.5) and (6.6) provide a complete solution for the case of a single line and, in combination with (4.94), for the random model. A solution for the Elsasser model can be obtained along the lines of § 4.4.1. Using (6.4) to give $\tau_{\nu-i\delta}(1,2)$, there results

$$\bar{T}(1,2) = \int\limits_{-1/2}^{+1/2} \prod\limits_{i=-\infty}^{i=+\infty} \left(\frac{(x-i)^2 + y^2(2)}{(x-i)^2 + y^2(1)} \right)^{\tilde{u}/\xi} dx \tag{6.7}$$

$$= \int\limits_{-1/2}^{+1/2} \left(\frac{\cosh 2\pi y(2) - \cos 2\pi x}{\cosh 2\pi y(1) - \cos 2\pi x} \right)^{\tilde{u}/\xi} dx, \tag{6.8}$$

by the Mittag–Leffler theorem. A number of writers have solved (6.8) by numerical quadrature (see Bibliography).

Equation (6.8) has asymptotic forms closely analogous to those for the Elsasser function (§ 4.4.1). For $y(2) = 0$ the single-line limit (6.6) is reached, cf. (4.10). As $y(1) \to \infty$ the limit (4.65) is reached, pressure having no more effect in the limit of strongly overlapping lines. The equivalent to the strong-line limit is obtained by substituting $y = y(1)/2$ and $u = 2\tilde{u}$ in (4.67).

6.1.2. Scaling approximations

The usual aim of approximate solutions for non-homogeneous paths is to reduce the problem to that of a homogeneous path with amount $\tilde{a}$, temperature $\tilde{\theta}$, and pressure $\tilde{p}$. This presupposes that the superposed lines of a particular profile but different widths can somehow be represented by a single line with the same profile. Clearly this cannot be done precisely, and the literature on the subject is often confused by a failure to recognize this assumption explicitly. Moreover, it is common practice to assume additionally, and without further proof, that $\tilde{p}$ and $\tilde{\theta}$ can be assigned fixed standard values and that the problem consists solely in determining the optimum form for $\tilde{a}$. Thus a quantity (the transmission) which requires three parameters to define its magnitude for a homogeneous path is treated as a function of a single variable for a non-homogeneous path. This variable ($\tilde{a}$) is the *scaled amount* of matter, and such an approximation can be referred to as a one-parameter *scaling approximation*.

The necessary and sufficient condition for the existence of a scaling approximation is that the absorption coefficient can be factored in the following way

$$k_\nu(p,\theta) = \phi(p,\theta)\psi(\nu). \tag{6.9}$$

The optical path is then

$$\begin{aligned}
\tau_\nu &= \int \psi(\nu)\phi(p,\theta)\,da \\
&= \psi(\nu)\int \phi(p,\theta)\,da \\
&= k_\nu(\tilde{p},\tilde{\theta})\int \frac{\phi(p,\theta)}{\phi(\tilde{p},\tilde{\theta})}\,da \\
&= k_\nu(\tilde{p},\tilde{\theta})\tilde{a}, \tag{6.10}
\end{aligned}$$

where $\tilde{p}$ and $\tilde{\theta}$ are fixed standards of pressure and temperature, both of which can be assigned at will. If the optical path can be expressed in terms of $\tilde{a}$ only so can all other absorptive properties.

As far as temperature is concerned, the most important factor is the Boltzmann term in the line intensity. This suggests that we may write

$$\phi(p,\theta) = S(\theta)P(p). \tag{6.11}$$

Early experimental work indicated that, for limited ranges of amount and pressure in the laboratory

$$\frac{\partial \ln \bar{T}}{\partial \ln p} \simeq n\frac{\partial \ln \bar{T}}{\partial \ln a}, \tag{6.12}$$

requiring

$$P(p) \propto p^n. \tag{6.13}$$

The scaled amount is then

$$\tilde{a} = \int \frac{S(\theta)p^n}{S(\tilde{\theta})\tilde{p}^n}\,da. \tag{6.14}$$

Equation (6.14) has been widely used, discussion having centred upon the alternatives $n \simeq \frac{1}{2}$ (as indicated by some laboratory data) or $n = 1$ (as required in the strong-line limit). Rarely is it pointed out that the equation is based upon two false premises, since neither (6.9) nor (6.11) is generally correct for the Lorentz line shape. If all lines in the spectrum are strong, then according to the results derived in Chapter 4, absorption is a function of the product $a.p$ and (6.12) is correct for $n = 1$.

If the lines are weak, however, absorption does not depend upon the pressure, and hence we must set $n = 0$. In atmospheric computations there is evidence for the importance of lines of intermediate strength, for which the uncertainty in $\tilde{a}$ is presumably of the order of $(p/\tilde{p})^{1/2}$. If we are dealing with stratospheric events, using data obtained at 1 atmosphere pressure, this uncertainty may be as great as a factor 10, rendering the results meaningless. If laboratory data are taken at pressures close to those for which computations are made, however, the error will be reduced.

Much of the work in atmospheric radiation over the last two decades has been based on the scaling approximation. Despite the unsatisfactory nature of the basic premises, this work can be of value. Under some circumstances the scaling approximation is probably sufficiently accurate; unfortunately, we have practically no objective estimates of errors, apart from a single comparison of flux computations with $n = \frac{1}{2}$ and $n = 1$ (see Table 6.5).

6.1.3. *Two-parameter approximations*

A more hopeful approach is to employ two disposable parameters in attempting to simulate a non-homogeneous by a homogeneous path. One possibility is to use a one-parameter scaling approximation in two different ways: once for the strong-line region, with $n = 1$; and once for the weak-line region with $n = 0$. The problem is then to determine the point of changeover from strong to weak lines. For a single spectral region to which a model and model parameters can be assigned, this presents no difficulty. Near to the mid-point both asymptotic forms will be in error, and this region of uncertainty will increase with the spread of line strengths involved (see Fig. 4.2), and for a whole band area the spread is so great that the approximation is hard to justify. The whole band data discussed in Chapter 5, for example, are rarely in one asymptotic region or the other. With careful experimental precautions whole-band data for the *weak-line* approximation have been recorded, but vibration-rotation bands have no minimum line intensity, and even for long atmospheric paths the effect of lines with incompletely absorbed centres is sufficient to prevent the band as a whole from following the *strong-line* approximation with any precision. The two-parameter scaling approximation should consequently be restricted to relatively narrow spectral ranges and should not be used for band areas.

The most useful two-parameter method which has been proposed (the Curtis–Godson approximation) assigns to a non-homogeneous path a temperature-scaled amount of absorbing matter, at a definable mean pressure. This again implies that a single line absorbing along a non-homogeneous path can be simulated by a line absorbing along a homogeneous path, but now both intensity and width can be adjusted, rather than the intensity alone, as for the scaling approximation. The choice of width and intensity is made to give the correct absorption in the strong- and weak-line regions; the accuracy in the transition region can then be tested by objective numerical methods. The Curtis–Godson approximation is the highest in which homogeneous and non-homogeneous paths can usefully be compared; further refinement will require that this kind of approximation be abandoned.

Consider the mean transmission over a spectral region $\Delta\nu$, wide compared to the equivalent widths of the individual lines (which are assumed to have a Lorentz profile),

$$\bar{T} = \frac{1}{\Delta\nu}\int_{\Delta\nu} d\nu \exp\left(-\int da \sum_i \frac{\alpha_i S_i}{\pi\{(\nu-\nu_i)^2+\alpha_i^2\}}\right). \tag{6.15}$$

The integration with respect to a is for an atmospheric path along which both α_i and S_i may vary, and the summation is over all lines in the frequency interval $\Delta\nu$. Consider this equation in the strong- and weak-line limits. In the strong-line limit we may neglect α_i in the denominator; instead, and with greater accuracy, we will replace it by a mean value $(\tilde{\alpha}_i)$, which does not vary along the path

$$\bar{T} \simeq \frac{1}{\Delta\nu}\int_{\Delta\nu} d\nu \exp\left(-\int da \sum_i \frac{\alpha_i S_i}{\pi\{(\nu-\nu_i)^2+\tilde{\alpha}_i^2\}}\right). \tag{6.16}$$

In the weak-line limit we can expand the exponential and retain only the first two terms

$$\bar{T} \simeq \frac{1}{\Delta\nu}\int_{\Delta\nu} d\nu\left(1-\int da \sum_i \frac{\alpha_i S_i}{\pi\{(\nu-\nu_i)^2+\tilde{\alpha}_i^2\}}\right).$$

Since $\Delta\nu \gg \tilde{\alpha}_i$ by hypothesis, the limits of the frequency integration are effectively infinite for each individual line; thus

$$\bar{T} \simeq 1 - \frac{1}{\Delta\nu}\int da \sum_i \int_{-\infty}^{+\infty} \frac{\alpha_i S_i\, d\nu}{\pi\{(\nu-\nu_i)^2+\tilde{\alpha}_i^2\}}$$

$$= 1 - \frac{1}{\Delta\nu}\int da \sum_i \frac{\alpha_i}{\tilde{\alpha}_i} S_i. \tag{6.17}$$

Returning to (6.15) we can obtain the weak-line limit directly

$$\bar{T} = 1 - \frac{1}{\Delta\nu} \int da \sum_i S_i. \tag{6.18}$$

Since $\alpha_i/\tilde{\alpha}_i = p/\tilde{p}$, (6.17) and (6.18) are consistent if

$$\tilde{p} = \frac{\int da\, p \sum_i S_i}{\int da \sum_i S_i}$$

$$= \frac{\int p\sigma\, da}{\int \sigma\, da}, \tag{6.19}$$

where σ is the mean line intensity. This is the Curtis–Godson approximation for the effective width for a Lorentz line and a non-homogeneous path. We will demonstrate later that the result is not restricted to Lorentz lines but we must first consider the optimum value to choose for the scaled amount.

Returning to (6.15), we now have an exact analogy to a homogeneous path if we write

$$\bar{T} \simeq \frac{1}{\Delta\nu} \int_{\Delta\nu} d\nu \exp\left(-\tilde{a} \sum_i \frac{\tilde{\alpha}_i \bar{S}_i}{\pi[(\nu-\nu_i)^2 + \tilde{\alpha}_i^2]}\right), \tag{6.20}$$

where $\bar{S}_i$, by hypothesis, can be evaluated at any specified temperature ($\tilde{\theta}$) and from (6.16) and (6.19)

$$\tilde{a} = \frac{\int \sum_i f_i(\tilde{\alpha}_i, \nu - \nu_i) S_i\, da}{\sum_i f_i(\tilde{\alpha}_i, \nu - \nu_i) \bar{S}_i}. \tag{6.21}$$

$f(\tilde{\alpha}_i, \nu - \nu_i)$ is the line-profile parameter $k_{\nu,i}/S_i$, see (3.48). Up to this point we have been completely general. To interpret (6.21) usefully, however, we must be more specific. For a regular band model S_i is constant over a spectral interval small compared with the band width. Then

$$\tilde{a} = \frac{\int \sigma\, da}{\tilde{\sigma}}. \tag{6.22}$$

Alternatively, for a random model, S_i is uncorrelated with α_i and ν_i, as assumed in the development of the model; then $\sum_i f_i S_i$ can be replaced by $(1/N) \sum_i f_i \sum_i S_i$, where N is the number of lines in the spectral interval. Again, we obtain (6.22). Finally, if f_i is independent of temperature and does not vary from line to line, (6.22) is again valid. The relation holds under such a variety of circumstances that it should be a close approximation for all.

Equations (6.19) and (6.22) together constitute the Curtis–Godson approximation. They define an effective amount ($\tilde{a}$) and an effective pressure ($\tilde{p}$), with reference to a standard temperature ($\tilde{\theta}$), which give the optimum simulation of a non-homogeneous by a homogeneous path. Since the equations are valid without reference to particular optical properties, they apply to any array with a definable mean line intensity,

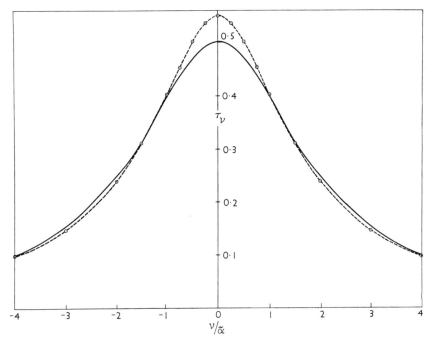

FIG. 6.1. Optical depths for a layer in a mixed atmosphere with $p_1 = 3p_2$.
The broken line is an exact computation of the optical depth following (6.4) and the full line uses the Curtis–Godson approximation.

even if the array includes a whole band. The approximation is not therefore restricted to narrow spectral regions.

The reason for the effectiveness of the Curtis–Godson approximation for a Lorentz line shape is illustrated in Fig. 6.1, which shows exact and approximate computations of τ_ν for a layer in a mixed atmosphere, with a pressure ratio from base to top of 3:1. The two profiles coincide in the wings, and the areas are the same; the agreement in detail is not perfect, but is sufficiently good for interpolation between the exact strong- and weak-line limits.

Although we have used the Lorentz profile as a basis for our derivation, the equations are of somewhat greater validity. Let us suppose that a scaling approximation is valid, and that

$$k_\nu = \psi(\nu)p^n S(\theta), \tag{6.23}$$

where $\psi(\nu)$ is a function of ν only. For a non-homogeneous atmosphere

$$\tau_\nu = \psi(\nu) \int p^n S(\theta)\, da. \tag{6.24}$$

Using (6.23), (6.19), and (6.22), we find for the Curtis–Godson approximation

$$\tau_\nu \simeq \psi(\nu)\left(\int S(\theta)\, da \right)^{1-n}\left(\int pS(\theta)\, da \right)^n. \tag{6.25}$$

Equations (6.24) and (6.25) are identical if $n = 1$ or 0. The former covers all types of pressure-broadening in the strong-line approximation (see § 3.6.5 viii); the latter covers any profile in the weak-line approximation, and Doppler-broadened lines under all conditions. Restrictions on the use of the Curtis–Godson approximation are therefore few.

TABLE 6.1

Single-line transmission between the base of a mixed atmosphere and pressure levels p/p_0 for $\tilde{u} = 1$ and a vertical path

After Kaplan (1959)

$$\frac{W(p, p_0)}{\pi \alpha_L(0)}$$

p/p_0	Exact	1st diff.	Curtis–Godson	1st diff.
0	1·0000		1·0476	
		0·0100		0·0262
0·1	0·9900		1·0214	
		0·0300		0·0418
0·2	0·9600		0·9796	
		0·0500		0·0591
0·3	0·9100		0·9205	
		0·0700		0·0744
0·4	0·8400		0·8461	
		0·0900		0·0933
0·5	0·7500		0·7528	
		0·1100		0·1117
0·6	0·6400		0·6411	
		0·1300		0·1307
0·7	0·5100		0·5104	
		0·1500		0·1504
0·8	0·3600		0·3600	

A further test is shown in Table 6.1. The quantity computed is the equivalent width of a single line in a non-dimensional form, for $\tilde{u} = 1$ when errors are a maximum. The path concerned starts vertically from the ground, and the calculation is therefore relevant to the first term

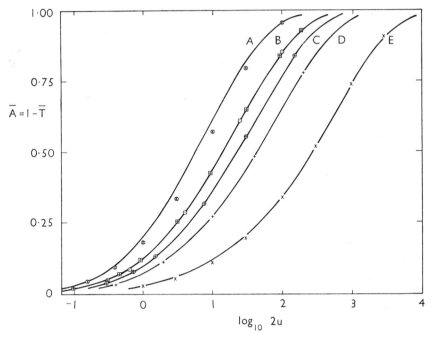

F‌IG. 6.2. A comparison between numerical computations and the Curtis–Godson approximation for a statistical model with exponential distribution of line intensities.

The full lines are for homogeneous paths with the Curtis–Godson mean pressure and scaled amount, and the points are from numerical computations. B, C, D, and E use values of y which are characteristic of water vapour. Distribution A and the values of y employed are more characteristic of ozone. In each description the first part defines the full line (homogeneous path) and the second the conditions under which the points were computed. In the case of curve B two distributions are computed which, according to the approximation, should lie on the same curve. After Curtis (1952).

Curve A. 9·6 μ ozone band with $p = 0\cdot05$ atm: $\otimes$, two slabs of absorbing matter with $a/9$ at 0·25 atm and $8a/9$ at 0·025 atm. *Curve B.* Water vapour with $p = 0\cdot9$ atm: $\boxplus$, constant mixing ratio over range $e:1$ in pressure having $\tilde{p} = 0\cdot9$ atm; $\odot$, two slabs of absorbing matter $a/9$ at 0·1 atm and $8a/9$ at 1 atm. *Curve C.* Water vapour with $p = 0\cdot7$ atm: $\oplus$, mixing ratio proportional to p over the whole atmosphere, with $\tilde{p} = 0\cdot7$ atm. *Curve D.* Water vapour with $p = 0\cdot5$ atm: $+$, constant mixing ratio over the whole atmosphere with $\tilde{p} = 0\cdot5$ atm. *Curve E.* Water vapour with $p = 0\cdot2$ atm: $\times$ as for $\otimes$ but pressures multiplied by 4.

on the l.h.s. of (2.79), giving the contribution to the intensity from a boundary. Fluxes are closely related to intensities (see, for example, (2.112) for the Eddington approximation); the entries in Table 6.1 have therefore been differenced in order to assess the accuracy of flux divergence computations using the Curtis–Godson approximation. Bearing

R

in mind that $\tilde{u} = 1$ gives maximum errors, the approximation is satisfactory for this particular example if the pressure range is less than $5:1$.

Tests of the accuracy of the Curtis–Godson approximation for other distributions of absorbing matter are shown in Fig. 6.2 and Table 6.2. The conclusions from Fig. 6.2 are again self-evident: that the approximation is poor if most of the matter is at the lower pressures (e.g. for ozone in the lower stratosphere), but good in the opposite circumstance (e.g. for water vapour and carbon dioxide). The same conclusion can be reached from Table 6.2. Only for Model L, where most of the matter is at the lowest pressure, is the agreement poor.

TABLE 6.2

Computations for three-layer atmospheres using the Elsasser model

After Godson (1953). T (exact) is the result of numerical computation without approximation. Columns 2, 3, and 4 indicate the amounts of matter ($2\pi y u = Sa$) in each of those layers whose pressures are in the ratios $5:3:1$. T (app.) is obtained from the Curtis–Godson approximation.

	$2\pi y u$			T	
	$2\pi y = 0{\cdot}5$	$2\pi y = 0{\cdot}3$	$2\pi y = 0{\cdot}1$	*exact*	*app.*
A	1·0	0·0	3·6	0·367	0·369
B	0·0	1·4	3·6	0·382	0·383
C	1·0	0·6	0·0	0·433	0·435
D	1·0	0·047	0·0	0·508	0·508
E	1·0	0·0	0·12	0·508	0·509
F	0·0	1·4	0·12	0·524	0·524
G	0·0	0·047	3·6	0·543	0·544
H	0·0	0·6	0·12	0·681	0·677
I	0·0096	0·6	0·0	0·691	0·692
J	0·0	0·047	0·12	0·895	0·910
K	0·0	0·047	0·036	0·933	0·940
L	0·0096	0·0	0·036	0·961	0·980
M	1·0	0·3	0·12	0·428	0·429
N	1·0	0·047	0·12	0·501	0·502
O	1·0	0·047	0·036	0·506	0·506

A further simplification to the Curtis–Godson approximation is valid when the central region of a line is opaque for moderate paths; i.e. $u \gg 1$ when estimated on the basis of the total amount of matter above the level concerned and the line width at that level. For a mixed atmosphere this also corresponds to $\tilde{u} \gg 1$. Under this condition, the lines are strong for moderate atmospheric paths and α_i^2 in the denominator of (6.15) can be neglected. But heating rates and fluxes involve integration over all distances, and for short paths lines are weak and the strong-line approximation fails. For these short paths, however, the pressure is close to that at the level of computation. The modified strong-line approximation therefore uses the pressure at the level of computation in place of the mean pressure (6.19). According to unpublished work by Curtis, for $\tilde{u}$ of the order of 10^3 (e.g. for strong lines in the ν_2 CO_2 band) it is not possible to distinguish between results based upon the two approximations.

6.2. Diffuse radiation

Before discussing the details of some computational schemes, we will briefly consider a relatively minor problem, which has been the subject of many theoretical papers.

According to (2.96) the problem of computing the flux of long-wave radiation in a stratified atmosphere is that of evaluating an integral of the type

$$F(z) = \int_0^\infty F_\nu(z)\, d\nu = \pm \int_0^\infty d\nu \int^{z'=z} \pi J_\nu(z')\, d2E_3\{\tau_\nu(z, z')\}. \qquad (6.26)$$

In order to use the methods described in Chapter 4, we may write

$$F = \sum_i F_i = \pm \sum_i \Delta\nu_i \int^{z'=z} \pi J_i(z')\, d\left(\frac{\int_i 2E_3\{\tau_\nu(z, z')\}\, d\nu}{\int_i d\nu}\right), \qquad (6.27)$$

where the ith range is sufficiently narrow for J_ν to be replaced by its average value J_i, but wide enough to contain many lines.

In the following sections we will discuss methods of evaluating an integral of the type

$$F_i = \pm \int^{z'=z} \pi J_i(z')\, \Delta\nu_i\, d\bar{Z}_i(z, z'), \qquad (6.28)$$

where $\bar{Z}_i$ has to be related to the transmission function $\bar{T}_i$ if we wish to make direct use of laboratory data and the theory of Chapter 4. Three courses appear to be open:

(i) We may write exactly

$$\bar{Z}_i(z, z') = \frac{\int_i 2E_3\{\tau_\nu(z, z')\}\, d\nu}{\int_i d\nu}. \qquad (6.29)$$

Expanding E_3 we find

$$\bar{Z}_i(z, z') = \int_0^1 2\xi \bar{T}_i(z, z'; \xi)\, d\xi, \qquad (6.30)$$

where

$$\bar{T}_i(z, z'; \xi) = \int_i e^{-\tau_\nu(z, z')/\xi}\, d\nu. \qquad (6.31)$$

Given $\bar{T}_i(a, p, \theta)$ a relatively simple numerical quadrature then leads to the required function $\bar{Z}_i$. This method has been used by Möller (see § 6.8.1).

(ii) $2E_3(x)$ behaves somewhat similarly to $\exp(-rx)$, where r is a numerical factor. Many workers have therefore attempted to write

$$\bar{Z}_i(x) \simeq \bar{T}_i(rx), \qquad (6.32)$$

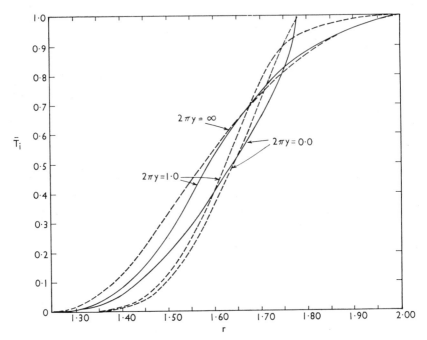

Fig. 6.3. Behaviour of the diffusivity factor r, as a function of $\bar{T}_i$ and $2\pi y$ for regular (full lines) and random spectral models (broken lines).
After Godson (1954).

where r is known as the *diffusivity factor*. Estimates of r have varied between 1 and 2. Elsasser (1942) showed that the best value for the regular model in the strong-line approximation is about 1·66. However, if we regard (6.32) as a definition of r, it is not a constant; while not a function of transmission alone, it appears in practice to be fairly well represented as a function of this one variable (Fig. 6.3). We note that if we adopt (6.32), $\bar{Z}_i(x)$ will correctly tend to 1 or 0 as x tends to 0 or ∞ regardless of the value of r. However, if we demand that the gradient of $\bar{Z}_i(x)$ tends correctly to 2 as $x \to 0$ (this is necessary if we wish to have the right contribution to heating from nearby layers), then we must insist that $r = 2$ at the origin for any empirical $(r \sim \bar{T})$ relation that is used.

(iii) A third possibility is to evaluate I_i rather than F_i. From (2.86)

$$I_i = \pm \int J_i(z') \, \Delta \nu_i \, d\bar{T}_i(z, z'; \xi). \qquad (6.33)$$

We can now evaluate (6.33) for a series of values of ξ and hence the flux from (2.94) and (2.136) by approximate numerical quadrature,

$$F_i^{+,-} = (+, -) 2\pi \int I_i^{+,-} \xi \, d\xi \qquad (6.34)$$

$$\simeq (+, -) 2\pi \sum_{j=1}^{m} a_j \xi_j I_i^{+,-}(\xi_j). \qquad (6.35)$$

If we choose to employ Gauss's method, the appropriate values of a_j and ξ_j are given by Table 2.2. We are, in effect, approximating

$$E_3(x) \simeq \sum_{j=1}^{m} a_j \xi_j e^{-x/\xi_j}. \qquad (6.36)$$

Equation (6.36) implies

$$\frac{dE_3(x)}{dx} \simeq - \sum_{j=1}^{m} a_j e^{-x/\xi_j}. \qquad (6.37)$$

In the limit of small x we have, from (2.137) (which refers to twice the number of ordinates considered here)

$$\lim_{x \to 0} \frac{dE_3(x)}{dx} = - \sum_{j=1}^{m} a_j = -1. \qquad (6.38)$$

This is a correct result (Appendix 8) and therefore (6.35) gives the correct contribution to the heating rate from short paths, regardless of the degree of approximation. The greatest likelihood of error is in $E_3(0)$, which should equal 0·5. From Table 2.2 and (6.36) we have for one ordinate $E_3(0) = 0\cdot5773$, for two ordinates $E_3(0) = 0\cdot5213$, and for three ordinates $E_3(0) = 0\cdot5098$.

In the following sections we will generally use a diffusivity factor to compute diffuse transmission. Most published work follows this approximation but, fortunately, its value does not depend on the validity of the approximation. If we redefine the optical path any computing scheme which yields fluxes approximately using a diffusivity factor will give exact results for the intensity. The flux can then be evaluated by procedure (iii) to any required degree of accuracy. Even without this saving factor, there is evidence to indicate that the approximation will not give rise to serious error in atmospheric computations. Hitschfeld and Houghton (see § 6.10 for details) have computed ozone heating rates through the stratosphere using (iii) with a three-ordinate numerical quadrature. The resulting heating rates differed by less than 5 per cent, at all levels and less than 1 per cent at most levels from those which can be obtained with a diffusivity factor $r = 1.667$. Robinson (1947) has presented observational evidence at ground level in favour of the same factor. Unless exceptional accuracy is required, procedure (ii) is probably sufficient for atmospheric computations.

6.3. Solar radiation

In all existing treatments of heating by solar radiation the source function is neglected; thermal radiation because it is small at wavelengths for which the solar flux is large (see § 2.2.1); scattered radiation for reasons which are rarely stated explicitly, but which may well represent a desire to avoid a particularly difficult problem. In some special circumstances a case for neglecting scattered radiation might be made. For example, where a very strong absorption is concerned (e.g. the centre of the Hartley bands) solar depletion will take place high in the atmosphere and scattering above the level will be negligible. Furthermore, the high absorption coefficient imparts a very small albedo to the lower atmosphere, and back scattering will also be negligible. Neither of these arguments holds for water vapour bands or for weak ozone bands, and it must be borne in mind that computations of heating in the troposphere and lower stratosphere may be seriously in error on this account.

Since no other treatments exist however we will proceed on the assumption that the source function can be neglected. Then from (2.103) and (2.6) we can write for the flux equation

$$F_\nu(z) = \xi_\odot f_\nu(\infty) e^{-|\tau_\nu(z,\infty)|/|\xi_\odot|},$$

or

$$F_\nu(z) = \xi_\odot f_\nu(\infty) \tau_\nu(z, \infty; \xi_\odot). \tag{6.39}$$

Hence

$$\frac{F(z)}{\xi_\odot} = \sum_i \frac{F_i(z)}{\xi_\odot} = \sum_i f_i(\infty) \overline{T}_i(z, \infty; \xi_\odot). \tag{6.40}$$

Equation (6.40) can be evaluated without any particular difficulty. If two gases with overlapping bands are important (e.g. oxygen and ozone near 70 km) then $\overline{T}_i$ can be factorized into two components, making use of the multiplication property of band transmission (§ 4.1).

If the transmission functions in (6.40) can be expressed in terms of a single variable (e.g. a scaled amount $\tilde{a}$), further simplification is possible. We can write

$$\frac{F(\tilde{a})}{\xi_\odot} = f(\infty) \overline{\overline{T}} \left(\frac{\tilde{a}}{\xi_\odot} \right), \tag{6.41}$$

where $f(\infty)$ is the solar constant, and

$$\bar{\bar{T}}(\tilde{a}/\xi_\odot) = \frac{\sum_i f_i(\infty)\bar{T}_i(\tilde{a}/\xi_\odot)}{\sum_i f_i(\infty)}, \qquad (6.42)$$

can be evaluated once and for all. It only remains to identify $\tilde{a}$ with a particular value of z and the flux is determined.

Before accepting this approximation, we must briefly review the conditions involved. The most important is that (6.9), defining a one-parameter scaling approximation, should be valid. The scaled amount $(\tilde{a})$ is defined by (6.10) in terms of the temperature variation of line intensity. This may sometimes be similar for all lines in one spectral interval, but not for all intervals in a vibration-rotation band, for in band wings absorption increases with temperature while band centres give the opposite result; $\tilde{a}$ should therefore be written with suffix i. For (6.42) to be valid we not only require that $\bar{T}_i$ can be expressed as a function of $\tilde{a}_i$, but also that all $\tilde{a}_i$ are equal. This is practically equivalent to assuming *no temperature variation of absorption* at all. This problem will recur in the next section.

One further assumption required to make (6.42) valid is either that bands of individual gaseous absorbing components do not overlap or, if they do, that the relative concentrations do not vary. In the former case, each gas can be treated independently; in the latter, the mixture can be treated as a single, complex gas.

The above assumptions can be justified for the stratosphere and mesosphere. Here electronic absorptions of oxygen and ozone govern the depletion of solar radiation, and since their absorption is thought to be independent of temperature or pressure, the amount of gas uniquely defines the absorption. At levels where oxygen and ozone are simultaneously of importance their relative concentration is constant, at least while the sun is shining and photochemical equilibrium prevails (see § 1.3). In the troposphere on the other hand conditions are far less favourable.

The sum in (6.42) can be evaluated from laboratory and theoretical data or alternatively, over part of the range of $\tilde{a}$, from direct observation. The method is illustrated in Fig. 6.4 in which the solar irradiation at ground level is plotted against wavelength, and a background curve interpolated above the absorption bands (a process which is difficult to perform with any real certainty). The shaded area is

$$\sum_i f_i\{1-\bar{T}_i(\tilde{a}/\xi_\odot)\} = f\{1-\bar{\bar{T}}(\tilde{a}/\xi_\odot)\}. \qquad (6.43)$$

Since f is the area under the background curve the required quantity $f\overline{\overline{T}}(\tilde{a}/\xi_{\odot})$ can be determined. By observing throughout the day, a wide range of the argument $(\tilde{a}/\xi_{\odot})$ can be obtained, but it cannot be lower than for mid-day on an occasion when gaseous densities are small.

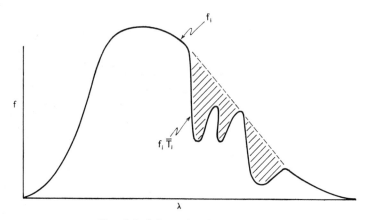

FIG. 6.4. Schematic solar spectrum.

Many attempts have been made to express $\overline{\overline{T}}(x)$ for water vapour in a simple algebraic form. One of the earliest gives

$$f\{1-\overline{\overline{T}}(x)\} = 0 \cdot 172 x^{0 \cdot 303} \text{ cal cm}^{-2} \text{ min}^{-1}, \qquad (6.44)$$

where x is in g cm^{-2}. A recent revision, using better data gives

$$\log_{10} f\{1-\overline{\overline{T}}(x)\}$$
$$= -0 \cdot 740 + 0 \cdot 347 \log_{10} x - 0 \cdot 056 (\log_{10} x)^2 - 0 \cdot 006 (\log_{10} x)^3, \quad (6.45)$$

where x is in g cm^{-2} and f in cal cm^{-2} min^{-1}. Despite its form, (6.45) is not based on a one-parameter scaling approximation, but on the Curtis–Godson approximation. Because the relative concentrations of water vapour at any two levels tend to be rather similar from one day to another it was found empirically that curves of $\overline{\overline{T}}(x)$ as a function of x calculated on the basis of the Curtis–Godson approximation did not greatly differ from one day to another, justifying the use of a mean curve. Without this simplification the use of a two-parameter approximation requires either a graphical or a numerical integration (see Bibliography for references to work of this kind).

The same techniques can be applied without modification to carbon dioxide or other atmospheric gases (see, for example, Karandikar, 1946).

6.4. Radiation charts

If the transmission can be expressed in terms of one variable, (6.28) can be evaluated graphically by means of *radiation charts*. This was first pointed out by Mügge and Möller (1932) and has been developed by many other workers.

The procedure is as follows. An atmospheric *ascent curve* defines a relationship between p, θ, and gas densities, from which both J_i and $\bar{Z}_i$ in (6.28) can be determined. In practice this is done using curved and non-linear scales of θ and $\tilde{a}$ which correspond to linear, orthogonal scales of $\pi J_i \Delta \nu_i$ and $\bar{Z}_i$ or equivalent transformations. The ascent data, when plotted with these non-linear axes, define a curve; according to (6.28), the area under this curve is a flux.

Considering the relatively straightforward problem involved here, the proliferation of radiation charts in the literature may seem surprising. It may help to clarify the situation if we summarize points of similarity and difference between them before considering details.

All derivations start from (6.28). In this equation, the assumption of a stratified atmosphere is implicit. Here the main problem arises with broken clouds, and no objective criterion of accuracy has been developed. Cloud decks can be taken to be two-dimensional if we consider only a long time-average, and provided that the level of computation is not in amongst the clouds.

All the methods under discussion treat cloud and ground as black bodies radiating at the surface temperature. We will see in Part 2 that this can be a poor assumption for clouds. Radiation charts can also be used with a more general boundary condition of the type (2.99), provided that the boundary radiation can be approximately simulated by black-body radiation. The quantity of radiation incident upon the surface can be determined in a separate calculation. The only attempt to discuss this problem appears to have been by Brooks (1952), who assigned a reflectivity of 0·09 to a soil surface. It will be shown in § 6.6 that the problem is of relatively small importance at ground level, but could be more serious where cloud surfaces are involved.

The most drastic approximation, common to all methods, is the assumption that absorption can be defined in terms of a single scaled amount. If we consider a single spectral range, the objections are those which apply to a one-parameter scaling approximation. But, as for the work discussed in § 6.3, a simple computing scheme demands that the transmission can still be expressed in terms of a single parameter after averaging over all spectral ranges. It has already been pointed out that this is tantamount to assuming that transmission is independent of temperature. This, at least, is a sufficient condition, and it will also be taken to be necessary when we require an explicit statement of the approximation.

The above are the main points in common between the available radiation diagrams. Points of difference are as follows. Most obvious, but of no physical significance, are the various equal-area transformations which can be made of (6.28); the possibility of differentiating to give a direct evaluation of heating rates; and the alternatives of graphical and numerical methods. Relatively unimportant are the differences between treatments of the diffuse field; unless otherwise stated we will use (6.32). The pressure-scaling factor in (6.14) differs from one writer to another, although $n = 1$ is most commonly employed for water vapour. The effect of overlapping carbon dioxide and water-vapour bands can be handled in three different ways, depending upon whether the $15\,\mu$ carbon dioxide band is assumed to be completely opaque, and whether overlap is taken fully into account. Slight differences also result from choices between alternative sets of laboratory data. Finally, we may or may not insist that the equations be stated in terms of the emissivity, (5.11), which is a directly measurable quantity.

It is appropriate to consider here another quantity which occurs naturally in the equations:

$$\epsilon_c^* = \frac{\sum\limits_i (1-\overline{T}_i)(dB/d\theta)_i \, \Delta\nu_i}{\sum\limits_i (dB/d\theta)_i \, \Delta\nu_i}. \tag{6.46}$$

While (6.46) bears no direct relationship to the emissivity, nevertheless it does not differ greatly from it. $dB_\nu(\theta)/d\theta$ is a function with a single maximum, similar in shape to $B_\nu(\theta)$ itself. For example, at 250° K the maximum of $dB_\nu(\theta)/d\theta$ is near $15\,\mu$ and the half-maximum points lie near $8\,\mu$ and $35\,\mu$, while for $B_\nu(\theta)$ the same points lie at $20\,\mu$, $10\,\mu$, and $50\,\mu$ respectively. Clearly (5.11) and (6.46) cannot greatly differ. If we assume that $\partial\epsilon/\partial\theta = 0$, and recall that we have also implicitly assumed that $\partial\overline{T}_i/\partial\theta = 0$, it can be shown that ϵ and ϵ^* are identical.

Extensive computations of ϵ^* for water vapour, ozone and carbon dioxide between $-80°$ C and $+40°$ C are given by Elsasser (1961).

Assuming the transmission in all spectral ranges to be determined by a single parameter, making use of a diffusivity factor of 1·66 to treat the diffuse field and adopting the Planck function as a source function,

(6.28) becomes

$$F = \sum_i F_i = \pm \sum_i \int_{\tilde{a}=\infty}^{\tilde{a}=0} \pi B_i(\theta)\, \Delta \nu_i \, d\overline{T}_i(1\cdot 66\tilde{a}). \qquad (6.47)$$

The limits have been placed at $\tilde{a} = 0$ and ∞ on the understanding that the limit of infinite opacity is approached along the curve $\theta = \theta^*$, where θ^* is the temperature, or effective emission temperature, of a relevant boundary (see § 2.3.2).

From (5.11) and (5.12)

$$B(\theta)\frac{\partial \epsilon_{\mathrm{s}}(\tilde{a}, \theta)}{\partial \tilde{a}} = B(\theta)\frac{\partial \epsilon_{\mathrm{c}}(1\cdot 66\tilde{a}, \theta)}{\partial \tilde{a}} = -\sum_i \frac{d\overline{T}_i(1\cdot 66\tilde{a})}{d\tilde{a}} B_i(\theta), \quad (6.48)$$

the total differential having been introduced since, by assumption, $\overline{T}_i$ is a function of $\tilde{a}$ only.

Hence

$$F = \pm \int_{\tilde{a}=\infty}^{\tilde{a}=0} \pi B(\theta)\frac{\partial \epsilon_{\mathrm{s}}(\tilde{a}, \theta)}{\partial \tilde{a}}\, d\tilde{a} = \pm \int_{\tilde{a}=\infty}^{\tilde{a}=0} \pi B(\theta)\frac{\partial \epsilon_{\mathrm{c}}(1\cdot 66\tilde{a}, \theta)}{\partial \tilde{a}}\, d\tilde{a}. \quad (6.49)$$

This is the basic equation used in connexion with empirical emissivity data. It can be employed in this form without modification, plotting the ascent data as a function of $\tilde{a}$ and $B(\theta)\partial \epsilon_{\mathrm{s}}/\partial \tilde{a}$. However, the diagram is awkward near $\tilde{a} = 0$ and transformations are used in practice.

The flux equation can also be expressed in terms of ϵ^*. Integrating (6.47) by parts, substituting from (6.46) and using the positive sign for convenience

$$F = \sum_i \pi B_i(\theta)\Delta \nu_i\, \overline{T}_i(1\cdot 66\tilde{a}) \Big|_{\tilde{a}=\infty}^{\tilde{a}=0} - \sum_i \int_{\tilde{a}=\infty}^{\tilde{a}=0} \Delta \nu_i\, \overline{T}_i(1\cdot 66\tilde{a})\, d\pi B_i(\theta)$$

$$= \pi B(\tilde{a} = 0) - \sum_i \int_{\tilde{a}=\infty}^{\tilde{a}=0} \Delta \nu_i\, \overline{T}_i(1\cdot 66\tilde{a})\frac{d\pi B_i(\theta)}{d\theta}\, d\theta$$

$$= \pi B(\tilde{a} = 0) - \int_{\tilde{a}=\infty}^{\tilde{a}=0} \{1 - \epsilon_{\mathrm{c}}^*(1\cdot 66\tilde{a}, \theta)\}\, d\pi B(\theta). \qquad (6.50)$$

Little purpose would be served in detailed discussion of the many different treatments of (6.49) and (6.50). Instead, in § 6.5 we will discuss one chart in detail, the Yamamoto chart. In § 6.6 this chart will be used to demonstrate how computations are made in practice. In § 6.7 we will examine in some detail the only recorded attempt to develop a simple computational method without assuming a one-parameter scaling approximation. While this has had only limited success it exposes many of the problems involved in improving existing methods, and illustrates the advantages of numerical techniques. In § 6.8, we will briefly indicate the characteristics of other computing methods, and give some comparative results. §§ 6.9 and 6.10 will deal with even more simple techniques on the one hand, and more complex techniques using electronic computers on the other. Finally, in § 6.11 the methods will be generalized to include non-equilibrium source functions.

6.5. Yamamoto's flux chart

This chart was developed for use in the troposphere, and takes account of water vapour and carbon dioxide but not ozone whose effect is negligible at low levels. Initially we will discuss water vapour alone and subsequently extend the treatment to include carbon dioxide. The chart (Fig. 6.5) uses (6.50) in unmodified form. The two axes are linear in $\pi B(\theta)$ and $1-\epsilon_s^*(\tilde{a}, \theta) = 1-\epsilon_c^*(1\cdot66\tilde{a}, \theta)$. Since πB is a function of θ only, this axis can be marked off with a non-linear temperature scale and *isotherms* will be vertical, straight lines. *Isopleths*, or lines of constant $\tilde{a}$, can be drawn by marking each isotherm with a non-linear scale in $\tilde{a}$ and joining the points thus constructed. Between 200° K and 310° K emissivity varies little with temperature, and the isopleths are close to horizontal, straight lines. Between these temperatures the peak of the Planck curve lies somewhere in the window between the ν_2 and rotation bands of water vapour ($8\,\mu$ to $20\,\mu$), and the average transmission, and hence ϵ^* does not greatly vary with temperature. Below 200° K the peak of the Planck curve approaches the strong intensities near the maximum of the rotation band, and hence ϵ^* approaches unity as $\theta \to 0$; the isopleths therefore curve downwards at low temperatures.

In Fig. 6.5 the upper, heavy lines are ascent curves, which will be discussed in § 6.6. The form of (6.50) shows that the flux is given by the area to the left of the ascent curves. In the case of the curve III, for example, the flux is the area between the $\tilde{a} = 0$ isopleth and the ascent curve; since the curve applies to the atmosphere above the point under consideration, the flux is downward and negative in sign as a matter of definition. Similarly, the upward flux (which is positive) is given by the area to the left of curve II. The net flux (in this case also positive) is then the area between curves II and III.

Up to this point we have discussed water vapour alone, but one of the most important features of Yamamoto's chart is the relatively sophisticated treatment of the interaction between water vapour and carbon dioxide. In the derivation of (6.50) it was only necessary to assume ϵ^* to be a function of one variable in order to obtain a two-dimensional diagram, and it is permissible to write

$$\epsilon_s^* = \epsilon_s^*(\tilde{a}_w, \tilde{a}_c, \theta), \qquad (6.51)$$

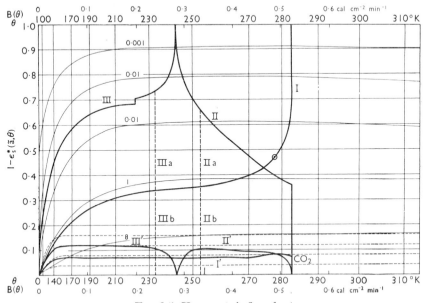

Fig. 6.5. Yamamoto's flux chart.

The curved, light lines are isopleths for water vapour (amounts in g cm^{-1}). The heavy, full lines are ascent curves (I, II, III), defined below and discussed in § 6.6. The lower, heavy lines (I', II', III') are ascent curves for the carbon dioxide corrections. The chain dashed lines are lines of constant $\Delta\epsilon^*$ (see text).

I. Ascent curve for *downward* flux at ground level. The point marked ($\odot$) is the first datum point above ground level (0·833 km, see discussion in § 6.5). II. Ascent curve for *upward* flux at 6·667 km. III. Ascent curve for *downward* flux at 6·667 km. II*b* and III*b* are the modifications to I given by *cloud bases* at 5 and 8·333 km respectively. II*a* II*b* is the modification to II for a *cloud top* at 5 km. III*a* III*b* gives the modification to III for a *cloud base* at 8·333 km. After Yamamoto (1952).

where $\tilde{a}_w$ and $\tilde{a}_c$ are reduced amounts of water vapour and carbon dioxide respectively. Since individual spectral ranges obey the multiplication property

$$\overline{T}_i(\tilde{a}_w, \tilde{a}_c, \theta) = \overline{T}_i(\tilde{a}_w, \theta)\overline{T}_i(\tilde{a}_c, \theta). \qquad (6.52)$$

The difference between (6.51) and the emissivity of water vapour alone is

$$\Delta\epsilon_s^* \equiv \epsilon_s^*(\tilde{a}_w, \tilde{a}_c, \theta) - \epsilon_s^*(\tilde{a}_w, \theta) = \frac{\sum_i (dB/d\theta)_i \Delta\nu_i \overline{T}_i(\tilde{a}_w, \theta)\{1 - \overline{T}_i(\tilde{a}_c, \theta)\}}{\sum_i (dB/d\theta)_i}. \qquad (6.53)$$

Since all $\overline{T}_i$ are less than unity, (6.53) is essentially positive, and $(1 - \epsilon^*)$

is over-estimated if only water vapour is considered. Yamamoto ingeniously incorporates this correction by plotting $\Delta\epsilon_s^*$ as a function of $B(\theta)$ at the base of the diagram, and excluding the area under this curve when measuring net flux.

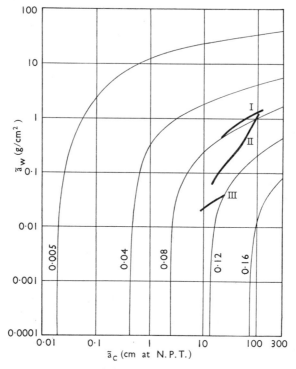

FIG. 6.6. $\Delta\epsilon_s^*$ as a function of $\tilde{a}_c$ and $\tilde{a}_w$.

The curves marked I, II, and III are defined in Fig. 6.5. After Yamamoto (1952).

Lines of constant $\Delta\epsilon_s^*$ (broken) are shown in Fig. 6.5. Since it is a small quantity it has only been evaluated at one temperature and the broken lines are horizontal except near $\theta = 0$ where $\Delta\epsilon_s^* \to 1$; this feature is incorporated approximately by curving down the lines below $150°$ K.

It remains to estimate $\Delta\epsilon_s^*$. This is performed with the help of an ancillary diagram (Fig. 6.6) which has lines of constant $\Delta\epsilon_s^*$ on a diagram whose axes are $\tilde{a}_c$ and $\tilde{a}_w$.

6.6. Examples of the use of a radiation chart

The use of a radiation chart now turns on the appropriate $\tilde{a} \sim \theta$ relationship for any particular atmospheric situation. Specimen computations are laid out in Table 6.3 (a) and (b).

TABLE 6.3 (a)

Specimen computation of ā for water vapour

After Möller (1943)

1	2	3	4	5	6	7	8	9†	10	11	12
z	θ	f	$\rho_w \times 10^6$	p	$\sqrt{\left(\dfrac{p}{1000}\right)}$	$\bar{\rho}_w \times 10^6$	$\tilde{\rho}_w \times 10^6$	Δz	$\Delta \bar{a}_w$	$\bar{a}_w(0)$	$\bar{a}_w(6\cdot7)$
(km)	(°C)	(%)	(g cm^{-3})	(mb)		(g cm^{-3})	(g cm^{-3})	(cm)	(g cm^{-2})	(g cm^{-2})	(g cm^{-2})
0	+10	70	6·58	1000	1·000	6·58	5·56	8·33×10⁴	0·453	0·0	1·338
0·833	+5	70	4·76	910	0·954	4·54	3·805	8·33×10⁴	0·317	0·453	0·884
1·667	0	70	3·40	816	0·903	3·07	2·56	8·33×10⁴	0·2135	0·770	0·568
2·500	−5	70	2·29	736	0·858	2·05	1·695	8·33×10⁴	0·1413	0·984	0·354
3·333	−10	70	1·65	660	0·812	1·340	1·1025	8·33×10⁴	0·0919	1·125	0·213
4·167	−15	70	1·12	596	0·772	0·865	0·7045	8·33×10⁴	0·0587	1·217	0·1208
5·000	−20	70	0·748	529	0·727	0·544	0·3725	1·667×10⁵	0·0621	1·275	0·0621
6·667	−30	70	0·310	421	0·649	0·201	0·1356	1·667×10⁵	0·0226	1·338	0·0
8·333	−40	70	0·122	332	0·576	0·0703	0·0460	1·667×10⁵	0·0077	1·360	0·0226
10·000	−50	70	0·0426	258	0·508	0·0216	—	—	0·0094	1·368	0·0303
∞	−50	—	—	—	—	—	—	—		1·377	0·0397

† If no height data are available, column 9 can be derived from the hydrostatic approximation $dz = -dp/\rho_a g$, where g is the gravitational acceleration.

TABLE 6.3 (b)

Specimen computation of ã for carbon dioxide

After Möller (1943). ρ_c/ρ_a is assumed to be $4\cdot59\times10^{-4}$, corresponding to a volume proportion of 3×10^{-4}. The data are consistent with Table 6.3 (a).

1	2	3	4	5	6	7	8	9
z	ρ_a	ρ_c	$\left(\dfrac{p}{1000}\right)^{0\cdot73}$	$\tilde{\rho}_a$	$\tilde{\rho}_c$	$\Delta\tilde{a}_c$	$\tilde{a}_c(0)$	$\tilde{a}_c(6\cdot7)$
(km)	(g cm^{-3})	(g cm^{-3})		(g cm^{-3})	(g cm^{-3})	(cm s.t.p.)	(cm s.t.p.)	(cm s.t.p.)
0	$1\cdot231\times10^{-3}$	$5\cdot65\times10^{-7}$	1·0	$5\cdot65\times10^{-7}$			0	107·4
					$5\cdot27\times10^{-7}$	22·2		
0·833	1·140	5·23	0·934	4·88			22·2	85·2
					4·50	19·0		
1·667	1·042	4·78	0·862	4·12			41·2	66·2
					3·82	16·1		
2·500	0·957	4·39	0·799	3·51			57·3	50·1
					3·24	13·6		
3·333	0·874	4·01	0·739	2·96			70·9	36·5
					2·75	11·6		
4·167	0·805	3·70	0·685	2·53			82·5	24·9
					2·32	9·8		
5·000	0·728	3·34	0·628	2·10			92·3	15·1
					1·79	15·1		
6·667	0·604	2·77	0·532	1·47			107·4	0
					1·25	10·5		
8·333	0·497	2·28	0·447	1·02			117·9	10·5
					0·86	7·2		
10·000	0·403	1·85	0·372	0·69			125·1	17·7
					—	13·2		
∞	—	—	—	—			138·3	30·9

In Table 6.3 (a), the ascent data are listed in columns 2, 3, and 5, i.e. temperature (θ), relative humidity (f, alternatively it might be frost- or dew-point) and pressure (p). The height (z) is a derived parameter and is irrelevant except for comparison with other phenomena. The relative humidity is constant through the troposphere for this particular example. ρ_w (water-vapour density) is a tabulated function of f and θ (see Appendix 6). In columns 6 and 7 the pressure-scaling factor ($n = \tfrac{1}{2}$) is introduced to give $\tilde{\rho}_w = \rho_w\sqrt{(p/1000)}$. Column 8 is interpolated, and the product of columns 8 and 9 gives $\Delta\tilde{a}_w$. The integral in (6.14) is approximated by a summation. By definition, $\tilde{a} = 0$ at the level of computation and columns 11 and 12 are constructed by setting $\tilde{a} = 0$ at two chosen levels (0 and 6·667 km) and adding the $\Delta\tilde{a}$ from this point. The evaluation of $\tilde{a}_c$ for carbon dioxide proceeds along similar lines (Table 6.3 (b)), except that a pressure-scaling factor $n = 0\cdot73$ is employed. This corresponds to one particular set of laboratory data, as interpreted by means of the scaling approximation (cf. (6.12) and (6.13)); the data in Table 5.17, for example, indicate a $p^{0\cdot88}$ scaling factor.

In Fig. 6.5, the full lines marked I, II, and III represent the data from columns 11 and 12. The area to the left of curve I is the downward flux from the whole atmosphere, and has a negative sign (the sign is, of course, not determined by the diagram). To understand curve II, we must recall the convention which utilizes an appropriate constant value of the source function after a boundary has been reached. How this source function is assigned is a matter for the individual research worker. Owing to the complex structure of the atmosphere near the surface of both land and sea, it is by no means clear how the many possible definitions of surface temperature relate to the effective emission temperature. Normal practice is to use the 'screen temperature' over land. According to Houghton (1958), the emission temperature lies between 'screen temperature' and 'surface temperature', and these two temperatures can differ by as much as 6° K. According to Robinson (1947), a colourless, liquid-in-glass thermometer, lying on the ground, gives an accurate measure of the emission temperature for the closely cropped grass surfaces used by him.

The problem is complicated by the partial reflectivity of the surface; fortunately, the results obtained from a radiation chart at ground level are not too sensitive to this parameter. From Fig. 6.5, curve I, we calculate that the downward flux at ground level is about 0.75 times the upward flux $(= \pi B_\nu^*$, i.e. the area under the $283°$ K isotherm). Thus, from (2.99) there will be a reflected intensity, $0.75 a_\nu B_\nu^*$, in addition to the thermal emission, $(1-a_\nu)B_\nu^*$. Combining these two contributions, we find for the effective source function of the lower boundary

$$J^* = 0.75\, a_\nu B_\nu^* + (1-a_\nu)B_\nu^*$$
$$= B_\nu^*(1-0.25 a_\nu). \tag{6.54}$$

The albedo of natural surfaces for thermal radiation is variable. Brooks (1952) assumes $a_\nu = 0.09$ for a soil surface; for water surfaces the reflectivity is generally smaller. In view of the uncertain value of B_ν^*, a value of $a_\nu < 0.1$ can be neglected in (6.54). In general therefore we may replace the earth's surface by an equivalent black body without serious error.

In the ensuing discussion we will follow published work and replace clouds also by a black body at the surface temperature; but the procedure is questionable, and the above argument for the earth's surface is weaker in two respects when applied to a cloud.

In the first place the cloud albedo may be greater than 0.1. The individual particles in a typical water cloud may have an albedo near

$10\,\mu$ as high as 0·5, although the albedo of the entire cloud will be less, depending upon its optical thickness.

Secondly, comparing the area to the left of curve III in Fig. 6.5 with that to the left of the 243° K isotherm, the downcoming radiation at 6·667 km is only about 0·35 of that from a black body at the ambient temperature. This value is typical for the upper surface of a medium-level cloud, and with $a = 0·25$ the upward radiation is

$$J^* \simeq B_\nu(6·7)(1-0·65 \times 0·25)$$

$$= 0·840 B_\nu(6·7).$$

To replace J^* by $B_\nu(6·7)$ in this example could lead to errors of the order of 16 per cent, somewhat larger than is normally tolerated.

Since existing radiation charts only permit black-body boundary conditions, they cannot readily be adapted to a more sophisticated treatment of the cloud problem. In the event that a_ν varies little with frequency and little from cloud to cloud, an emission temperature might be assigned as a function of surface temperature and incident intensity. However, neither condition is met in the atmosphere and a marked improvement of this particular feature is difficult within the framework of existing radiation charts.

Bearing in mind the above criticisms, we will follow standard practice and incorporate the effect of the earth's surface emission in curve II (Fig. 6.5) by passing from $\tilde{a} = 1·338$ g cm^{-2} to $\tilde{a} = \infty$ along the 283° K isotherm. Similarly, the effect upon II of a cloud *top* at 5 km (253° K) will be to terminate the curve along the 253° K isotherm.

On curve III (the downward radiation at 6·667 km) the kink represents the tropopause. A *cloud base* above at 8·333 km (233° K) is incorporated by terminating the curve along III (a) III (b); similarly the modification to II caused by a *cloud top* at 5 km is given by II (a) II (b). Note that, for normal tropospheric ascent curves, cloud either above or below the level under consideration greatly decreases the net flux of radiation.

A point of fundamental importance with respect to the overall accuracy of radiation chart computations is illustrated by curve I. The first datum point in Table 6.3 is at 0·833 km, with $\tilde{a} = 0·453$ g cm^{-2}, and this is marked on the curve. Approximately 0·8 of the area is beneath that part of the ascent curve which joins $\tilde{a} = 0$ to this first datum point; most of the flux therefore comes from a part of the atmosphere lying between two successive datum levels.

Another unsatisfactory feature is illustrated by a computation of the net radiation at ground level, i.e. the area between the 283° K isotherm and curve I. The downward flux originating between the ground and the first datum point is almost exactly cancelled by a similar upward flux from the ground. Thus a large part of the diagram is being used for an unsatisfactory evaluation of two terms which largely cancel.

These factors suggest that such charts are not ideal for solving the flux equation, and that they do not make the most efficient use of the available data.

A practical limitation of a chart arises when fluxes are small compared with the emission from a black body at 300° K; areas are then too small to be determined with accuracy. Consequently, charts designed for the troposphere are usually unsuitable for stratospheric problems.

To determine a heating rate from a flux chart requires numerical differentiation with respect to the height. Net fluxes are estimated at two close pressure levels. This gives two close curves like II, and two close curves like III, with long, narrow areas between the two sets. If temperature decreases monotonically with height (as in the troposphere), both upward and downward components of the flux decrease in magnitude as height increases; consequently these two areas partially cancel giving rise to an awkward problem in the estimation of area. The change of independent variable from pressure to height can be made with the help of the hydrostatic equation. Thus

$$h = -\mathbf{\nabla}.\mathbf{F} = -\frac{dF}{dz} = \rho_a g \frac{dF}{dp}.$$ (6.55)

6.7. Curtis's two-parameter method

6.7.1. *General considerations*

The work discussed in this section represents an attempt to develop a computational method for the 15μ CO_2 band having the flexibility of the methods described in § 6.5 without some of the major limitations. The work, as presented, concentrates upon three major sources of error:

(i) the use of a one-parameter scaling approximation for the pressure;
(ii) the numerical or graphical differentiation of imprecise empirical data to obtain heating rates;
(iii) the large contributions to the flux from close layers.

In place of (i) the Curtis–Godson approximation is used initially, and is later replaced for large $\tilde{u}$ by the 'modified strong line' technique described in § 6.1.3. (iii) is treated analytically on the assumption that

fifth and higher derivatives of the data (viz. gaseous densities and temperatures) with respect to height can be neglected. The improvement
of (ii) is based upon consideration of the fundamental physics of gaseous
absorption to obtain accurate derivatives of absorption data.

These and other improvements can readily be incorporated if an
electronic digital computer is available; the interest in Curtis's work
lies in its attempt to do the same for a manual computer. The investigation is experimental in the sense that non-essential approximations
have been made in order to come to grips with these three points. In
particular: an assumption of non-overlapping lines restricts the results
to levels greater than about 30 km; only one gaseous distribution
(corresponding to carbon dioxide) is considered; the effect of temperature on line intensity is sketchily treated. While there are objections
to the general applicability of Curtis's work, it provides a valuable basis
for further development, besides giving good numerical results in the
special case for which it was designed (i.e. for the $15\,\mu$ carbon dioxide
bands in the mesosphere).

The fundamental equations are stated in terms of the heating function H_i (2.21), using the dimensionless quantity

$$K_i(z',z) = \frac{\int_i e_\nu(z)E_2\{|\tau_\nu(z',\infty)-\tau_\nu(z,\infty)|\}\,d\nu}{\int_i e_\nu(z)\,d\nu}. \tag{6.56}$$

Substituting (6.56) in (2.97) we find

$$H_i(z) = \int_{K_i^{(1)}=0}^{1} \{J_i(z')-J_i(z)\}\,dK_i^{(1)} + \int_{K_i^{(2)}=0}^{1} \{J_i(z')-J_i(z)\}\,dK_i^{(2)}, \tag{6.57}$$

where $K_i^{(1)}$ is defined for $z' < z$, and $K_i^{(2)}$ is defined for $z' > z$. The
conventions with regard to the boundaries must now be interpreted as
follows (cf. (2.81)). For $z' < z$, and values of $K_i^{(1)}$ less than $K_i(0,z)$,
$J_i(z')$ is taken to equal J_i^*; for $z' > z$, and values of $K_i^{(2)}$ less than
$K_i(z,\infty)$, $J_i(z')$ is taken to be zero.

6.7.2. The method of computation

The atmosphere is divided into datum levels separated by equal
intervals of $\log_{10}\alpha_L$.† The interval chosen is $0{\cdot}2$ and the levels are
numbered by consecutive integers (r,s), equal to the integral values of
$-5\times\log_{10}\alpha_L$. Given $\log_{10}\alpha_L$ at one particular pressure (normally 1

---

† α_L, the Lorentz width, is being used here as a label. Its precise meaning is
unimportant at this stage as long as it is proportional to the pressure and has a value
of the order of $0{\cdot}1$ cm^{-1} at s.t.p.

atmosphere), the ascent can be represented by a series of numbers z_s, θ_s, B_s, ρ_s interpolated from the reference levels. (6.57) can now be transformed into a sum, using a suitable formula for numerical integration:

$$H_i^{(r)} = \sum_{s=0}^{35} A_i^{(r,s)} J_i^{(s)}, \tag{6.58}$$

where contributions from levels at which the Lorentz width would be $< 10^{-7}$ cm^{-1} ($s > 35$ or pressure less than about 1 dyne cm^{-2}) are neglected.

We have seen that the heating rate at the level r depends strongly upon the data at levels $s = r-1$, r, and $r+1$. Contributions to the integrals from this region are computed analytically assuming the source function at $(r-2)$, $(r-1)$, r, $(r+1)$, and $(r+2)$ to be fitted by a quartic curve. This gives an objective method of treating the data in a region to which the final answer is very sensitive, and the cancellation near to the level of evaluation, which plagues graphical methods, is automatically incorporated in the computed matrix elements.

To obtain the matrix element $A^{(r,s)}$ for a level s, outside the range $(r-1)$, r, $(r+1)$, it is assumed that the source functions at any three levels can be fitted by a quadratic curve. Owing to these conventions, which automatically fit quartic and quadratic curves to the data, the $A^{(r,s)}$ do not have a simple significance in the sense of relating physically the heating at r to the Planck function at s. The relation is formal and depends upon the method of integration.

We will not discuss the actual computation of the matrix $A^{(r,s)}$ given the function K, since this is a straightforward, albeit complex problem in numerical methods. The matrix is laid out so that the coefficients required to compute the heating at one particular level are in a column. The column is labelled with the value of r at the level of computation. If we wish to compute at a level for which $\log_{10} \alpha_L = -3.2$ we choose the column labelled $r = 16$ {$= -5 \times (-3.2)$}. Then list the $J_i^{(s)}$, starting at the top with the ground emission J_i^*. The row at which this first entry is made depends upon the Lorentz width at ground level. For example, if $\log_{10} \alpha_L = -0.8$ at ground level the first entry is placed in the row $s = 4$ and the matrix elements for $s < 4$ can be added and assigned to this row.

Table 6.4 gives part of a column of matrix elements for the strong lines in the 15μ carbon dioxide bands, under the assumption that lines do not overlap, and that the line intensity multiplied by the mixing ratio does not change with height (cf. § 6.1.1). The general pattern of the matrix elements is typical. All are positive except for the one at the level of computation, which is negative and numerically the largest, and the elements decrease as $|s-16|$ increases, except at the lower boundary.

TABLE 6.4

Curtis's computing scheme (example)

The level under consideration has $\log_{10} \alpha_{\mathrm{L}} = -3\cdot2$ $(r = 16)$

s	$A_i^{(r,s)}$	$\pi J_i^{(s)}$	$A_i^{(r,s)} \times \pi J_i^{(s)}$
0	0·0000	—	0
1	0·0000	—	0
2	0·0000	—	0
3	0·0000	—	0
4	+0·0499	$\pi J_i(-0\cdot8)$	+...
5	+0·0230	$\pi J_i(-1\cdot0)$	+...
6	+0·0364	$\pi J_i(-1\cdot2)$	+...
..	..	..	+...
..	..	..	+...
14	+2·0142	$\pi J_i(-2\cdot8)$	+...
15	+5·8745	$\pi J_i(-3\cdot0)$	+...
16	−23·3593	$\pi J_i(-3\cdot2)$	−...
17	+3·7395	$\pi J_i(-3\cdot4)$	+...
18	+0·8185	$\pi J_i(-3\cdot6)$	+...
..	..	..	+...
..	..	..	+...
23	+0·0113	$\pi J_i(-4\cdot6)$	+...
..	..	..	+...
..	..	..	+...
Space	−7·9228	0	0

The last entry marked *space* is equal to $K_i(z, \infty)$ and has already been included in the large negative term at $s = 16$; to avoid including it twice it is automatically multiplied by zero. If multiplied by $\pi J_i(-3\cdot2)$ it gives the contribution to $\pi H_i(-3\cdot2)$ caused by direct radiation to space, which is a large part of the net cooling at mesospheric levels. Since the magnitude of this term depends only upon the temperature at the level of computation, important simplifications can be made when it is large (see § 6.11).

6.7.3. Calculation of K for thermal radiation

Any two-parameter method of treating non-homogeneous paths (§ 6.1.3) will give a different $K(z', z)$ for each relationship between $\tilde{a}$ and $\tilde{p}$, and the computed matrix applies to one distribution of absorbing matter only. For a single line we may choose to define the distribution in terms of the parameter

$$\tilde{u}(z) = \frac{S_m(z)c(z)}{2\pi g} \frac{dp}{d\alpha_{\mathrm{L}}}, \tag{6.59}$$

generalizing (6.3) to include the possibility that mixing ratio and line intensity both vary with height. Two auxiliary functions, analogous to the optical path between z and z', are defined for a single line

$$
\tau_{0,j}(z, z') = \left| \int_1^{\alpha_L(z')/\alpha_L(z)} \tilde{u}_j(z'') \, d\!\left(\frac{\alpha_L(z'')}{\alpha_L(z)}\right) \right|
$$

$$
= \frac{\int_0^{a(z,z')} S_j(z'') \, da(z, z'')}{2\pi\alpha_L(z)} , \tag{6.60}
$$

where the suffix j indicates the jth line in an array. The second function is

$$
\tau_{1,j}(z, z') = 2\left| \int_1^{\alpha_L(z')/\alpha_L(z)} \tilde{u}(z'') \frac{\alpha_L(z'')}{\alpha_L(z)} \, d\!\left(\frac{\alpha_L(z'')}{\alpha_L(z)}\right) \right|
$$

$$
= \frac{\int_0^{a(z,z')} S_j(z'')\alpha_L(z'') \, da(z, z'')}{\pi\alpha_L^2(z)} . \tag{6.61}
$$

If $y_j(z, z')$ and $u_j(z, z')$ are the functions defined in (4.8), but employing the parameters $\tilde{a}$ and $\tilde{\alpha}$ appropriate to the Curtis–Godson approximation ((6.19) and (6.22)), we find

$$
u_j(z, z') = \frac{2\tau_{0j}^2}{\tau_{1j}} = \tau_{0j}\, q, \tag{6.62}
$$

where

$$
q = \frac{2\tau_{0j}}{\tau_{1j}} = \frac{\alpha_L(z)}{\tilde{\alpha}_L} . \tag{6.63}
$$

First consider the function $K_{L,i}(z, z')$ for the ith spectral range, in which there is assumed to be a random array of Lorentz lines. From (4.94), (4.10), and (6.56) it can be shown that

$$
K_{L,i}(z, z') = \frac{1}{\sum_j S_j(z)} \sum_j \int_1^\infty \frac{d\xi}{\xi^2} S_j(z)\phi_{L,i}\!\left(\frac{u_j(z, z')}{\xi}\right), \tag{6.64}
$$

where the sums are over the spectral region designated by i and

$$
\phi_{L,i} = \bar{T}_{L,i}\, e^{-u_j/\xi}|I_0(u_j/\xi) + (q_j - 1)I_1(u_j/\xi)|, \tag{6.65}
$$

where I_0 and I_1 are Bessel functions of the first kind with imaginary arguments.

The sums and integrals are still difficult to evaluate and, to proceed, we may introduce the following assumptions, which are arbitrary and could be avoided if so desired:

(i) the line-strength histogram follows the exponential law (4.22)

$$p(S)\, dS = \frac{dS}{\sigma} e^{-S/\sigma};\tag{6.66}$$

(ii) the 'modified strong line approximation' is valid, i.e. $\tilde{\alpha}_L$ is replaced by $\alpha_L(z)$ or, in the terms of the symbols used here, $q = 1$;

(iii) the atmosphere is transparent and lines do not overlap to any important degree, which requires the mean transmission to be near unity; thus we set $\overline{T}_{L,i} = 1$.

With assumptions (i) and (iii) it can be shown that

$$K_{L,i}(z,z') = \int_1^\infty \frac{d\xi}{\xi^2} \frac{1+qu(z,z')/\xi}{\{1+2u(z,z')/\xi\}^{3/2}},\tag{6.67}$$

where u is defined as for u_j, but with σ in place of S_j. We may note that the introduction of assumption (i) has made it impossible to consider temperature effects on individual lines except through the variation of σ.

This computation scheme was developed for high altitudes, where Doppler effects are important. For a single, isolated Doppler-shaped line $K_{D,i}(z,z')$ is given by the same expression (6.64) but with $\phi_{D,i}$ in place of $\phi_{L,i}$ and w_j in place of u_j where

$$w_j = \frac{\int S_j\, da}{\pi^{1/2}\alpha_{D,j}},\tag{6.68}$$

$\alpha_{D,j}$ = Doppler line width (assumed constant at all levels),

$$\phi_D(X) = \pi^{-1/2} \int_{-\infty}^{+\infty} e^{-t^2 - Xe^{-t^2}}\, dt.\tag{6.69}$$

With assumption (i) above, we find

$$K_{D,i} = \pi^{-1/2} \int_{-\infty}^{+\infty} e^{-t^2}\gamma(w_j e^{-t^2})\, dt,\tag{6.70}$$

with
$$\gamma(X) = 1+2X\log\frac{X}{1+X} + \frac{X}{1+X}.\tag{6.71}$$

Equations (6.68) and (6.71) were joined by assuming that
$$(1-K) = (1-K_L)(1-K_D),\tag{6.72}$$
which gives the correct asymptotic behaviour for very low and very high pressures, but which is probably inaccurate in the transition region $(K_L \simeq K_D)$.

6.7.4. *Summation of frequency ranges*

If the effect of temperature upon the function K can be neglected, the summation over all the frequency ranges can be performed conveniently as follows. Between 200° K and 300° K $B_i(0)$ can be represented to a remarkably high degree of accuracy by a quartic in θ; for example, at 667 cm^{-1}

$$B_{667}(\theta) \simeq \sum_0^4 b_k(\theta-250)^k, \qquad (6.73)$$

with

$$\pi b_0 = 244\cdot25, \qquad\qquad \pi b_3 = -5\cdot50,$$
$$\pi b_1 = 306\cdot25, \qquad\qquad \pi b_4 = -2\cdot62,$$
$$\pi b_2 = 98\cdot60,$$

the remainder being always less than one part in 3000. In general the ith range can be specified in terms of the set of coefficients $b_{k,i}$. Then in place of the matrix $A_i^{(r,s)}$ we form a new matrix

$$C_k^{(r,s)} = \sum_i b_{k,i}\, A_i^{(r,s)}\beta_i^{(s)}, \qquad (6.74)$$

where

$$\beta_i^{(s)} = \frac{\int\limits_i k_\nu^{(s)}\, d\nu}{\int\limits_0^\infty k_\nu^{(s)}\, d\nu}. \qquad (6.75)$$

We may now write, using (2.21), (6.74), and (6.58)

$$H^{(r)} = \frac{h^{(r)}}{2\pi \int\limits_0^\infty k_\nu^{(r)}\, d\nu} = \sum_{k=0}^4 \sum_{s=0}^{35} C_k^{(r,s)}(\theta^{(s)}-250)^k. \qquad (6.76)$$

The computational problem does not therefore increase indefinitely with the number of spectral ranges; with the approximations used here it is sufficient to sum four terms only.

6.7.5. *Extension of the method*

The functions τ_0 and τ_1, and all others derived therefrom, depend upon the variation of $u(z)$ with z. Thus the matrix coefficients $A^{(r,s)}$ depend upon the density and temperature distributions. Dependence upon the density distribution means that the matrix has to be recomputed whenever the density varies. Dependence upon the temperature distribution imposes the same restriction with respect to temperature, and $A^{(r,s)}$ encompasses no variability at all. If the matrix must be recomputed for each situation no purpose is served by the computing scheme, whose only advantage relative to a straightforward machine computation lies in the possibility of using it for different ascent data.

The position is, however, not quite so bad as this may suggest. The matrix coefficients have been computed for a model of the strong lines of the $15\,\mu$ CO_2 bands, with and without a temperature effect on σ, using the relation (6.22). The difference was shown to be small, suggesting that the effect of temperature on absorption might be included in a suitable linear approximation. For example, it might partly be included in the multiplying factor which converts H to h, and partly by computing a separate set of matrix elements for each of a number of standard atmospheres, and interpolating linearly between them.

Variations of density distribution can probably be treated similarly. Matrices can be computed for a number of standard density distributions. Additional linear effects of density upon heating rates, as for a shallow layer with humidity differing greatly from that above and below, can probably be included in the transformation of H to h.

These questions have not yet been seriously studied. They are fundamental to the idea of a *radiation chart* or similar device. If they cannot be solved in an approximation which can be shown numerically to be adequate, then the idea that fluxes and heating rates can be computed by simple numerical processes must be abandoned. As yet no alternative has been presented, except for the unwieldy one of detailed numerical computation from first principles. Three such methods will be described in § 6.10, but they are not suitable for routine use, and the search for a simple, but accurate technique will probably continue.

6.8. Other computational techniques

6.8.1. *The Mügge–Möller chart*

This was the first radiation chart. A scaling approximation was used, and the tabular computation in Fig. 6.3 follows one of Möller's earlier papers. At one stage Möller computed diffuse transmission functions (§ 6.2 (i)), but in later papers he appears to be content with a diffusivity factor of 1·66. In order to obtain a convenient chart, the following substitutions are made in (6.49),

$$X(\tilde{a}) = \pi B(\tilde{\theta})\epsilon_c(1\cdot 66\tilde{a}, \tilde{\theta}),\tag{6.77}$$

$$Y(\theta, \tilde{a}) = \frac{\pi B(\theta)\partial\epsilon_c(1\cdot 66\tilde{a}, \theta)/\partial\tilde{a}}{\pi B(\tilde{\theta})\partial\epsilon_c(1\cdot 66\tilde{a}, \tilde{\theta})/\partial\tilde{a}},\tag{6.78}$$

where $\tilde{\theta}$ is an arbitrary, standard temperature. There results

$$F = \pm \int_{X=\pi B(\tilde{\theta})}^{X=0} Y(\theta, \tilde{a})\, dX(\tilde{a}).\tag{6.79}$$

In the Mügge–Möller chart the abscissa is X with a non-linear scale in $\tilde{a}$. Since X is not a function of the variable θ, but only of the fixed $\tilde{\theta}$, isopleths are vertical, straight lines. The ordinate, Y, is not strongly dependent upon θ, and the isotherms are therefore approximately horizontal.

The effect of carbon dioxide is included by assuming that emission in the range $13 \cdot 5\,\mu < \lambda < 16 \cdot 5\,\mu$ is by the ν_2 band alone, while water vapour alone emits at all other wavelengths. Equations (6.77) and (6.78) are evaluated independently for the two gases, using the appropriate emissivities and scaled amounts of matter. The limits of the integral (6.79) are $X = 0$ and $X = \pi B(\tilde{\theta})\epsilon_c(\tilde{a} = \infty, \tilde{\theta})$, but since the total emissivity has an upper limit of unity, we cannot permit both $\epsilon_c(\tilde{a}_c)$ and $\epsilon_c(\tilde{a}_w)$ to have this upper bound.

To preserve our assumption of independence, we must write

$$\left.\begin{array}{l} \overline{T}_i(\tilde{a}_c = \infty) = 0 \\ \overline{T}_i(\tilde{a}_w = \infty) = 1 \end{array}\right\} \quad (13 \cdot 5\,\mu < \lambda < 16 \cdot 5\,\mu),$$

$$\left.\begin{array}{l} \overline{T}_i(\tilde{a}_c = \infty) = 1 \\ \overline{T}_i(\tilde{a}_w = \infty) = 0 \end{array}\right\} \quad (\lambda < 13 \cdot 5\,\mu; \lambda > 16 \cdot 5\,\mu), \tag{6.80}$$

and hence from (5.11)

$$\epsilon_c(\tilde{a}_c = \infty, \tilde{\theta}) = \frac{\displaystyle\sum_{13 \cdot 5\mu < \lambda < 16 \cdot 5\mu} \pi B_i(\tilde{\theta})}{\pi B(\tilde{\theta})}, \tag{6.81}$$

$$\epsilon_c(\tilde{a}_w = \infty, \tilde{\theta}) = \frac{\displaystyle\sum_{\lambda < 13 \cdot 5\mu, \lambda > 16 \cdot 5\mu} \pi B_i(\tilde{\theta})}{\pi B(\tilde{\theta})}. \tag{6.82}$$

Möller finds that these sums do not strongly depend upon the temperature, and that at $273°$ K

$$\epsilon(\tilde{a}_c = \infty) = 0 \cdot 146,$$

$$\epsilon(\tilde{a}_w = \infty) = 0 \cdot 854.$$

The abscissae of the carbon dioxide and water vapour diagrams are therefore made proportional to these two numbers.

Details of an early chart by Dmitriev (1940) are not available in the English language, but it appears to be similar to the Mügge–Möller chart in all important respects.

6.8.2. *The Elsasser chart*

The Elsasser radiation chart is the most widely used in the English literature. In its most recent form it is a transform of the Yamamoto chart. Equation (6.50) is transformed by the substitutions

$$X = \chi\theta^2,$$

$$Y = \frac{1}{2\chi\theta} \{1-\epsilon_c^*(1\cdot66\tilde{a}, \theta)\} \frac{d\pi B}{d\theta}, \tag{6.83}$$

where χ is a scaling factor which can be varied at will to give a convenient diagram. The integral in (6.50) is then of the form

$$F = \int Y \, dX. \tag{6.84}$$

The most important practical advantage of this transformation is that isopleths are approximately straight. Since X is a function of θ only the isotherms are vertical, and necessarily straight. As $\tilde{a} \to \infty$, $\epsilon^* \to 1$, and $Y \to 0$. The bottom of the diagram is therefore the straight, horizontal isopleth $\tilde{a} = \infty$. When $\tilde{a} = 0$, $\epsilon^* = 0$, and $Y = (2\sigma/\chi)\theta^2 = 2\sigma X/\chi^2$. The isopleth $\tilde{a} = 0$ is therefore a straight line, with slope $Y/X = 2\sigma/\chi^2$. Thus restrained at the two limits, all intermediary isopleths are close to straight lines. The diagram takes the form of a triangle with the isopleths all converging on $\theta = 0°$ K, for the same reason that they also do so in the Yamamoto chart.

Elsasser's original treatment of carbon dioxide divided the wavelength range in a similar way to Möller, except that $13\cdot1\,\mu < \lambda < 16\cdot9\,\mu$ was assigned to carbon dioxide, with a black-body emission equal to $18\cdot4$ per cent of the total at $273°$ K. Inside this range carbon dioxide was assumed to be completely opaque ($\overline{T}_i = 0$) under all conditions, which has the effect of adding $0\cdot184\pi B(\theta)$ to both upward and downward fluxes, giving no net flux and no radiational heating for this gas. A later version of Elsasser's chart follows Yamamoto's treatment.

6.8.3. *The Kew chart*

The isopleths in Fig. 6.5 are close to horizontal straight lines for temperatures greater than $200°$ K. Near to ground level, emission from temperatures less than $200°$ K can usually be neglected. For some purposes therefore it is sufficient to approximate the diagram by one in which isotherms and isopleths are orthogonal.

This approximation implies that ϵ^* is not a function of temperature. Taken together with the implicit assumption that $\overline{T}_i$ does not depend upon temperature this requires that ϵ also is independent of temperature. Equation (6.49) now becomes

$$F = \pm \int\limits_{}^{\tilde{a}=0} \pi B(\theta) \, d\epsilon_s(\tilde{a}). \tag{6.85}$$

With suitable laboratory empirical data, such as those given in Table 5.13, the evaluation of (6.85) is a simple matter. If observations of the flux for known $\theta \sim \tilde{a}$ relationships are available, (6.85) can be inverted by trial and error to give $\epsilon_s(\tilde{a})$. The data in Table 5.13 were partly derived on this semi-empirical basis.

Carbon dioxide emission in the Kew chart is incorporated as for the earlier Elsasser chart, but a more elaborate treatment, equivalent to that of Yamamoto, has been developed by Shekhter (1953) and by Kondratiev and Niilisk (1960). As with the Mügge–Möller chart, two diagrams are used, one for water vapour alone, and one for the range of wavelengths $12\,\mu$ to $18\,\mu$ in which both gases are important. The emissivity of this latter range can be evaluated as in (6.53), and the result transferred to a rectangular diagram of the Kew type.

A chart similar to the Kew chart has been given by F. A. Brooks (1952).

6.8.4. *Bruinenberg's numerical method*
Bruinenberg (1946) first drew attention to two important points, namely the superiority of numerical methods over graphical methods for this problem, and that where heating rates are required it is better to compute the flux divergence directly rather than to take finite differences of fluxes.

In differentiating the flux equation we can treat either $\tilde{a}$ or θ as an independent variable; Bruinenberg chooses to use θ. Let us write

$$R(\tilde{a}, \theta) = \pi B(\theta)\epsilon_s(\tilde{a}, \theta). \tag{6.86}$$

From (6.49), re-introducing dependence upon z and z', we have

$$F(z) = \pm \int_{\tilde{a}=\infty}^{\tilde{a}=0} \frac{\partial R\{\tilde{a}(z, z'), \theta(z')\}}{\partial \tilde{a}(z, z')} \, d\tilde{a}(z, z'). \tag{6.87}$$

To obtain the flux divergence (6.87) must be differentiated with respect to z, while $\tilde{a}(z, z')$ is held constant. This means that z' varies simultaneously with z and that variation in the integral (6.87) comes from the change in $\theta(z')$. Some confusion may arise because $d\tilde{a}(z, z')$ in (6.87) indicates the result of changing z' while holding z constant; it is not a complete differential in a strict sense. With the same convention for forming differentials both with regard to θ and $\tilde{a}$ (i.e. permitting z' to vary but not z), we find

$$\frac{\partial F(z)}{\partial z} = \pm \frac{\partial \tilde{a}}{\partial z} \int \frac{\partial^2 R(\tilde{a}, \theta)}{\partial \tilde{a} \partial \theta} \, d\theta. \tag{6.88}$$

The multiplying factor before the integral depends upon the definition of $\tilde{a}$. If we employ (6.14) we have

$$\frac{\partial \tilde{a}}{\partial z} = \pm \rho(z) \, \frac{S\{\theta(z)\} p^n(z)}{S(\tilde{\theta}) \tilde{p}^n},$$
(6.89)

the sign depending upon whether z is the upper or lower limit of (6.14). The integral has to be evaluated twice, once for upward and once for downward fluxes, and the sign can be chosen on the obvious grounds that exchange with a hotter layer gives rise to heating and vice versa. The quantity $\partial^2 R / \partial \tilde{a} \partial \theta$ can be evaluated either directly from empirical data or by summing over spectral ranges, using transmission functions.

The integral with respect to θ in (6.88) is performed by numerical quadrature

$$\frac{\partial F(z)}{\partial z} = \pm \frac{\partial \tilde{a}}{\partial z} \sum_j \left(\frac{\partial^2 R}{\partial \tilde{a} \partial \theta} \right)_j \Delta \theta_j,$$
(6.90)

where the subscript j signifies an atmospheric layer which is assumed homogeneous.

Bruinenberg's computational scheme derives directly from the Mügge–Möller chart, and has the same fundamental strengths and weaknesses. Carbon dioxide flux divergence is evaluated by assigning the wavelength range $13 \cdot 5 \, \mu < \lambda < 16 \cdot 5 \, \mu$ to carbon dioxide alone and performing a separate set of computations, in a manner analogous to the Mügge–Möller chart.

6.8.5. *D. L. Brooks's numerical method*

This is based on the same assumptions as the Kew chart, namely that $\overline{T}_i$ and ϵ are independent of temperature. We may write (6.85) in the form

$$F = \pm \left\{ \pi B(z) - \int_{\tilde{a} = \infty}^{\tilde{a} = 0} \epsilon_s(\tilde{a}) \, d\pi B(\theta) \right\}.$$
(6.91)

Taking $\tilde{a}$ to be the independent variable, we have

$$\frac{\partial F}{\partial z} = \pm \frac{\partial \tilde{a}}{\partial z} \int_{\tilde{a} = \infty}^{\tilde{a} = 0} \epsilon_s'(\tilde{a}) \, d\pi B(\theta).$$
(6.92)

Equation (6.92) has been used to evaluate heating rates by D. L. Brooks (1950). Consistent with the assumptions of the Kew chart, the flux divergence of carbon dioxide is neglected.

6.8.6. *Flux divergence chart of Yamamoto and Onishi*

If we use (6.50) as a starting-point and take $\tilde{a}$ to be the independent variable, we find

$$\frac{\partial F}{\partial z} = \frac{\partial \tilde{a}}{\partial z} \int_{\tilde{a}=\infty}^{\tilde{a}=0} \frac{\partial \epsilon_s^*(\tilde{a}, \theta)}{\partial \tilde{a}} \, d\pi B(\theta). \tag{6.93}$$

The integral in (6.93) can be evaluated graphically in a manner analogous to the Yamamoto chart, but with a different ordinate. However, this is inconvenient for the troposphere because large contributions to the integral come from small $\tilde{a}$, and because there is usually some cancellation between heating from below and cooling from above. In an attempt to rectify the first difficulty, Yamamoto and Onishi (1953) make the following transformations

$$r(\tilde{a}, B) = \frac{(\partial \epsilon_s^*/\partial \tilde{a})^{1/2}}{W(B)},$$

$$d\Theta = 2W^2(B) \, d\pi B, \tag{6.94}$$

as a result of which

$$\int \partial \epsilon_c^*/\partial \tilde{a} \, d\pi B = \int \tfrac{1}{2} r^2 \, d\Theta. \tag{6.95}$$

Heating rates are then proportional to areas on a polar diagram with angular coordinate Θ and radius r. The function $W(B)$ is arbitrary and can be used to distort the diagram to any convenient shape. For small amounts of water vapour $W(B) = e^{\gamma B}$ was used, where γ is a constant, while for carbon dioxide and large amounts of water vapour $W(B) = 1$ was used. Thus three separate charts are employed, and, unlike the Yamamoto flux chart, the interaction between water vapour and carbon dioxide is not taken into account; instead a method equivalent to that of the Mügge–Möller chart is employed.

6.8.7. *Comparison of techniques*

Table 6.5 shows computations of the net flux at altitudes of 0, 3, and 8 km, for average, cloudless conditions in seven latitude zones of the northern hemisphere. Seven different methods are used, and two kinds of pressure-scaling approximation; the square root of pressure for the Elsasser, Kew, Shekhter, F. A. Brooks, and Möller charts, the first power for Yamamoto and Dmitriev. In order to assess the importance of this difference of procedure, Yamamoto's computations are repeated with a square-root scaling approximation.

In Fig. 6.7 a similar comparison is made of radiative heating rates (or rather potential rate of change of temperature $\partial \theta/\partial t = h/\rho_a c_p$). The results attributed to Yamamoto and Elsasser were obtained from finite flux differences; those of Brooks and Yamamoto and Onishi directly from a flux-divergence computation.

TABLE 6.5

Comparative computations of the net flux of terrestrial radiation
($cal\ cm^{-2}\ min^{-1}$)

After Kondratiev and Niilisk (1960)

$$
\begin{matrix}
\text{E} & = & \text{Elsasser (§ 6.8.2)} \\
\text{M} & = & \text{Möller (§ 6.8.1)} \\
\text{Sh} & = & \text{Shekhter (§ 6.8.3)} \\
\text{B} & = & \text{F. A. Brooks (§ 6.8.3)} \\
\text{K} & = & \text{Kew (§ 6.8.3)} \\
\text{Y*} & = & \text{Yamamoto (§ 6.5)}
\end{matrix}
\left. \right\} \text{Pressure-scaling factor, } n = \tfrac{1}{2}.
$$

$$
\begin{matrix}
\text{Y} & = & \text{Yamamoto (§ 6.5)} \\
\text{D} & = & \text{Dmitriev (§ 6.8.1)}
\end{matrix}
\left. \right\} \text{Pressure-scaling factor, } n = 1.
$$

Latitude zone	E	M	Sh	B	K	Y*	Y	D
				$z = 0$ km				
0°–10° N	0·110	0·124	0·096	0·090	0·142	0·118	0·125	0·107
20°–30° N	0·123	0·147	0·117	0·125	0·154	0·137	0·141	0·135
40°–50° N	0·112	0·125	0·117	0·118	0·142	0·129	0·132	0·128
60°–70° N	0·110	0·109	0·124	0·127	0·142	0·128	0·131	0·131
				$z = 3$ km				
0°–10° N	0·180	0·218	0·189	0·187	0·204	0·217	0·225	0·235
20°–30° N	0·198	0·221	0·211	0·213	0·228	0·238	0·245	0·254
40°–50° N	0·175	0·189	0·195	0·199	0·205	0·209	0·218	0·229
60°–70° N	0·150	0·159	0·181	0·184	0·180	0·182	0·190	0·204
				$z = 8$ km				
0°–10° N	0·275	0·302	0·300	0·307	0·296	0·334	0·349	0·353
20°–30° N	0·289	0·315	0·333	0·330	0·309	0·347	0·358	0·378
40°–50° N	0·254	0·278	0·301	0·303	0·269	0·297	0·306	0·334
60°–70° N	0·202	0·220	0·253	0·253	0·220	0·240	0·251	0·272

The differences between results in Fig. 6.7 and Table 6.5 are difficult to rationalize. The safest conclusion is that the spread of the results (of the order of 50 per cent both in fluxes and heating rates) gives the order of accuracy of these computational techniques. Below 8 km differences between the two types of scaling approximation appear to be small. The large discrepancy between Elsasser and all others above 13 km (Fig. 6.7) might be partly attributable to the difference in pressure-scaling factor but Brooks's results do not support this conclusion.

An interesting comparison in Fig. 6.7 is between Yamamoto and Yamamoto and Onishi. Here, the only major difference is in the form of the computation, although the carbon-dioxide–water-vapour inter-

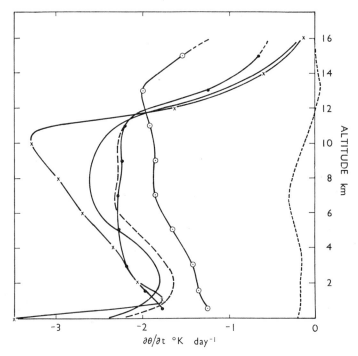

FIG. 6.7. Comparative computations of heating by terrestrial radiation. The ascent data are the same as those used in Table 6.5 for the latitude zone 0–10° N, and are taken from London (1952).

	Author	Ref.	Gas	Pressure-scaling factor, n
——————	Yamamoto and Onishi	§ 6.8.6	$CO_2 + H_2O$	1
— — —	Yamamoto and Onishi	§ 6.8.6	H_2O	1
– – – – –	Yamamoto and Onishi	§ 6.8.6	CO_2	1
×—×—×	D. L. Brooks	§ 6.8.5	H_2O	$\frac{1}{2}$
•—•—•—	Yamamoto	§ 6.5	$H_2O + CO_2$	1
⊙–⊙–⊙	Elsasser	§ 6.8.2	H_2O	$\frac{1}{2}$

action is also treated slightly differently. Differences between the two sets of results probably illustrate the magnitude of random computing errors. Above 15 km, where the radiative heating decreases rapidly in magnitude, the percentage error can be very large. This illustrates an unavoidable difficulty in atmospheric computations, namely that since the atmosphere does not differ greatly from its equilibrium temperature (approximately 246° K, see § 1.1), the flux divergence is a small residual between large, nearly compensating components.

Fig. 6.7 shows a tendency for carbon dioxide to have an increasing influence as the height increases. Between 12 and 16 km there is a slight radiational heating by carbon dioxide, which is of some theoretical interest (see Chapter 8). In the mesosphere carbon dioxide is probably of considerably greater importance than any other atmospheric gas.

6.9. Simplified methods

In view of the lack of precise agreement between radiation charts, it is but a short step to seek a simpler, semi-empirical approach. A method in common use assumes that the radiation flux (the methods described here refer only to the downward component) depends upon a limited number of variables rather than the entire atmospheric structure. The following relationships have been proposed for a cloudless sky (θ, p, and e are the screen temperature in °K, air pressure and water-vapour partial pressure in millibars, all at the observing site):

$$F^- = -\sigma\theta^4(a - b10^{-\gamma e}) \quad \text{(Ångström, 1915, 1929),} \quad (6.96)$$

$$F^- = -\sigma\theta^4(a - b\sqrt{e}) \quad \text{(Brunt, 1932),} \quad (6.97)$$

$$F^- = -\sigma\theta^4(a + b\log e) \quad \text{(Elsasser, 1942),} \quad (6.98)$$

all of which apply at $p = 1013$ mb only, and

$$F^- = -\sigma\theta^4 \frac{(ap + be)}{\theta} \quad \text{(Robitzsch, 1926),} \quad (6.99)$$

which is not so restricted.

The coefficients a and b can be found from observational data by partial regression, but the values vary from site to site, as is illustrated for (6.97) in Table 6.6. The average correlation coefficient for these data is about 0·7, and Brunt's formula therefore generally accounts for about 0·5 of the total observed variance.

In Ångström's formula, constants have been quoted in the ranges $a = 0.71$ to 0.80, $b = 0.24$ to 0.325, $\gamma = 0.40$ to 0.74. Robitzsch has given $a = 0.135$, $b = 6.0$ for (6.99). Elsasser has suggested $a = 0.21$, $b = 0.22$ for (6.98).

If the effective ground emission temperature equals the screen temperature,

$$F^+ = \sigma\theta^4,$$

and for clear skies

$$F = F^+ + F^- = \sigma\theta^4(1 + F^-/\sigma\theta^4), \tag{6.100}$$

where F^- can be found from any one of (6.96), (6.97), (6.98), or (6.99).

TABLE 6.6

Coefficients in Brunt's empirical relation for the downward flux of terrestrial radiation (6.97)

After Elsasser (1942)

Observer	Place	a	b	r†
Dines	Benson, England	0·53	0·065	0·97
Askloef	Uppsala, Sweden	0·43	0·082	0·83
Ångström	Bassour, Algeria	0·48	0·058	0·73
Ångström	Mt. Whitney, California, U.S.A.	0·50	0·032	0·30
Bontario	Montpellier, France	0·60	0·042	—
Kimball	Washington, D.C., U.S.A.	0·44	0·061	0·29
Kimball	Mt. Weather, Virginia, U.S.A.	0·52	0·066	0·84
Kimball	Various places in U.S.A.	0·53	0·062	0·88
Eckel	Kanzelhoehe, Austria	0·47	0·063	0·89
Ramanathan and Desai	Poona, India	0·55	0·038	0·63
Ramanathan and Desai	Poona, India	0·62	0·029	0·68

† Correlation coefficient between predicted and observed downward fluxes.

The presence of cloud decreases the net flux at ground level. An empirical correction for an overcast sky can be made by writing

$$F_{ov} = \lambda F, \tag{6.101}$$

where the factor λ varies with the height of cloud base. Clearly, $\lambda = 0$ if the cloud is at ground level. The values shown in Table 6.7 for other altitudes have been quoted for use with Ångström's formula (6.96). For high clouds, discordant values have been quoted (Elsasser, 1942) suggesting that the principle of the method is not sound.

TABLE 6.7

The coefficient λ as a function of cloud-base heights

After Phillips (1940)

Height km:	2	5	8
λ	0·17	0·38	0·45

A valuable numerical approach to the problem has been made by Möller (1954). An empirical procedure was developed and tested against a radiation chart. A thick atmospheric layer lying between 1000 and

400 mb was considered, and the flux at the upper and lower boundaries was found statistically to be closely related to the temperature and frost-point temperature (θ and θ_f respectively, both expressed in °K) at the nearest boundary. The following relations were found for a cloud-free atmosphere:

$$F(400) = [121\cdot4 - 4\cdot187\{\theta_f(400) - 245\} + 1\cdot25 \times 10^{-4}\{\theta_f(400) - 245\}^3 +$$
$$+ 3\cdot9\{\theta(400) - 222\}^{1\cdot105}] \times 10^{-3} \text{ cal cm}^{-2} \text{ min}^{-1}, \quad (6.102)$$

$$F(1000) = [115\cdot0 - 3\cdot24\{\theta_f(1000) - 273\} - 0\cdot0185\{\theta_f(1000) - 273\}^2 +$$
$$+ 3\cdot0\{\theta(1000) - 273\}] \times 10^{-3} \text{ cal cm}^{-2} \text{ min}^{-1}. \quad (6.103)$$

The influence of cloud is of paramount importance, and if we adopt (6.101) it is not sufficient to express λ as a function of cloud base height alone. If θ_t and θ_b are cloud top and base temperatures, and if f is the relative humidity, suitable empirical relations were found to be

$$\lambda(400) = 0\cdot585 + 0\cdot0116\{\theta_t - \theta(400)\} - 4 \times 10^{-5}\{\theta_t - \theta(400)\}^2 \quad (6.104)$$
and
$$\lambda(1000) = 0\cdot043\{\theta_b - \theta(1000)\}^{0\cdot7}\left(1 + \frac{0\cdot15f(1000)}{100}\right) \times$$
$$\times [1 - 0\cdot0027\{\theta(1000) - 273\}]. \quad (6.105)$$

If a fraction (q) of the sky is overcast the effective λ is assumed to be

$$\lambda_{\text{eff}} = \{1 - q(1 - \lambda)\}. \quad (6.106)$$

These formulae give an adequate substitute for the Mügge–Möller radiation diagram for the particular problem under consideration. In 18 cases the computed values of $F(400) - F(1000)$ were compared with those derived from the radiation chart, and a root-mean-square difference between the two methods of 6·5 per cent was found, which, in the light of other uncertainties, is probably negligible. For use with routine weather forecasts therefore detailed chart evaluations can probably be replaced by this simpler technique.

6.10. Machine methods

Plass (1956 a and b), Kaplan (1959), and Hitschfeld and Houghton (1961) have described numerical techniques requiring the use of electronic digital computers, but still not entirely free from approximations.

Plass computes the cooling rate at all levels up to 70 km, and the downward flux at the ground, for both the 9·6 μ ozone band and the 15 μ carbon dioxide band, using empirical laboratory data to the maximum extent possible. In the former case the computations are for three temperature distributions and three density distributions; in the latter for three density distributions but only one temperature distribution. The effect of temperature on the population of the rotational levels for the (000–010) CO_2 transition was included approximately in the computation, although the effect on the band intensity of the flanking upper-state bands was not. For ozone the effect of temperature upon absorption was neglected. Equation (6.28) was evaluated directly by numerical integration, using variable height intervals, close enough to avoid rounding-off errors. At high levels the flux divergence is very small but, by retaining sufficient significant figures in the computation, differentiation by finite differences was possible. Above 70 km sufficient accuracy could no longer be obtained by this method.

A two-parameter scaling approximation was used to allow for atmospheric inhomogeneity. Empirical data on the 15 μ carbon dioxide band were divided into 1 μ ranges, inside which the variation of line intensity is limited in extent; consequently errors associated with the doubtful transition region between strong and weak lines are minimized. The ozone band, however, was treated as a whole, as a result of which the asymptotic strong- and weak-line regions are practically never achieved. It was further assumed that bands do not overlap, an assumption which is probably adequate above the tropopause, and that the diffusivity factor depends only on transmission.

Kaplan's method is intended to provide accurate fluxes in the troposphere, where water vapour, carbon dioxide, and clouds are the most important constituents. The method was designed for application to numerical weather forecasting, which assumes a small number of thick, homogeneous atmospheric layers. The aim of the computation is to find the downward flux at the ground and the upward flux at the upper surface of a thick layer, assuming a linear vertical temperature gradient.

With this model Kaplan is able to discuss a computation which is largely free from approximations. Pressure effects are included by the use of the Curtis–Godson approximation, whose accuracy is tested for water vapour and carbon dioxide in the troposphere. Detailed data on line widths and intensities are employed in conjunction with the Lorentz line profile; temperature effects are included exactly for each line individually; line absorptions are combined either according to (4.94) for the statistical model, or using the Elsasser model and the multiplication property for overlapping sub-bands, whichever is appropriate. Since the model is defined by few parameters it proves possible to compute fluxes for every reasonable situation and to store the results for subsequent incorporation in numerical weather prediction schemes.

Hitschfeld and Houghton have attempted to integrate the flux equation directly for the $9 \cdot 6 \mu$ ozone band between $9 \cdot 5$ and $32 \cdot 5$ km. They do not use a spectral model and their treatment of pressure effects is exact; it is instructive to see how far one can proceed with this degree of rigour. A diffusivity factor was employed and the atmosphere divided into layers 1 km thick, each assumed to be homogeneous. The Lorentz profile was used, even up to 10 cm^{-1} from line centres, and the smallest frequency interval chosen (near to line centres) was equal to the line width at 5 mb pressure; thus the computation could not be extended in its present form below a pressure of about 10 mb. The theoretical line strengths of Kaplan, Migeotte, and Neven (see § 5.6.3) were normalized upon Walshaw's laboratory data (also see § 5.6.3.) It may be recalled that this theory takes no account of the severe Coriolis interactions.

With these data theoretical spectra can be evaluated in the form of a frequency-height matrix, and stored on tape. Because Hitschfeld and Houghton assume that line strengths do not vary with temperature, this matrix can be re-used with different ozone density distributions to yield a transmission matrix. Given the transmission matrix fluxes and heating rates follow from straightforward numerical procedures. The assumption of line intensity independent of temperature is, however, unsatisfactory, and should be eliminated, as the authors apparently contemplate for future occasions.

The main limitation encountered was the amount of the spectrum which could be treated in such detail. Only three ranges, averaging a span of $0 \cdot 4$ cm^{-1} and containing 8 lines each could be handled in the time available on an electronic digital computer, and results for the entire band had to be inferred from these three ranges together with a few computations on single lines. Given the best available computer the authors estimate that fifteen $0 \cdot 4$ cm^{-1} ranges could be computed in one hour, for a single ascent curve. These figures illustrate clearly the limitations of the direct approach to radiative computations, even if the spectral data are entirely satisfactory.

6.11. Non-equilibrium source functions

For the purpose of this section we will assume the existence of an adequate procedure for computing the heating function $H_i(z)$ for given values of the source function $J_i(z')$, of the form

$$H_i(z) = \mathscr{L}_z\{J_i(z')\}, \tag{6.107}$$

where according to (2.97), $\mathscr{L}$ in a linear operator. Let us indicate by

$H_i^{(0)}$ the value of H_i computed with the use of the source function for thermodynamic equilibrium. Then

$$H_i^{(0)}(z) = \mathscr{L}_z\{B_i(z')\}.$$

If the band of wavelengths designated by (i) is sufficiently wide to embrace the whole of a vibration-rotation band, then according to (2.73)

$$H_i(z) = \mathscr{L}_z\left(B_i(z') + \frac{\lambda_i(z')H_i(z')}{2\phi_i}\right)$$

$$= H_i^{(0)}(z) + \mathscr{L}_z\left(\frac{\lambda_i(z')H_i(z')}{2\phi_i}\right). \qquad (6.108)$$

Equation (6.108) is a linear, operational equation from which we can compute H_i by iteration, given the operator $\mathscr{L}_z$. The iteration has been

TABLE 6.8

Computations on the 15 μ carbon dioxide band

After Curtis and Goody (1956)

$\log_{10} p$	z	πB	$-\pi H^{(0)}$	$-\pi H^{(1)}$	$-\pi H$	πJ
(dyne cm^{-2})	(km)					
3·0	49·2	329·1	5·4	5·4	5·4	329·1
2·8	52·9	317·3	5·0	5·0	5·0	317·3
2·6	56·5	291·5	4·2	4·2	4·2	291·5
2·4	60·1	252·0	3·2	3·2	3·2	252·0
2·2	63·4	208·4	2·2	2·2	2·2	208·2
2·0	66·6	169·4	1·4	1·4	1·4	169·2
1·8	69·8	139·0	0·7	0·7	0·7	138·9
1·6	72·7	121·6	0·6	0·6	0·8	121·2
1·4	75·6	111·6	1·3	1·3	2·0	110·2
1·2	78·5	105·2	2·3	2·2	3·8	101·0
1·0	81·3	101·5	4·2	3·7	5·9	91·2
0·8	84·2	99·9	6·8	5·2	7·3	79·6
0·6	87·1	101·5	12·2	7·0	7·7	67·6
0·4	90·1	106·4	21·6	7·9	7·2	56·4
0·2	93·1	113·1	35·0	7·3	6·0	47·1
0·0	96·3	121·9	54·2	6·1	4·7	39·7
−0·2	99·6	132·9	(73·8)	(4·5)	(3·7)	(31·4)
−0·4	103·0	144·5	(94·1)	(3·2)	(2·7)	(24·9)
−0·6	106·5	159·2	(116·9)	(2·2)	(2·0)	(19·0)
−0·8	110·1	185·7	(149·0)	(1·7)	(1·5)	(15·1)
−1·0	114·1	227·3	(194·5)	(1·3)	(1·2)	(11·8)
−1·2	118·4	299·0	(268·0)	(1·1)	(1·0)	(9·1)

B is the black-body intensity at 667 cm^{-1}. B, $H^{(0)}$, $H^{(1)}$, and J are in ergs cm^{-2} sec^{-1} ster^{-1} per cm^{-1}. Values in parentheses depend upon the doubtful assumption that CO_2 has a constant mixing ratio at and above 120 km.

studied by numerical methods, and there are difficulties with its convergence as $\lambda \to \infty$ (i.e. as $z \to \infty$) unless an adequate first approximation is used with the correct asymptotic properties for large z. This may be obtained as follows.

At low pressures the term representing cooling to space (see § 6.7.2) tends to dominate all others. As a first approximation we will neglect all other terms. Thus, in terms of Curtis's computing scheme, we assume that as $z \to \infty$

$$\mathscr{L}_z\{J_i(z')\} \to -K_i^{(2)}(z, \infty)J_i(z),\qquad(6.109)$$

and hence,

$$H_i^{(1)}(z) = H_i^{(0)}(z)\left(1 + \frac{\lambda_i(z)K_i^{(2)}(z, \infty)}{2\phi_i}\right)^{-1}.\qquad(6.110)$$

$H_i^{(1)}(z)$ has the correct asymptotic behaviour that $H_i^{(1)}(z) \to 0$ as $z \to \infty$, and if used as a first approximation the iteration of (6.108) can be made to converge satisfactorily.

To illustrate the relationship between B, J, $H^{(0)}$, $H^{(1)}$, and H, a specimen computation for an idealized model of the $15\,\mu$ CO_2 band is shown in Table 6.8. For this band $\lambda_i/\phi_i \simeq 1$ at 75 km. The comparatively small difference between $H^{(0)}$ and H for λ_i/ϕ_i close to unity is caused by the near black-body conditions in the band centre, even at these high altitudes. At higher levels where $H^{(0)}$ and H differ greatly, $H^{(1)}$ gives a good first approximation.

BIBLIOGRAPHY

6.1. Transmission along non-homogeneous paths

6.1.1. *An exact solution*

The theory for a single line is by

PEDERSEN, F., 1942, 'On the temperature-pressure effect on absorption of long-wave radiation by water vapour', *Met. Ann. (Norwegian Met. Inst.)* **1**, No. 6; and

STRONG, J., and PLASS, G. N., 1950, 'The effect of pressure broadening of spectral lines on atmospheric temperature', *Astrophys. J.* **112**, p. 365.

The integral (6.8) has been evaluated, for certain discrete values of $y(1)$ and $y(2)$, by

KAPLAN, L. D., 1952, 'On the pressure dependence of radiative heat transfer in the atmosphere', *J. Met.* **9**, p. 1.

A graphical representation for $y(2) = 0$ is given by

GOODY, R. M., 1952, *The physics of the stratosphere*. Cambridge University Press.

6.1.2. *Scaling approximations*

The origin of the scaling approximation is obscure. As a desperate measure for securing a quick result it must have been re-discovered many times. The experimental evidence most commonly appealed to by the earlier workers is that of

SCHNAIDT, F., 1939, 'Über die Absorption von Wasserdampf und Kohlendioxyd und ihre Druck- und Temperaturabhängigkeit', *Beitr. Geophys.* **54**, p. 203.

This was interpreted in terms of the scaling approximation by ELSASSER (1942) (see § 1.1) and in a slightly more complex manner by

MÖLLER, F., 1943, 'Das Strahlungsdiagram', *Reichsampt für Wetterdienst* (Luftwaffe).

6.1.3. *Two-parameter approximations*

The two-parameter scaling approximation has been exploited in machine computations by

PLASS, G. N., 1956a, 'The influence of the 9·6 μ ozone band on the atmospheric infra-red cooling rate', *Quart. J. R. Met. Soc.* **82**, p. 30; and
—— 1956b, 'The influence of the 15 μ carbon-dioxide band on the atmospheric infra-red cooling rate', ibid. p. 310.

The concept of assigning a mean pressure to a non-homogeneous path has been suggested independently by a number of writers. The first explicit statement known to the author is by

VAN DE HULST, H. C., 1945, 'Theory of absorption lines in the atmosphere of the earth', *Annales d'astrophys.* **8**, p. 21.

Application to atmospheric heat-transfer problems was first proposed by

CURTIS, A. R., 1952, 'Discussion of *A statistical model for water-vapour absorption* by R. M. Goody', *Quart. J. R. Met. Soc.* **78**, p. 638;

and independently by

GODSON, W. L., 1953, 'The evaluation of infra-red radiative fluxes due to atmospheric water vapour', ibid. **79**, p. 367.

Table 6.1 follows

KAPLAN, L. D., 1959, 'A method for calculation of infrared flux for use in numerical models of atmospheric motion', in *The atmosphere and the sea in motion*, Rockefeller Inst. Press, p. 170.

6.2. Diffuse radiation

Many writers have had something to say on the question of diffuse radiation. ELSASSER (1942) (§ 1.1) introduced the diffusivity factor 1·66, although the idea of a diffusivity factor is of much earlier origin. Observational evidence in its favour was given by

ROBINSON, G. D., 1947, 'Notes on the measurement and estimation of atmospheric radiation', *Quart. J. R. Met. Soc.* **73**, p. 127.

Fig. 6.3 follows GODSON (1954) (§ 4.2).

The ideas suggested in (iii) may have been utilized by A. A. DMITRIEV (1940), see

KONDRATIEV, K. Y., and NIILISK, H. J., 1960, 'Comparison of radiation charts', *Geofisica pura e applicata*, **46**, p. 231.

Details of Dmitriev's paper are not available in the English language.

6.3. Solar radiation

A review of methods of computing solar heating is given by

FRITZ, S., 1951, 'Solar radiant energy and its modification by the earth and its atmosphere', *Compendium of meteorology*, ed. T. F. Malone, Am. Met. Soc., p. 13.

Equations (6.44) and (6.45) were given, respectively, by

MÜGGE, R., and MÖLLER, F., 1932, 'Zur Berechnung von Strahlungsströmen und Temperaturänderungen in Atmosphären von beliebigem Aufbau', *Z. für Geophysik*, **8**, p. 53; and

KORB, G., MICHALOWSKY, J., and MÖLLER, F., 1956, *Investigations of the heat balance of the troposphere*. Air Force Cambridge Research Center, TN–56–881.

A graphical, two-parameter method of evaluating heating rates has been proposed by

YAMAMOTO, G., and ONISHI, G., 1952, 'Absorption of solar radiation by water vapour in the atmosphere', *J. Met.* **9**, p. 415.

A numerical computation by

ROACH, W. T., 1961, 'The absorption of solar radiation by water vapour and carbon dioxide in a cloudless atmosphere', *Quart. J. R. Met. Soc.* **87**, p. 364, employs the laboratory data of HOWARD *et al.* (1956) (see § 5.4) and the Curtis–Godson approximation.

An early computation of heating by all atmospheric components is by

KARANDIKAR, R. V., 1946, 'Radiation balance of the lower stratosphere. Part I. Height distribution of solar energy absorption in the atmosphere', *Proc. Ind. Acad. Sci.* **23**, p. 70.

6.4. Radiation charts

The first radiation chart was proposed by MÜGGE and MÖLLER (1932) (§ 6.3) although the idea was not widely popularized until the publication of a practical chart by MÖLLER (1943) (§ 6.1.2). Almost simultaneously appeared the well-known work by ELSASSER (1942) (see § 1.1). This has recently been improved by

ELSASSER, W. M., and CULBERTSON, M. F., 1961, *Atmospheric radiation tables*. Meteor. Monographs **4**, No. 23, Am. Met. Soc.

According to KONDRATIEV and NIILISK (1960) (§ 6.2), DMITRIEV appears to have published a practical chart in the Soviet Union as early as 1940. It is unfortunate that this work was not known to Western meteorologists.

A discussion of the boundary condition at the earth's surface is given by

BROOKS, F. A., 1952, 'Atmospheric radiation and its reflection from the ground', *J. Met.* **9**, p. 41.

6.5. Yamamoto's chart

This section follows

YAMAMOTO, G., 1952, 'On a radiation chart', *Science Rept. Tohoku University, Series 5, Geophysics*, **4**, No. 1, p. 9.

6.6. Examples of the use of a radiation chart

The data follow MÖLLER (1943) (§ 6.1.2). Other references are to ROBINSON (1947) (§ 6.2) and to

HOUGHTON, J. T., 1958, 'The emissivity of the earth's surface', *Quart. J. R. Met. Soc.* **84**, p. 448.

6.7. A two-parameter method

The only published reference to this work is by

CURTIS, A. R., 1956, 'The computation of radiative heating rates in the atmosphere', *Proc. Roy. Soc.* A, **236**, p. 145.

The text of this section is based upon unpublished manuscripts by the same author.

6.8. Other computational techniques

The Mügge-Möller chart is described by MÖLLER (1943) (§ 6.4), and the Elsasser chart by ELSASSER (1942) (§ 1.1) and ELSASSER and CULBERTSON (1961) (§ 6.4). The Kew chart is by

ROBINSON, G. D., 'Notes on the measurement and estimation of atmospheric radiation—2', *Quart. J. R. Met. Soc.* **76**, p. 37.

The work of DMITRIEV (1940) and SHEKHTER (1953) is discussed by KONDRATIEV and NIILISK (1960) (see § 6.2), and a carbon dioxide flux diagram is given by

KONDRATIEV, K. Y., and NIILISK, H. I., 1960, 'On the question of carbon dioxide heat radiation in the atmosphere', *Geophysica pura e applicata*, **46**, p. 216.

Other references are to BROOKS (1952) (§ 6.4),

BRUINENBERG, A., 1946, *A numerical method for the calculation of temperature changes by radiation in the free atmosphere.* K. Ned. Met. Inst. de Bilt, Med. en ver., serie B, Deel 1, Nr. 1;

BROOKS, D. L., 1950, 'A tabular method for the computation of temperature change by infrared radiation in the free atmosphere', *J. Met.* **7**, p. 313; and

YAMAMOTO, G., and ONISHI, G., 1953, 'A chart for the calculation of radiative temperature changes', *Science Rept., Tohoku Univ., Series 5, Geophysics*, **4**, No. 3, p. 108.

The comparisons of methods for determining fluxes and heating rates are taken from YAMAMOTO and ONISHI (1953) and KONDRATIEV and NIILISK (1960) (§ 6.2).

The data are from

LONDON, J., 1952, 'The distribution of radiational temperature change in the Northern hemisphere during March', *J. Met.* **9**, p. 145.

6.9. Simplified methods

For details see ELSASSER (1942) (see § 1.1);

ÅNGSTRÖM, A., 1915, 'A study of radiation in the atmosphere', *Smithsonian Misc. Coll.* **65**, No. 3;

—— 1929, 'On the variation of atmospheric radiation', *Beitr. Geophys.* **21**, p. 145;

BRUNT, D., 1932, 'Notes on radiation in the atmosphere—1', *Quart. J. R. Met. Soc.* **58**, p. 389;

ROBITZSCH, M., 1926, 'Strahlungsstudien', *Arb. preuss. aero. Obs.* **15**, p. 194;

PHILLIPS, H., 1940, 'Zur Theorie der Wärmestrahlung in Bodennähe', *Beitr. Geophys.* **56**, p. 229; and

MÖLLER, F., 1954, 'Ein Kurzverfahren zur Bestimmung der langwelligen Ausstrahlung dicker Atmosphärenschichten', *Arch. für Met. Geophys. u. Biokl.*, Ser. A, **7**, p. 158.

6.10. Machine methods

See PLASS (1956a, b) (§ 6.1.3), KAPLAN (1959) (§ 6.1.3); and

HITSCHFELD, W., and HOUGHTON, J. T., 1961, 'Radiative transfer in the lower stratosphere due to the 9·6 micron band of ozone', *Quart. J. R. Met. Soc.* **87**, p. 562.

6.11. Non-equilibrium source functions

The presentation follows CURTIS and GOODY (1956) (see § 2.2).

7

EXTINCTION BY MOLECULES AND DROPLETS

7.1. The problem in terms of the electromagnetic theory

THE formal theory developed in Chapter 2 assumed the Stokes parameters to be additive. The sufficient condition for additivity is that
the radiation fluxes in the atmosphere shall have no *phase coherence*.
Thermal emission from independently excited molecules is necessarily
incoherent with respect to phase. Atmospheric scattering centres are
widely and randomly spaced, and they can be treated as independent
and incoherent scatterers. The situation differs, however, when we
consider details of the scattering process within a single particle, and
in order to derive the extinction coefficient and the scattering matrix
(see § 2.1.3) we must make use of a theoretical framework which involves
the phase explicitly.

The problem of extinction in electromagnetic theory is complex, and
is rendered the more difficult by preconceptions based on the approximations of elementary optics. A scholarly presentation of the known facts
by van de Hulst (1957) has recently provided us with a survey whose
breadth and precision cannot readily be bettered; this chapter will
therefore follow his treatment closely.

The geometry of the problem is illustrated in Fig. 7.1. An isolated
particle is irradiated by an incident, plane electromagnetic wave. The
plane wave preserves its character only if it propagates through a homogeneous medium; the presence of the scattering particle, with electric
and magnetic properties differing from those of the surrounding medium,
distorts the wave front. The disturbance has two aspects: firstly, the
plane wave is diminished in amplitude; secondly, at distances from the
particle large compared with the wavelength and particle size, there is
an additional outward-travelling spherical wave. The energy carried
by this spherical wave is the *scattered* energy; the total energy lost by
the plane wave corresponds to *extinction*; the difference is the *absorption*.

The properties of the spherical wave in one particular direction (the
line of sight) will be considered. This direction can be specified by the
scattering angle θ (see Fig. 7.1) in a plane containing both the incident
and scattered wave normals (*the plane of reference*), and the azimuth
angle ϕ between the plane of reference and a plane fixed in space. For

spherical particles, from symmetry considerations, scattering will be independent of the azimuth angle.

An electromagnetic wave is characterized by electric and magnetic vectors **E** and **H** which form an orthogonal set with the direction of propagation of the wave (i.e. the direction of the wave normal). In any one medium $|\mathbf{E}|$ and $|\mathbf{H}|$ are related and, since we will examine the

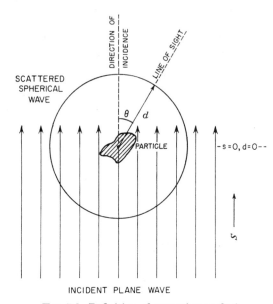

FIG. 7.1. Definition of scattering angle θ.

properties of the radiation in air surrounding a scattering particle, we may without loss of generality speak in terms of the electric vector only. The *direction of polarization* is defined as the direction of the electric vector.

Let **l** and **r** be two unit vectors which form an orthogonal set with the direction of propagation, respectively parallel to and perpendicular to the *plane of reference*. Note that this defines two directions *which depend upon the direction of the line of sight and are not fixed in space.* If $E^{(l)}$ and $E^{(r)}$ are the complex amplitudes of the parallel and perpendicular components then

$$\mathbf{E} = E^{(l)}\mathbf{l} + E^{(r)}\mathbf{r}, \tag{7.1}$$

and, for a general (elliptically polarized) wave

$$E^{(l)} = |E^{(l)}|e^{-i\delta}e^{-2\pi is/\lambda + 2\pi i\nu t},$$

$$E^{(r)} = |E^{(r)}|e^{-2\pi is/\lambda + 2\pi i\nu t}, \tag{7.2}$$

where δ is the phase difference between the two components and s the distance along the direction of propagation. λ varies from medium to medium and may be written $\lambda_0/\tilde{m}$ where $\tilde{m}$† is the refractive index. In general $\tilde{m}$ is complex

$$\tilde{m} = \tilde{n} - i\tilde{n}', \qquad (7.3)$$

where n and n' are both real. From (7.3) and (7.2) the complex index leads to an attenuation factor $e^{-2\pi\tilde{n}'s/\lambda_0}$ in the amplitude of the electric vector.

The energy carried by an electromagnetic wave is measured by the Poynting vector, directed along the wave normal, whose magnitude is

$$N = |\mathbf{E}|^2$$
$$= (|E^{(l)}|^2 + |E^{(r)}|^2). \qquad (7.4)$$

Since it is irrelevant to our discussion, no constant of proportionality is included in (7.4), thus avoiding problems associated with the choice of electromagnetic units.

It is a straightforward matter to show that there is a vector $(N^{(1)}, N^{(2)}, N^{(3)}, N^{(4)})$ corresponding to $(I, Q, U, V)\,d\omega$ where

$$N^{(1)} = E^{(l)}E^{(l)*} + E^{(r)}E^{(r)*},$$
$$N^{(2)} = E^{(l)}E^{(l)*} - E^{(r)}E^{(r)*},$$
$$N^{(3)} = E^{(l)}E^{(r)*} + E^{(r)}E^{(l)*}, \qquad (7.5)$$
$$N^{(4)} = i(E^{(l)}E^{(r)*} - E^{(r)}E^{(l)*}),$$

where the star denotes the complex conjugate.

If we apply these definitions to the elliptically polarized wave (7.2), we obtain

$$N^{(1)} = |E^{(l)}|^2 + |E^{(r)}|^2,$$
$$N^{(2)} = |E^{(l)}|^2 - |E^{(r)}|^2,$$
$$N^{(3)} = 2|E^{(l)}||E^{(r)}|\cos\delta, \qquad (7.6)$$
$$N^{(4)} = 2|E^{(l)}||E^{(r)}|\sin\delta.$$

If the direction of polarization happens to be $\mathbf{l}$ or $\mathbf{r}$ then $|E^l|$ or $|E^r|$ is zero, and $N^{(3)} = N^{(4)} = 0$. Natural or unpolarized light can be looked upon as an incoherent sum of two beams polarized at right angles. We may, for convenience, take these two beams to be polarized in the $\mathbf{l}$ and $\mathbf{r}$ directions. If there is no phase coherence, the Stokes parameters add, and $N^{(2)}$, $N^{(3)}$, and $N^{(4)}$ are therefore all zero for natural light.

† The tilde denotes the optical properties of the particle. Properties of the entire medium consisting of small particles in space or suspended in a gas will be designated m, n, and n'.

7.2. Scattering functions

Let the components of the electric vector be $E_0^{(l)}$ and $E_0^{(r)}$ for the incident wave and $E^{(l)}$ and $E^{(r)}$ for the scattered spherical wave. For distances (d) large compared with the wavelength and the particle size $E^{(l)}$ and $E^{(r)}$ fall off as d^{-1}. Moreover the scattered wave will possess a phase difference $2\pi i(d-s)/\lambda$ from the incident wave (see Fig. 7.1). When writing down a formal relationship between incident and scattered amplitudes we may therefore take out a factor

$$\left(\frac{2\pi id}{\lambda_0}\right)^{-1} e^{-2\pi i(d-s)/\lambda_0}.$$

According to Maxwell's equations a linear relationship exists between scattered and incident amplitudes. We can write in matrix form:

$$\begin{pmatrix} E^{(l)} \\ E^{(r)} \end{pmatrix} = \left(\frac{2\pi id}{\lambda_0}\right)^{-1} e^{-2\pi i(d-s)/\lambda_0} \begin{pmatrix} S_2 & S_3 \\ S_4 & S_1 \end{pmatrix} \begin{pmatrix} E_0^{(l)} \\ E_0^{(r)} \end{pmatrix} \tag{7.7}$$

The matrix $\begin{pmatrix} S_2 & S_3 \\ S_4 & S_1 \end{pmatrix}$ is known as the *amplitude scattering matrix*; the unconventional numbering of elements follows established usage. For homogeneous, spherical scatterers (the only case which we will consider in detail) S_3 and S_4 vanish. Our discussion will therefore be limited to S_2 and S_1 which, for spherical scatterers, are functions of scattering angle (θ) only.

A particularly important role is played in the theory by the matrix coefficients for $\theta = 0$. Consider a thin slab of material of thickness ds containing N scattering centres per cm^3, and a wave incident from one side. The electric field on the far side can be found by compounding the incident wave with all scattered waves, taking due account of the phase. Consider a scalar wave, with a one-component scattering matrix. The resultant amplitude E_0' can be shown to be

$$E_0' = E_0(1 - \lambda_0^2 N \, ds \, S(0)/2\pi). \tag{7.8}$$

Looking at the problem from a different viewpoint we may now suppose the slab to have a complex refractive index $m = n - in'$. The amplitude can be written (7.2)

$$\frac{E_0'}{E_0} = e^{-2\pi ids(m-1)/\lambda_0} = 1 - \frac{2\pi i \, ds(m-1)}{\lambda_0}, \tag{7.9}$$

since ds is infinitesimally small. Comparing (7.8), (7.9), and (7.3), we find

$$n = 1 + 2\pi N(2\pi/\lambda_0)^{-3} \mathscr{I}\{S(0)\}, \tag{7.10}$$

$$n' = 2\pi N(2\pi/\lambda_0)^{-3} \mathscr{R}\{S(0)\}, \tag{7.11}$$

where $\mathscr{R}$ and $\mathscr{I}$ denote real and imaginary components. Relations similar to (7.10) and (7.11) can be shown to hold for vector waves, but, in the case of spherical particles, it is clear from symmetry considerations that for $\theta = 0$ both states of polarization will be similarly affected, and therefore $S_1(0) = S_2(0)$. (7.10) and (7.11) therefore also apply to spherical particles. Since intensity is proportional to the square of the amplitude, it involves an attenuating factor $e^{-4\pi n' ds/\lambda_0}$. From the definition of extinction coefficient (2.14) it follows that

$$e_v = \frac{4\pi n'}{\lambda_0}. \tag{7.12}$$

Since the theory of Chapter 2 is given in terms of the phase matrix, we must show how this quantity can be related to the amplitude scattering matrix. From (7.5) and (7.7) we can derive a linear relation between scattered and incident Poynting vectors. Extracting a factor $(\lambda_0/2\pi d)^2$ we may write

$$N^{(j)} = \left(\frac{\lambda_0}{2\pi d}\right)^2 F_{ij}(\theta, \phi) N_0^{(i)}, \tag{7.13}$$

where the sum convention for repeated indices is employed. Equation (7.13) can be compared with (2.28) and (2.29) if we bear in mind that (7.13) applies to a single particle only. Make the transformation

$$N \to I\,d\omega, \tag{7.14}$$

and take account of the spherical nature of the scattered wave by writing $d\omega_d = d^{-2}$. There results

$$\left(\frac{\lambda_0}{2\pi}\right)^2 F_{ij}(\theta, \phi) \equiv \frac{s_n}{4\pi} P_{ij}(\mathbf{s}, \mathbf{d}), \tag{7.15}$$

where s_n is the scattering coefficient per particle (see Appendix 3).

The scattering coefficient can be evaluated independently by applying the first law of thermodynamics to the intensity. Let us suppose that the incident light is unpolarized $\{(N_0^{(2)}, N_0^{(3)}, N_0^{(4)}) = 0\}$ and inquire about the scattered *intensity* $(N^{(1)})$. From (7.13)

$$N^{(1)} = \left(\frac{\lambda_0}{2\pi d}\right)^2 F_{11}(\theta, \phi) N_0^{(1)}.$$

Now integrate over the surface of a sphere of radius d to discover the total scattered component of the Poynting vector. The fraction scattered by a single particle can be equated to s_n, and hence

$$s_n = \int \left(\frac{\lambda_0}{2\pi}\right)^2 F_{11}(\theta, \phi)\,d\omega. \tag{7.16}$$

s_n has the dimensions of an area, and is conveniently made non-dimensional by dividing by the cross-sectional area of the particle (πa^2, where a = radius of the particle). The result is a *scattering efficiency factor*

$$Q_s = \int \frac{F_{11}(\theta, \phi)\, d\omega}{\pi x^2}, \qquad (7.17)$$

where $x = 2\pi a/\lambda_0$. Similarly, from (7.12) and (7.11), we can define an *extinction* efficiency factor

$$Q_e = \frac{e_n}{\pi a^2} = \frac{e_v}{N \pi a^2} = \frac{4}{x^2} \mathscr{R}\{S(0)\}. \qquad (7.18)$$

The difference

$$Q_a = Q_e - Q_s, \qquad (7.19)$$

is the *absorption efficiency factor*.

Now let

$$i_1 = |S_1|^2,$$
$$i_2 = |S_2|^2, \qquad (7.20)$$

and let δ be the phase difference between S_1 and S_2 (both complex in the general case). Substituting (7.7) in (7.5), with $S_3 = S_4 = 0$, we find

$$N^{(1)} = \left(\frac{\lambda_0}{2\pi d}\right)^2 \{\tfrac{1}{2}(i_1+i_2)N_0^{(1)} + \tfrac{1}{2}(i_2-i_1)N_0^{(2)}\},$$

$$N^{(2)} = \left(\frac{\lambda_0}{2\pi d}\right)^2 \{\tfrac{1}{2}(i_2-i_1)N_0^{(1)} + \tfrac{1}{2}(i_1+i_2)N_0^{(2)}\},$$

$$N^{(3)} = \frac{\sqrt{(i_1 i_2)}}{(\lambda_0/2\pi d)^2} (N_0^{(3)}\cos\delta - N_0^{(4)}\sin\delta), \qquad (7.21)$$

$$N^{(4)} = \frac{\sqrt{(i_1 i_2)}}{(\lambda_0/2\pi d)^2} (N_0^{(3)}\sin\delta + N_0^{(4)}\cos\delta).$$

The transformation matrix is therefore

$$F_{ij} = \begin{pmatrix} \tfrac{1}{2}(i_1+i_2) & \tfrac{1}{2}(i_2-i_1) & 0 & 0 \\ \tfrac{1}{2}(i_2-i_1) & \tfrac{1}{2}(i_1+i_2) & 0 & 0 \\ 0 & 0 & \sqrt{(i_1 i_2)}\cos\delta & -\sqrt{(i_1 i_2)}\sin\delta \\ 0 & 0 & \sqrt{(i_1 i_2)}\sin\delta & \sqrt{(i_1 i_2)}\cos\delta \end{pmatrix}. \qquad (7.22)$$

7.3. Rayleigh's solution for small particles

One confusing aspect of scattering theory is that a complete formal solution (Mie's theory) exists for homogeneous spheres, which sometimes seems to differ from approximate solutions applicable in certain limiting cases. Examples of such limiting cases are ray optics, Huygens's principle, Fresnel's theory of diffraction, and Rayleigh's theory of molecular scattering. Mie's theory contains all the diverse phenomena of classical optics, and is difficult to comprehend in simple terms. Thus despite its generality the complete theory will only be used when no simple limiting form is satisfactory. We will first consider the case of very small particles.

The phase change along the radius of a sphere is $2\pi a|\tilde{m}|/\lambda_0$. If this is small, i.e. if $|\tilde{m}x| \ll 1$ then the impressed electric field is constant throughout the particle and equal to $\mathbf{E}_0$. If the polarizability tensor is $\boldsymbol{\alpha}$, then the induced dipole moment is, by definition

$$\mathbf{M} = \boldsymbol{\alpha}\mathbf{E}_0. \tag{7.23}$$

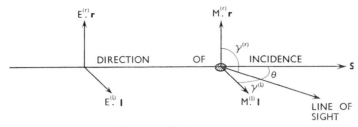

FIG. 7.2. Dipole scattering.
By definition, $\mathbf{l}$, $\mathbf{s}$, and the line of sight all lie in the plane of reference. From the definition of scattering angle, $\theta+\gamma^{(l)} = \tfrac{1}{2}\pi$ and $\gamma^{(r)} = \tfrac{1}{2}\pi$.

If we now add the emitted radiation from $\mathbf{M}$ to the incident plane wave we obtain a solution to Maxwell's equations consistent with the presence of a single scattering particle. The classical solution by Hertz for the wave propagated in space for $d \gg \lambda_0$ is

$$\mathbf{E} = \frac{1}{d\mathbf{c}^2}\frac{\partial^2\mathbf{M}}{\partial t^2}\sin\gamma$$

$$= (2\pi/\lambda_0)^2\frac{\mathbf{M}}{d}\sin\gamma, \tag{7.24}$$

where γ is the angle between $\mathbf{M}$ and the direction of observation, and we have replaced the operator $\partial/\partial t$ by $2\pi i\nu = 2\pi i\mathbf{c}/\lambda_0$.

Rayleigh's original theory assumed α to be a scalar, which is correct for a sphere. Then $\mathbf{M}$, $\mathbf{E}_0$, and $\mathbf{E}$ are parallel, or, in other words, cross terms in the scattering matrix are zero. Since $\mathbf{r}$ is, by definition, perpendicular to the plane of reference it follows that this component corresponds to viewing the induced dipole sideways on (see Fig. 7.2) and $\sin\gamma^{(r)} = 1$. For parallel polarization on the other hand

$$\sin\gamma^{(l)} = \cos\theta.$$

Combining (7.23) and (7.24)

$$E^{(r)} = \left(\frac{2\pi}{\lambda_0}\right)^2 \frac{\alpha}{d} E_0^{(r)},$$

$$E^{(l)} = \left(\frac{2\pi}{\lambda_0}\right)^2 \frac{\alpha}{d} E_0^{(l)} \cos\theta,$$

(7.25)

giving a scattering matrix

$$\mathbf{S} = i\left(\frac{2\pi}{\lambda_0}\right)^3 \alpha \begin{pmatrix} \cos\theta & 0 \\ 0 & 1 \end{pmatrix},$$

(7.26)

and a transformation matrix

$$F_{ij} = \left(\frac{2\pi}{\lambda_0}\right)^6 |\alpha| \begin{pmatrix} \frac{1}{2}(1+\cos^2\theta) & -\frac{1}{2}\sin^2\theta & 0 & 0 \\ -\frac{1}{2}\sin^2\theta & \frac{1}{2}(1+\cos^2\theta) & 0 & 0 \\ 0 & 0 & \cos\theta & 0 \\ 0 & 0 & 0 & \cos\theta \end{pmatrix}.$$

(7.27)

An alternative form of the transformation matrix, often quoted in the literature, is

$$F'_{ij} = \left(\frac{2\pi}{\lambda_0}\right)^6 |\alpha| \begin{pmatrix} \cos^2\theta & 0 & 0 & 0 \\ 0 & 1 & 0 & 0 \\ 0 & 0 & \cos\theta & 0 \\ 0 & 0 & 0 & \cos\theta \end{pmatrix}.$$

(7.28)

This is the appropriate form if $(I^{(l)}, I^{(r)}, U, V)$ are chosen in place of (I, Q, U, V) as Stokes parameters. The apparent advantage of simplicity is illusory, but it makes more obvious the complete perpendicular polarization of scattered radiation for $\theta = \frac{1}{2}\pi$.

In Fig. 7.3 the results are shown graphically on a scattering diagram. Incidence is from left to right. The length of a radius vector at the scattering angle θ from the central point gives the scattered intensity. The scattering depends upon the polarization of the incident radiation. Three possibilities are shown: for polarization in the r- and l-directions; and for natural or unpolarized radiation. Note that the scale is not the same for all cases. According to (7.27) the only relevant matrix element for natural radiation ($Q = U = V = 0$) is $(2\pi/\lambda_0)^6 |\alpha|^2 \frac{1}{2}(1+\cos^2\theta)$, while according to (7.28) the matrix elements for parallel and perpendicular polarization are $(2\pi/\lambda_0)^6 |\alpha|^2 \cos^2\theta$ and $(2\pi/\lambda_0)^6 |\alpha|^2$ respectively. The factor $\frac{1}{2}$ means that, for natural light, half of the intensity is to be attributed to each state of polarization.

It is frequently and erroneously implied that for computing scattered intensity in the atmosphere natural light can be assumed. Using a prime to denote this special condition:

$$s_n \frac{P'(\cos\theta)}{4\pi} = \left(\frac{\lambda_0}{2\pi}\right)^2 F(\theta) = \left(\frac{2\pi}{\lambda_0}\right)^4 |\alpha|^2 \tfrac{1}{2}(1+\cos^2\theta). \qquad (7.29)$$

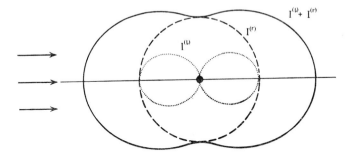

FIG. 7.3. Scattering diagram for small particles.
Full line $= I^{(r)}+I^{(l)}$; broken line $= I^{(r)}$; dotted line $= I^{(l)}$.

This is sufficient for primary scattering of solar radiation but not for secondary or higher order scattering. Multiplying (7.29) by $d\omega$ and integrating over all solid angles

$$s_n = \frac{1}{2}\left(\frac{2\pi}{\lambda_0}\right)^4 |\alpha|^2 \int (1+\cos^2\theta)\,d\omega$$
$$= \frac{8\pi}{3}\left(\frac{2\pi}{\lambda_0}\right)^4 |\alpha|^2. \qquad (7.30)$$

The scalar phase function for incident unpolarized radiation is therefore

$$P'(\cos\theta) = \tfrac{3}{4}(1+\cos^2\theta). \qquad (7.31)$$

According to (7.26) $\mathbf{S}$ is imaginary, and according to (7.11), m is real. Eliminating $S_1(0)$ or $S_2(0)$ between (7.10) and (7.26), we find

$$\alpha = \frac{m-1}{2\pi N}, \qquad (7.32)$$

and, for m close to unity, as for a gas

$$s_n = \frac{32\pi^3}{3\lambda_0^4}\frac{(m-1)^2}{N^2} \sim \frac{8\pi^3}{3\lambda_0^4}\frac{(m^2-1)^2}{N^2}. \qquad (7.33)$$

This is the usual form of Rayleigh's fourth-power law of scattering. Its important application is to molecular scattering. Only spherical top molecules have a scalar polarizability, however, and a small, but significant, correction term must therefore be included for non-spherical molecules. Let the tensor components of $\boldsymbol{\alpha}$, as referred to the three principal axes, be α_1, α_2, and α_3. We define

$$A = \tfrac{1}{15}(\alpha_1 \alpha_1^* + \alpha_2 \alpha_2^* + \alpha_3 \alpha_3^*),$$

$$B = \tfrac{1}{30}(\alpha_1 \alpha_2^* + \alpha_2 \alpha_3^* + \alpha_3 \alpha_1^* + \alpha_2 \alpha_1^* + \alpha_3 \alpha_2^* + \alpha_1 \alpha_3^*). \qquad (7.34)$$

Since in meteorological problems we are interested in the combined effect of many randomly oriented particles, it is permissible to take suitable averages of the transformation matrix. The matrix F'_{ij} (7.28) becomes

$$F'_{ij} = \left(\frac{2\pi}{\lambda_0}\right)^6 \begin{pmatrix} (2A+3B)\cos^2\theta + A - B & A - B & 0 & 0 \\ A - B & 3A + 2B & 0 & 0 \\ 0 & 0 & (2A+3B)\cos\theta & 0 \\ 0 & 0 & 0 & 5B\cos\theta \end{pmatrix}. $$

$$(7.35)$$

The most important difference between (7.28) and (7.35) lies in the polarization at a scattering angle $\theta = \tfrac{1}{2}\pi$. For incident natural light $(I_0^{(r)} = I_0^{(l)})$, (7.35) gives

$$\Delta = \frac{I^{(l)}(\theta = \tfrac{1}{2}\pi)}{I^{(r)}(\theta = \tfrac{1}{2}\pi)} = \frac{2A - 2B}{4A + B}, \qquad (7.36)$$

where Δ is known as the *depolarization factor*. It is a quantity which lends itself relatively easily to measurement in the laboratory.

For natural light (but not for other polarizations), the scattering coefficient and phase function can be specified in terms of Δ alone,

$$s_n = \frac{8\pi^3}{3\lambda_0^4} \frac{(m^2-1)^2}{N^2} \frac{3(2+\Delta)}{(6-7\Delta)}, \qquad (7.37)$$

$$P'(\cos\theta) = \frac{3}{2(2+\Delta)} \{1 + \Delta + (1-\Delta)\cos^2\theta\}. \qquad (7.38)$$

Depolarization factors of atmospheric constituents are as follow: $\Delta(O_2) = 0.054$; $\Delta(N_2) = 0.0305$; $\Delta(CO_2) = 0.0805$; $\Delta(A) = 0$; with an effective mean for dry air of $\Delta(\text{air}) = 0.0350$. At s.t.p. (7.37) and (7.38) give

$$s_n(\text{s.t.p.}) = 1.214 \times 10^{-37} \frac{(m^2-1)^2}{\lambda_0^4}, \qquad (7.39)$$

$$P'(\cos\theta) = 0.7629(1 + 0.932\cos^2\theta). \qquad (7.40)$$

Appendix 4 gives values of $m-1$ for dry air, $0.2\,\mu \leqslant \lambda \leqslant 20\,\mu$ for temperatures between $-30°$ C and $+30°$ C. The refractive index is dependent upon wavelength, varying from $m-1 = 3.4187 \times 10^{-4}$ at $\lambda_0 = 0.2\,\mu$ and $0°$ C to $m-1 = 2.8757 \times 10^{-4}$ at $\lambda_0 = 20\,\mu$ and $0°$ C. This gives rise to slight departures from a simple inverse fourth-power law for the wavelength dependence of the Rayleigh scattering coefficient. In Appendix 12 are given computations of s_n, s_m, s_v, and $P'(\cos\theta)$ for Rayleigh scattering at s.t.p.

In the following sections we will discuss the properties of single particles rather than an assembly of particles and the equations must be rewritten in terms of the optical properties of the particle itself. For a sphere the Lorentz relation states

$$\alpha = \frac{\tilde{m}^2-1}{\tilde{m}^2+2}\, a^3, \tag{7.41}$$

where a is the drop radius, and $\tilde{m}$ is complex and not close to unity. There results from (7.30)

$$Q_s = \frac{s_n}{\pi a^2} = \frac{8}{3}x^4 \left|\frac{\tilde{m}^2-1}{\tilde{m}^2+2}\right|. \tag{7.42}$$

From (7.18) and (7.26) we have

$$Q_e = \frac{4}{x^2}\,\mathcal{R}\left\{i\alpha\left(\frac{2\pi}{\lambda_0}\right)^3\right\}$$

$$= -4x\,\mathcal{I}\left\{\frac{\tilde{m}^2-1}{\tilde{m}^2+2}\right\}. \tag{7.43}$$

Equation (7.43) is paradoxical. It implies that if α is real, then $Q_e = 0$. But we expect $Q_e = Q_s$ if α is real and from (7.42) both must be non-zero. Radiation reaction is neglected in the Rayleigh theory and because of this the phase of the scattered wave is incorrect; as a result only the absorbed component is properly accounted for, and it can be shown that (7.43) gives Q_a—the absorption efficiency—and not the extinction efficiency.

Under certain conditions small elements of a large particle of arbitrary shape can be treated as independent Rayleigh scatterers whose amplitudes can be summed on the scattered wave front, provided due account is taken of the phase. van de Hulst calls this the Rayleigh–Gans approximation and shows the appropriate conditions to be

$$|\tilde{m}-1| \ll 1,$$

and

$$2x|\tilde{m}-1| \ll 1. \tag{7.44}$$

These conditions can be satisfied in the X-ray spectrum, but not in the visible spectrum for any circumstances encountered in the earth's atmosphere. While the approximation has no applications it is interesting to note some of the properties of Rayleigh–Gans scattering by a sphere, which are typical of large-particle scattering in general.

First, the scattering efficiency factor tends to (7.42) in the limit $x \ll 1$, as is expected. For $x \gg 1$, however, we have

$$Q_\mathrm{s} = 2x^2 |\tilde{m} - 1|. \tag{7.45}$$

In place of Rayleigh's fourth-power law, scattering now varies as λ_0^{-2}, i.e. the scattering is 'whiter'.

Secondly, owing to destructive interference, the phase matrix can have zeros for certain values of the scattering angle. The scattering diagram can therefore exhibit *lobes*.

Thirdly, for $\theta = 0$ the scattering is the same as for the Rayleigh case, but for all other scattering angles the intensity is diminished. The scattering therefore has a strongly forward component, and the scattering coefficient is less than predicted by the Rayleigh theory.

In subsequent sections we will find all three properties to be typical of scattering by large, dielectric spheres as opposed to Rayleigh's theory.

7.4. Large particles as $|\tilde{m}| \to 1$

The condition $x \gg 1$ does not by itself uniquely define an important asymptotic form of the exact electromagnetic theory. Clearly, there is some connexion with the laws of refraction and reflection of geometric optics, since these are usually effective for large surfaces. Also the concept of a *ray* becomes meaningful. If the particle is very much larger than the first few Fresnel zones, then we may look upon a ray as *localized* in these central zones. A complete wave-description then ceases to be necessary, although this does not mean that the wave character (i.e. the properties of amplitude and phase) can be neglected.

The concept of localization enables us to make a useful distinction between *diffraction* on the one hand, and *reflection* and *refraction* on the other. In the context of a complete electromagnetic theory the distinction is not meaningful, all phenomena being aspects of a single solution of Maxwell's equations. However, if we recognize the localization principle, we can refer to the *shadow area* of an obstacle and consider rays outside it to be *diffracted* while those inside are *reflected* or *refracted*. The distinction is, however, fraught with paradoxes. For example, for absorbing spheres with moderate values of x, Q_a can exceed unity. Now, since absorption can only occur if a quantum actually strikes the scattering particle, we have the situation whereby more quanta strike the particle than pass through the shadow area. A complete theory is, however, so complex that despite such difficulties we cannot pass over the opportunity of utilizing the large body of optical theory based on the localization principle, even if some risk of error is entailed.

We know from theory and observation that the *diffracted* light can be accounted for on the basis of Fresnel's theory. Narrow diffraction rings are observed, which become narrower as the particle size is increased. On a scattering diagram these diffraction rings would be portrayed as strong, forward lobes, but we may note that they will not necessarily be detectable experimentally. If the scattered light is viewed with an instrument of large angular dispersive power the diffraction rings will be resolved and distinguished from the incident beam. In this case a measurement of the extinction will include the diffraction term. If, however, we picture a small water drop close to, and in front of a pyrheliometer or other non-image forming device, it is obvious that the diffracted radiation cannot be distinguished from undisturbed incident radiation, and the Fresnel diffraction should not be included in an extinction computation.

A relationship between the two components is provided by Babinet's principle. Consider a beam of light falling on an opaque obstacle of shadow area G, and also consider the complementary experiment whereby all of the wave front is obscured *except* the area G. The amplitudes in the two cases must add up to the original, undisturbed wave front. It follows that the diffracted *wave amplitudes* are equal and opposite in the two cases; since intensity is proportional to the square of the amplitude, diffracted *intensities* are therefore the same in the two experiments. Now, in the case of the opening G in the otherwise opaque wave front, we know that all of the radiation is disturbed by diffraction to a greater or lesser extent. The same therefore applies to the opaque obstacle of area G, which must therefore have a diffraction cross-section G. Since it also has a cross-section G for interception of photons in the shadow zone, it follows that the total cross-section is $2G$. As has already been pointed out, the result of an experiment with limited angular resolution may be to measure G only, but the expected theoretical result is $Q_e \to 2$ as $x \to \infty$.

We will now state the familiar theory of Fresnel diffraction at an aperture (Fig. 7.4) in the formal language adopted in this chapter, in order that it can later be applied to the problem of a sphere with $(\tilde{m})$ close to unity. Fresnel diffraction involves no phase change and no change in the state of polarization. It follows that $S_1 = S_2$ and that both are real quantities. Moreover, foreseeing the result that diffraction effects are only important for small scattering angles, and considering distances large compared with the aperture, we may regard each element of the aperture G as an isotropic scatterer, contributing an intensity proportional only to its area. The only factor distinguishing

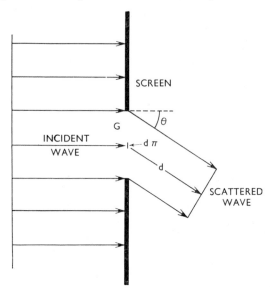

FIG. 7.4. Fresnel diffraction.

the scattering in different directions is that caused by destructive interference. Thus

$$\frac{S_{1,2}(\theta, \phi)}{S(0)} = \left| \frac{\int\limits_{G} e^{-2\pi i d/\lambda_0} \, d\pi}{\int\limits_{G} d\pi} \right| = D(\theta, \phi). \tag{7.46}$$

A little consideration shows that $D(\theta, \phi)$ is a function only of θ, ϕ, and G and not of d. For real $S(0)$ (7.18) can be written

$$Q_e = \frac{\lambda_0^2}{\pi G} S(0), \tag{7.47}$$

and we have already shown that $Q_e = 2$ for a large obstacle. By Babinet's principle, the *ratio* (7.46) is the same for obstacle and aperture; consequently for an obstacle

$$S_{1,2}(\theta, \phi) = \frac{2\pi G}{\lambda_0^2} D(\theta, \phi). \tag{7.48}$$

If G is circular, as for a sphere, (7.46) can be evaluated in terms of Bessel functions of integral order

$$S_{1,2}(\theta) = x^2 \frac{J_1(x \sin \theta)}{x \sin \theta}. \tag{7.49}$$

Let us briefly review the meaning of (7.49). It refers to the *diffraction component* of the amplitude scattering matrix (S_d) for a very large sphere, whether opaque or transparent. To obtain the complete matrix it must be added (taking account of phase) to the matrix resulting from rays in the shadow zone (S_s). The result has been obtained by a very loose argument based on Babinet's principle together with the conclusion that $Q_e = 2$ for an *opaque* aperture or screen. However, the localization principle permits us to distinguish between S_d and S_s, and therefore the result should be true for any kind of obstacle. For a translucent obstacle, however, we need not expect to find $Q_e = 2$; indeed in the limit $\tilde{m} = 1$ we have no screen at all and therefore Q_e must obviously be zero. For values of $\tilde{m}$ between 1 and ∞, some light will penetrate the sphere and interfere with the diffracted component, leading to a situation called *anomalous diffraction* by van de Hulst.

In one circumstance, namely $|\tilde{m}| \to 1$, the problem of anomalous diffraction can be treated very simply. Since we now have two parameters (viz. x and $\tilde{m}$) going to limits we must be careful to define their mutual behaviour. Let us assume $\tilde{m}$ to be real. The important parameter is then

$$\rho = 2x(\tilde{m}-1), \tag{7.50}$$

which is the phase lag of a ray passing through the centre of the sphere. We require that this parameter remain finite but it need no longer be small, as it was in the Rayleigh–Gans case. The importance of this condition is that the ray suffers a phase change as it passes through the sphere, but (since $\tilde{m}$ is small and real) is not reflected, refracted, or absorbed. The amplitude of a wave reaching $d\pi$ (Fig. 7.5) is the same as if the sphere were not there, but the phase is shifted by $\rho \sin \tau$.

From (7.47) we can write the diffraction component of the amplitude matrix in the form

$$S_d(0) = \frac{2\pi G}{\lambda_0^2} = \frac{2\pi}{\lambda_0^2} \int_G d\pi. \tag{7.51}$$

By Babinet's principle

$$S_s(0) = -\frac{2\pi}{\lambda_0^2} \int_G d\pi, \tag{7.52}$$

is the amplitude factor for the transmitted ray through a sphere with $\tilde{m}-1 = 0$ (i.e. an aperture). The sphere depicted in Fig. 7.5 introduces a phase shift $\rho \sin \tau$ for the ray passing through $d\pi$, and therefore we have to modify (7.52) to

$$S_s(0) = -\frac{2\pi}{\lambda_0^2} \int_G e^{-i\rho \sin \tau} \, d\pi. \tag{7.53}$$

Adding (7.52) and (7.53), setting $d\pi = 2G\cos\tau\sin\tau\,d\tau$, and performing the integration, we find

$$S(0) = x^2 K(i\rho),\tag{7.54}$$

where
$$K(w) = \frac{1}{2} + \frac{e^{-w}}{w} + \frac{e^{-w}-1}{w^2}.$$

Hence
$$Q_{\mathrm{e}} = \frac{4}{x^2}\mathscr{R}\{S(0)\} = 4\mathscr{R}\{K(i\rho)\}.\tag{7.55}$$

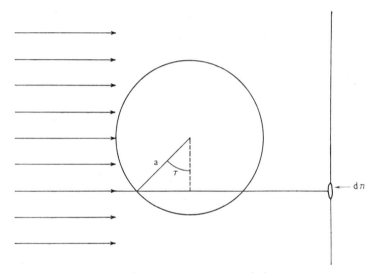

FIG. 7.5. Geometry of a sphere as $|\tilde{m}| \to 1$.

For real $\tilde{m}$ (7.55) is

$$Q_{\mathrm{e}} = 2 - \frac{4}{\rho}\sin\rho + \frac{4}{\rho^2}(1-\cos\rho).\tag{7.56}$$

Fig. 7.6 shows a comparison between this equation and exact computations for $\tilde{m} = 0\cdot8$, $0\cdot93$, $1\cdot33$, and $1\cdot5$ (see § 7.6). The agreement is remarkably good even though $\tilde{m}$ is far from unity. Maxima and minima appear at the predicted values of ρ, indicating that these features of the extinction curve are the result of interference of the transmitted and diffracted rays. If $\tilde{m} \neq 1$ curves have a fine structure, which is clearly not of fundamental importance in natural systems, for the inevitable mixture of particle sizes will blur over any such detail. There is a systematic increase in the heights of the maxima and minima as $\tilde{m}$ increases which is not given by the approximate theory. It has been proposed, as an empirical correction, that approximate extinction efficiencies should be multiplied by a factor

$$F(\tilde{m}, x) = \left(1 + \frac{3\tilde{m}+1}{2} x^{-2/3}\right). \tag{7.57}$$

We have already seen from the case of Rayleigh–Gans scattering how increase of particle size leads to more neutral scattering. Fig. 7.6 shows

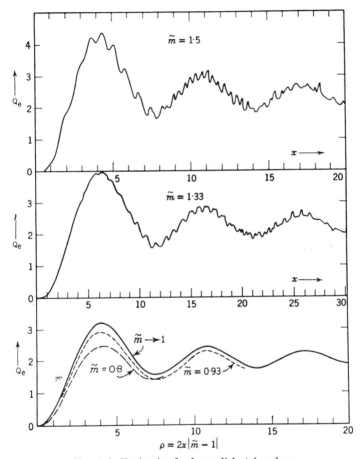

FIG. 7.6. Extinction by large dielectric spheres.

The full line in the lower panel follows the approximate equation (7.56). The other four curves are computed from Mie's theory (§ 7.6) for the values of $\tilde{m}$ shown. The accuracy of computation for $\tilde{m} = 0.8$ and 0.93 is somewhat lower than for $\tilde{m} = 1.5$ and $\tilde{m} = 1.33$. After van de Hulst (1957).

how this trend continues, and how Q_e hardly varies with λ for $\rho > 10$. Of considerable interest for atmospheric optical phenomena are the portions of the efficiency curve which slope downward to the right, particularly that section for $4.09 < \rho < 7.63$. In this region, close to an octave in extent, long wavelengths are scattered more strongly than short wavelengths. Such behaviour is occasionally observed for natural aerosols.

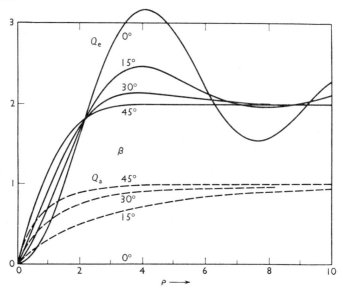

Fig. 7.7. Extinction and absorption efficiencies for a medium with refractive index $\tilde{m} = 1 + \epsilon - i\epsilon \tan \beta$ ($\epsilon \ll 1$).

After van de Hulst (1957).

For spheres with a complex refractive index, we may redefine ρ using only the real part of the refractive index and introduce the new variable

$$\tan \beta = \frac{\tilde{n}'}{\tilde{n} - 1}. \tag{7.58}$$

Thus $\beta = 0$ denotes a non-absorbing medium and $\beta = \frac{1}{4}\pi$ a *black* medium. Equation (7.56) becomes

$$Q_e = 2 - 4e^{-\rho \tan \beta} \frac{\cos \beta}{\rho} \sin(\rho - \beta) - 4e^{-\rho \tan \beta} \left(\frac{\cos \beta}{\rho}\right)^2 \cos(\rho - 2\beta) +$$

$$+ 4\left(\frac{\cos \beta}{\rho}\right)^2 \cos 2\beta. \tag{7.59}$$

Accepting the localization principle we know the ray paths through the drop, and can calculate the absorption from first principles. We find

$$Q_a = 2K(2\rho \tan \beta). \tag{7.60}$$

Equations (7.59) and (7.60) are illustrated in Fig. 7.7. We have seen that the maxima and minima on the curve $\beta = 0°$ owe their existence to interference of rays penetrating the sphere. As the absorption coefficient increases this interference decreases and the wave structure disappears while preserving the asymptotic value $Q_e = 2$ as $\rho \to \infty$.

Finally, we wish to know the angular distribution of the scattered light. This involves a straightforward application of the principles which we have already established. The integrals in (7.51) and (7.53) give the amplitude of the wave front in the area $d\tau$ of Fig. 7.5. The amplitude factor in a direction θ can now be found by vectorial addition as for (7.46).

Integrating over the wave front for a sphere with real refractive index, we find in a straightforward manner,

$$S_1(\theta) = S_2(\theta) = x^2 \int_0^{\frac{1}{2}\pi} (1 - e^{i\rho\sin\tau}) J_0(x\theta\cos\tau)\cos\tau\sin\tau\, d\tau$$

$$= x^2 A(\rho, x\theta). \tag{7.61}$$

The integral has to be computed by numerical quadrature, and the results are displayed in the form of an altitude chart in Fig. 7.8. In Fig. 7.9 a cross-section of this diagram, appropriate to water drops, is compared with exact computations, and also with the Fraunhofer diffraction pattern of an aperture.

The agreement between the approximation and exact computations shown in Fig. 7.9 is good qualitatively and quantitatively, except in one respect, namely that the approximation predicts no polarization effects at all. As regards the results displayed in Fig. 7.9, this is a good prediction, for the difference between S_1 and S_2 would not be visible on this scale. This is because the computations are restricted to scattering angles of 20° and less, but for larger scattering angles polarization is an important feature of the scattering pattern.

A solution has been proposed which reduces to that discussed in this section for $\theta = 0$, but which permits polarization effects at other angles. This is a mathematical approximation to the complete Mie theory (see § 7.6), which replaces the internal electromagnetic field in the droplet by the WKB approximation, and proceeds analytically from this point. The degree of polarization for large scattering angles is predicted with reasonable accuracy, but the absolute magnitude of the scattered intensities is less satisfactory.

7.5. Geometric optics

The elementary rainbow theory of Descartes and other early theories of droplet scattering are based upon the principles of *geometric optics*. This is an inclusive term for a number of assumptions of which the

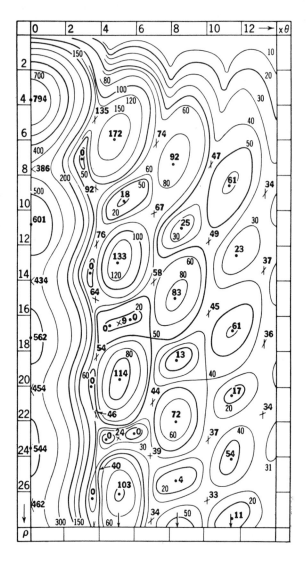

FIG. 7.8. Altitude chart of the amplitude function $|A| = |x^{-2}S|$.
After van de Hulst (1957).

localization principle is one, but not the only one. It is assumed in
addition that the disturbance is scalar and that the energy flux is
proportional to the density of rays. Thus, if the incident wave front is
represented by an equidistant set of rays, the scattering matrix is
proportional to the angular density of rays leaving the scattering
centre, the path of each having been traced by the laws of reflection

and refraction (an attenuating factor, dependent upon the path length, can be added if absorption has to be taken into account). As we have already discussed, the concept of a ray (i.e. the localization principle)

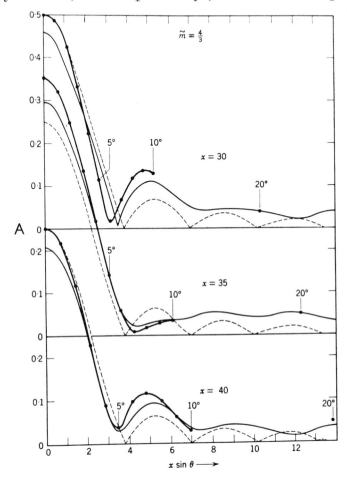

FIG. 7.9. Amplitude function $A = x^{-2}S$ for $\tilde{m} = 4/3$.

The points and heavy full lines are exact calculations. The less heavy full line is the approximation $\tilde{m} \to 1$. The broken line is the Fraunhofer diffraction pattern. Values of θ are given against some points. After van de Hulst (1957).

is sound, but rays can only be considered to be independent as long as two or more do not meet, when according to the assumptions of geometric optics, the intensity is infinite. Since the waves are now in a position to interfere the concept of a scalar ray is inadequate, and phase must be taken into account. At these singular points Fresnel's theory can be used to modify the simple concepts of geometric optics.

Applying the theory to the wave front before the focal point there results a system of diffraction rings. The focal point of principal interest is the final image, which is formed at infinity from an emergent parallel bundle of rays. Interest in intermediary foci is restricted to the phase changes which may take place; the diffraction blurring is too small to matter. It is a well-known result of Fresnel's theory that on passing

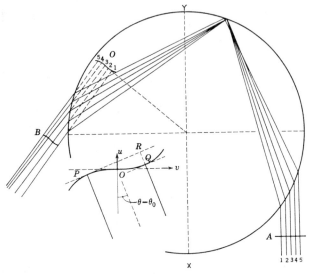

FIG. 7.10. Drawing in correct proportion of five equidistant rays which contribute to the first rainbow ($\tilde{m} = 4/3$).

Insert is a schematic drawing of the cubic wave front at O. After van de Hulst (1957).

through a focus there is a phase advance of $\frac{1}{2}\pi$. At a reflection there can be shown to be a phase shift of π, while refraction involves no phase shift. With these results we can determine the phase at any point on an emergent ray.

There are a variety of different situations in which the phase is important. Fig. 7.10 shows a familiar diagram illustrating the formation of the first rainbow. Because there is a ray whose deviation is a minimum (ray 3) there must be flanking rays which are parallel. These can interfere at the image formed in the eye or telescope. The nature of the image in the vicinity of this minimum deviation can only be found by constructing the wave front B and applying Fresnel's theory. Other less obvious singular points are when a ray intersects the line XY or runs parallel to it. Since the incident wave front is unlimited in size the net effect of wave fronts such as A is obtained by rotating the figure about the axis XY. A ray intersecting this axis will therefore intersect others, and a parallel ray ($\theta = 0$ or π) will form a parallel bundle.

It is scarcely necessary to add that geometric optics takes no account of diffraction of light which is not incident upon the scatterer. For large particles this diffracted component is all in a strongly forward direction. If it is not to interfere with transmitted rays these latter must be strongly deflected. The requirement is that ($\tilde{m}-1$) shall *not* be vanishingly small. Then we can consider the diffraction component separately and, if the particle is large enough, neglect its deviation entirely.

The most important application of geometric optics is to the problems of the *rainbow* and the *glory*, i.e. to scattering by a raindrop, which will typically have a value of $x = 5\times10^3$ in the visible spectrum. In the next section we will describe an exact solution which can be used conveniently for x up to about 30. The convergence of the series involved would be far too slow for the raindrop problem, even for the fastest electronic computers. In addition, the approximation $|\tilde{m}| \to 1$ is irrelevant for the present purpose, for it gives no solution for the phenomena under consideration. Only geometric optics, with all the associated doubts as to its validity, can be used flexibly, and it is of some interest to examine briefly the extensive work in this field. Some reviews of this popular topic are given in the Bibliography.

Except for the singular directions at $\theta = 0$ and π and the rainbow angles (at minimum or maximum deviation) geometric optics should be satisfactory without any corrections. Ignoring diffraction effects entirely, we obtain $Q_s = 1$ and a phase function which does not depend upon particle size. This phase function was first computed by Wiener in 1909 and is named after him.

A number of different rays have to be distinguished, identified by the ordinal numbers $p = 0, 1, 2, 3,\dots$. $p = 0$ is the externally reflected ray, which does not enter the drop. $p = 1$ is the ray which leaves the drop after two refractions, but no reflections. $p = 2$ is the ray which suffers one internal reflection, $p = 3$ suffers two and so on. In terms of the angles θ and τ (defined in Fig. 7.5) it is straightforward to show that

$$N^{(r)}(p,\tau) = \frac{a^2}{d^2} N_0^{(r)} \epsilon_r^2 \frac{\sin\tau\cos\tau}{\sin\theta\,|d\theta'/d\tau|}, \tag{7.62}$$

where

$$\epsilon_r = q_r \text{ for } p = 0,$$

$$= (1-q_r^2)(-q_r)^{p-1} \text{ for } p = 1, 2, 3,$$

and other symbols are defined below.

In (7.62) the factor d^{-2} accounts for the divergence of light leaving the raindrop. The factor ϵ_r involving the Fresnel reflection coefficient q_r accounts for the loss of light at the reflections and refractions which take place along the path. The index r refers to perpendicular polarization; the same expression with appropriate indices holds for the parallel case. Absorption in the drop has been neglected. Finally, the total deflexion θ' is defined to take account of the fact that the outgoing ray may have rotated through one or more multiples of 2π by means of internal reflections.

It is convenient to write

$$g^{(r)} = 4\epsilon_r^2 \frac{\sin\tau \cos\tau}{\sin\theta |d\theta'/d\tau|}, \qquad (7.63)$$

where $g^{(r)}$ is the gain relative to isotropic scattering. It is a well-known property of a spherical drop that, for $p \geqslant 2$, $d\theta'/d\tau = 0$ at certain angles, known as the *rainbow angles*. Thus, geometric optics gives $g^{(r)} = \infty$ at the rainbow, but if the gain factor is averaged over a small but finite angle, the result is finite. Table 7.1 shows rainbow angles

TABLE 7.1

Rainbow angles and degree of polarization

After Shifrin and Rabinovich (1957)

p	$\tilde{m}$	1·3200	1·3250	1·3300	1·3350	1·3400	1·3450
2	θ	135° 59′ 36″	136° 44′ 40″	137° 29′ 00″	138° 12′ 24″	138° 55′ 46″	139° 38′ 08″
	P	0·8995	0·9087	0·9185	0·9277	0·9364	0·9446
3	θ	132° 34′ 38″	131° 13′ 44″	129° 53′ 58″	128° 34′ 12″	127° 17′ 30″	126° 00′ 48″
	P	0·7832	0·7929	0·8017	0·8108	0·8193	0·8277
4	θ	46° 35′ 58″	44° 42′ 34″	42° 50′ 40″	41° 00′ 12″	39° 09′ 48″	37° 23′ 28″
	P	0·7573	0·7658	0·7752	0·7838	0·7922	0·8004

and degree of polarization for the first, second, and third rainbows ($p = 2$, 3, and 4 respectively, $p = 1$ gives no rainbow). The first rainbow has a minimum deviation, while the second and third rainbows are maxima. The dispersion of $\tilde{m}$ (see Appendix 13) gives different edges for each colour, leading to the familiar rainbow effect. The colour sequence in the second and third rainbows reverses that of the first. An extract from an extensive computation of gain factors is shown in Table 7.2. Also shown in Table 7.2 are the degree of polarization

$$P_N = \frac{g^{(r)} - g^{(l)}}{g^{(r)} + g^{(l)}}$$

TABLE 7.2

Gain factors and polarization for water drops ($n = 1 \cdot 3350$)

Extracted from the tables of Shifrin and Rabinovich (1957)

$\theta°$	$p = 0$		$p = 1$		$p = 2$		$p = 3$		$p = 4$		Total	
	$g^{(r)}$	$g^{(l)}$	$g^{(r)}$	$g^{(l)}$	$g^{(r)}$	$g^{(l)}$	$g^{(r)}$	$g^{(l)}$	$g^{(r)}$	$g^{(l)}$	$2g_N$	P_N
0	1·0000	1·0000	15·2339	15·2339	—	—	0·0003	0·0003	(∞)	(∞)	32·4684	0·0000
2	0·9246	0·8696	15·1147	15·0947	—	—	0·0003	0·0003	0·0102	—	32·0144	−0·00265
5	0·8211	0·7034	14·5377	14·6077	—	—	0·0003	0·0003	0·0052	—	30·6757	−0·00172
10	0·6746	0·4931	12·7009	12·9462	—	—	0·0003	0·0003	0·0028	—	26·8182	−0·00227
20	0·4582	0·2389	7·9051	8·4030	—	—	0·0003	0·0002	0·0026	—	17·0083	−0·0162
30	0·3152	0·1108	4·1517	4·7692	—	—	0·0003	0·0002	0·0055	—	9·3529	−0·0436
40	0·2208	0·0473	1·9692	2·5256	—	—	0·0004	0·0002	0·0933	0·0184	4·8752	−0·0631
50	0·1580	0·0171	0·8312	1·2320	—	—	0·0004	0·0002	—	—	2·2389	−0·1606
60	0·1158	0·0043	0·2787	0·4955	—	—	0·0005	0·0002	—	—	0·8950	−0·1173
70	0·0873	0·0002	0·0527	0·1169	—	—	0·0007	0·0002	—	—	0·2579	0·0911
80	0·0674	0·0005	0·0007	0·0016	—	—	0·0010	0·0001	—	—	0·0713	0·9383
90	0·0534	0·0028	—	—	—	—	0·0016	0·0001	—	—	0·0578	0·9031
100	0·0435	0·0059	—	—	—	—	0·0025	—	—	—	0·0519	0·7726
110	0·0364	0·0091	—	—	—	—	0·0053	—	—	—	0·0508	0·6417
120	0·0312	0·0121	—	—	—	—	0·0310	0·0000	—	—	0·0935	0·3305
130	0·0275	0·0146	—	—	—	—	—	0·0192	—	—	0·0421	0·3064
140	0·0247	0·0168	—	—	1·1239	0·1545	—	—	—	—	1·3199	0·7404
150	0·0228	0·0185	—	—	0·2983	0·2021	—	—	—	—	0·5417	0·1855
160	0·0217	0·0197	—	—	0·1133	0·0995	—	—	—	—	0·2542	0·0622
170	0·0208	0·0204	—	—	0·0839	0·0768	—	—	—	—	0·2019	0·0371
180	0·0206	0·0206	—	—	0·0795	0·0795	—	—	—	—	0·2002	0·0000

and gain factor (g_N) for incident natural (unpolarized) light. Calculations are made at integral numbers of degrees and not at the actual rainbow angles; therefore there are no infinities except for $\theta = 0, p = 4$.

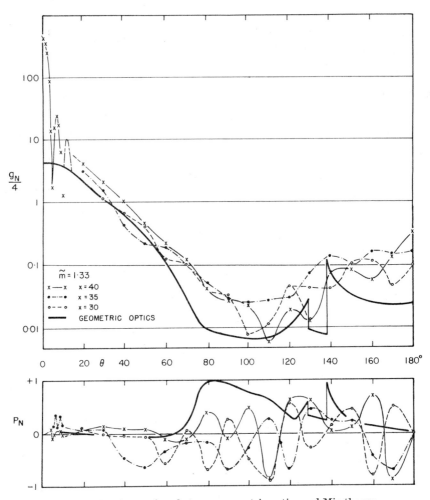

FIG. 7.11. Comparison between geometric optics and Mie theory. Note the discontinuities corresponding to the first and second rainbow (137° 29′ and 129° 54′). After Volz (1961).

Averaged over a degree, the contribution from this intensity is negligible and it is not included in g_N.

In Fig. 7.11 are shown somewhat similar data to those in Table 7.2, compared with exact computations for $x = 30$, 35, and 40. There is a general similarity between the scattering phase functions, which should improve with larger values of x. The degree of polarization, on the other hand, does not show any agreement.

We now have to consider the singularities in the light of Fresnel's theory of diffraction. At $\theta = 0$ we are concerned with the diffraction rings for rays passing close to, but not through, the raindrop. The theory for large x has been given in § 7.4 and the result for a sphere is given by (7.49). The Bessel function has zeros at $x \sin \theta = 3\cdot83$, $7\cdot02$, $13\cdot32$, etc., so that the diffraction rings close in as x increases. For $x = 10^4$ (a raindrop) the first zero is at $\theta = 1\cdot4'$, and no rings will be seen around a source as large as the sun. For mists of very small droplets the rings may sometimes be sufficiently far apart to be seen round either the sun or moon.

Near the rainbow angles a really satisfactory theory is lacking. Attempts to work from the full electromagnetic theory require various degrees of approximation (see van de Hulst, 1957, for details) and show little improvement over the Airy theory, which applies Fresnel's theory to the outgoing wave front assuming it to be a cubic function.

Airy's theory is illustrated in Fig. 7.10. The insert shows details of the virtual wave front at O, obtained by considering path lengths along the various rays. The fundamental assumption is that the equation of the wave front is approximately

$$u = \frac{hv^3}{3a^2},\qquad(7.64)$$

where the constant h is $4\cdot89$ for the first rainbow and $27\cdot86$ for the second, if $\tilde{m} = 1\cdot3333$. Applying Fresnel's principle at an angle $(\theta - \theta_0)$ to the rainbow direction gives an amplitude factor proportional to

$$\int_{-\infty}^{+\infty} e^{-2\pi iv(\theta-\theta_0)/\lambda + 2\pi ihv^3/3a^2\lambda}\, dv,$$

which can be written in the form of the Airy integral

$$f(Z) = \int_0^\infty \cos \tfrac{1}{2}\pi(Zt - t^3)\, dt,\qquad(7.65)$$

$$Z = (12/h\pi^2)^{1/2}x^{2/3}(\theta - \theta_0),$$

where θ_0 is the rainbow angle. The integral has been computed numerically and its square (which is proportional to the scattered intensity) is shown in Fig. 7.12. It has a series of maxima which, under ideal conditions, can be seen inside the first rainbow and outside the second.

Rain has a wide spectrum of drop sizes, which smooths out the diffraction pattern. Fig. 7.12 (*b*) illustrates three typical cases. The finite angular diameter of the sun leads to further averaging, and as a result it is rare that the diffraction rings (*supernumerary bows*) are seen in nature.

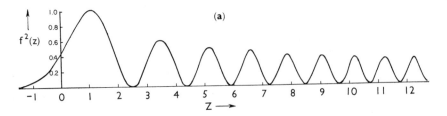

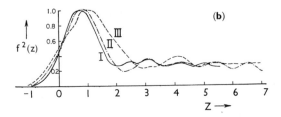

Fig. 7.12. The Airy rainbow integral for $\tilde{m} = 4/3$.

(*a*) $f^2(Z)$ from (7.65). (*b*) Averaged over naturally occurring drop-size distributions. I, light shower rain; II, warm front rain; III, drizzle. Z is computed on the assumption that $a = 0\cdot5$ mm, since the maximum contribution to the rainbow intensity comes from drops of approximately this size. After Volz (1961).

The glory is the least satisfactorily explained of all the raindrop diffraction phenomena. From Table 7.2 at $\theta = \pi$ we note there are zeros for $p = 1$, 3, and 4 and finite values for $p = 0$ and 2. Thus, the theory does not predict an infinity for backward scattering and it should therefore be correct; yet the existence of rings at $\theta = \pi$ is a well-known observed phenomenon. No satisfactory explanation of this paradox has been given, although various possibilities have been discussed by van de Hulst.

7.6. The Mie theory

The Mie theory is a complete, formal theory of the scattering of a plane wave by a dielectric sphere. Its results can be given in the form of a convergent series, but the convergence is very slow for large values of x. However, with the advent of electronic digital computers, this is no longer a serious defect and a large number of computations of scattering functions have now been made for real $\tilde{m}$, and a few for $\tilde{m}$ complex.

The derivation of the solution is a straightforward application of classical electrodynamics, and here we will only quote the results (see bibliography for further references). The two amplitude functions have the symmetrical form

$$S_1(\theta) = \sum_{n=1}^{\infty} \frac{2n+1}{n(n+1)} \{a_n \pi_n(\cos \theta) + b_n \tau_n(\cos \theta)\},$$

$$S_2(\theta) = \sum_{n=1}^{\infty} \frac{2n+1}{n(n+1)} \{b_n \pi_n(\cos \theta) + a_n \tau_n(\cos \theta)\}, \qquad (7.66)$$

where

$$\pi_n(\cos \theta) = \frac{1}{\sin \theta} P_n^1(\cos \theta),$$

$$\tau_n(\cos \theta) = \frac{d}{d\theta} P_n^1(\cos \theta),$$

P_n^1 is an associated Legendre polynomial, and the coefficients a_n and b_n are given by

$$a_n = \frac{\psi_n'(\tilde{m}x)\psi_n(x) - \tilde{m}\psi_n(\tilde{m}x)\psi_n'(x)}{\psi_n'(\tilde{m}x)\zeta_n(x) - \tilde{m}\psi_n(\tilde{m}x)\zeta_n'(x)},$$

$$b_n = \frac{\tilde{m}\psi_n'(\tilde{m}x)\psi_n(x) - \psi_n(\tilde{m}x)\psi_n'(x)}{\tilde{m}\psi_n'(\tilde{m}x)\zeta_n(x) - \psi_n(\tilde{m}x)\zeta_n'(x)},$$

where

$$\psi_n(x) = (\tfrac{1}{2}\pi x)^{1/2} J_{n+1/2}(x),$$

$$\zeta_n(x) = (\tfrac{1}{2}\pi x)^{1/2} H_{n+1/2}^{(2)}(x),$$

are Riccati–Bessel functions and $J_{n+1/2}$ and $H_{n+1/2}^{(2)}$ are spherical Bessel functions. The Bessel functions have zeros which increase in number with the size of the argument with the result that S_1 and S_2 can change rapidly for very small variations of x.

If extinction and scattering efficiencies alone are required these can be obtained from the expressions

$$Q_e = \frac{2}{x^2} \sum_{n=1}^{\infty} (2n+1)(a_n + b_n),$$

$$Q_s = \frac{2}{x^2} \sum_{n=1}^{\infty} (2n+1)(|a_n|^2 + |b_n|^2). \qquad (7.67)$$

van de Hulst has attempted to list all the calculations prior to 1957 in his monograph. Further computations have appeared since that date, and some of particular importance for water droplets in the visible and infra-red spectrum are noted in the Bibliography. Some results for real refractive indices, close to that of water, have already been given in Figs. 7.6 and 7.9. Further results for all scattering angles and

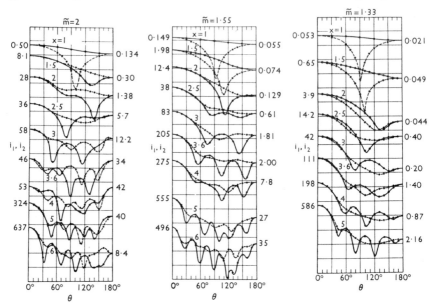

FIG. 7.13. Scattering diagrams from Mie theory.

The solid curves are for $i_1 = |S_1|^2$ and the broken curves for $i_2 = |S_2|^2$. The vertical scale is logarithmic, with 1 div = a factor 10. Values of i_1 and i_2 at $0°$ and $180°$ (where they are equal) are given beside the diagram. The results are taken from the tables of Lowan (1949). After van de Hulst (1957).

$\tilde{m} = 1·33$, $1·5$, and $2·0$ are shown in Fig. 7.13. These do not lend themselves to any simple discussion. The curves for $x = 1$ show some of the features of Rayleigh scattering with large, positive polarization (i.e. $i_1 > i_2$) near $90°$. For $x > 2$, however, both positive and negative polarizations occur, with changes from one to another which occur more frequently as x increases. For large particles the forward lobe often has a net negative polarization.

Some results for complex refractive indices are shown in Figs. 7.14 and 7.15. Fig. 7.14 shows almost no qualitative features which are not given by the approximate theory as $|\tilde{m}| \to 1$; in fact the quantitative agreement between comparable curves is also good if the empirical factor (7.57) is applied. The extinction maximum near $\rho = 4$ is damped out as the absorption increases and the asymptotic limits as $\rho \to \infty$ ($Q_e \to 2$, $Q_a \to 1$) are approached more rapidly. A weak maximum near $\rho = 1$ appears when absorption is large; this feature does not appear

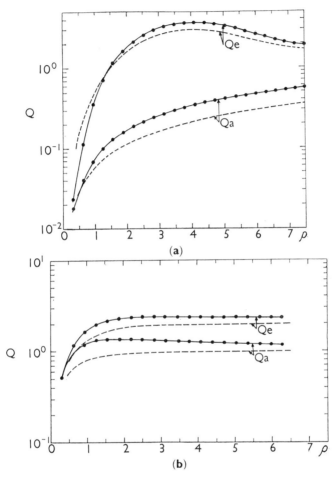

FIG. 7.14. Q_e and Q_a for $\tilde{n} = 1\cdot315$ according to the Mie theory (points and full line) and the approximate theory for $|\tilde{m}| \to 1$ (broken line).

(a) $\tilde{n}' = 0\cdot0143i$, (b) $\tilde{n}' = 0\cdot4298i$. After Deirmendjian *et al.* (1961).

in the approximate theory. Comparison with the curves in Fig. 7.6 shows that the fine structure is completely eliminated even for small absorptions.

The curves in Fig. 7.15 illustrate the large changes which result from a variation in the complex component of the refractive index.

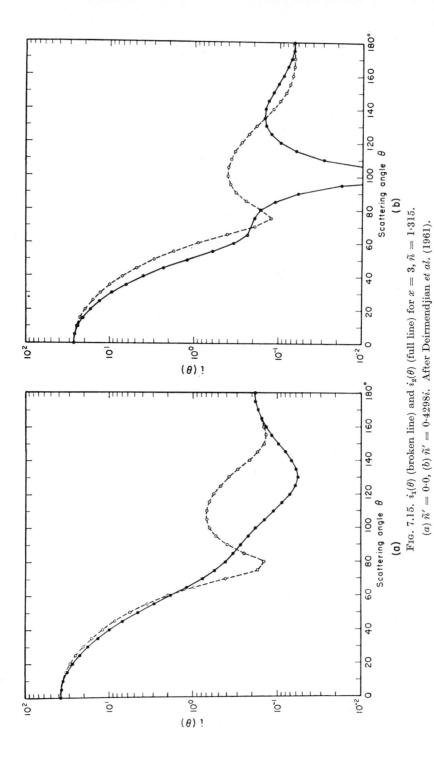

Fig. 7.15. $i_1(\theta)$ (broken line) and $i_2(\theta)$ (full line) for $x = 3$, $\tilde{n} = 1{\cdot}315$.

(a) $\tilde{n}' = 0{\cdot}0$, (b) $\tilde{n}' = 0{\cdot}4298i$. After Deirmendjian et al. (1961).

BIBLIOGRAPHY

7.1. The problem in terms of the electromagnetic theory
7.2. Scattering functions
7.3. Rayleigh's solution for small particles
7.4. Large particles as $|\tilde{m}| \rightarrow 1$
7.5. Geometric optics
7.6. The Mie theory

These sections follow closely the excellent text by

VAN DE HULST, H. C., 1957, *Light scattering by small particles*. New York: Wiley,

which contains a wealth of references. Two textbooks on general electromagnetic theory are by

STRATTON, J. A., 1941, *Electromagnetic theory*. New York: McGraw-Hill; and
ABRAHAM, M., and BECKER, R., 1932, *The Classical theory of electricity and magnetism*. London: Blackie.

A series of papers by Penndorf deals with numerical values of the scattering efficiency for a number of special cases:

PENNDORF, R. B., 1957a, 'Tables of the refractive index for standard air and the Rayleigh scattering coefficient for the spectral region between $0 \cdot 2 \, \mu$ and $20 \cdot 0 \, \mu$ and their application to atmospheric optics', *J. Opt. Soc. Am.* **47**, p. 176;
—— 1957b, 'Total Mie scattering coefficients for spherical particles of refractive index $n \sim 1 \cdot 0$', ibid., p. 603; and
—— 1957c, 'New tables of total Mie scattering coefficients for spherical particles of real refractive indexes $(1 \cdot 33 \leqslant n \leqslant 1 \cdot 50)$', ibid., p. 1010.

The empirical correction factor in (7.57) and a description of the WKB theory discussed in § 7.4 (this theory was developed by D. S. SAXON but remains unpublished) are given by

DEIRMENDJIAN, D., 1957, 'Theory of the solar aureole. Part I. Scattering and radiative transfer', *Annales de Géophysique* **13**, p. 286.

The problem of scattering by raindrops has recently been reviewed, with a copious bibliography by

VOLZ, F., 1961, 'Der Regenbogen', *Handbuch der Geophysik*, Band viii, p. 943. Berlin: Borntraeger; and
MEYER, R., 1956, 'Kränze, Glorien and verwandte Erscheinungen', ibid., p. 898. Berlin: Borntraeger.

Two earlier books of great interest are

MINNAERT, M., 1940, *Light and colour in the open air*. London: Bell; and
PERNTER, J. M., and EXNER, F. M., 1922, *Meteorologische Optik*. Vienna: Braumüller.

Tables 7.1 and 7.2 follow

SHIFRIN, K. S., and RABINOVICH, YU. I., 1957, 'The spectral indicatrices of large water drops and the spectral polarisation of rainbows', *Bull. Acad. Sci. U.S.S.R. Geophys.* **12**, p. 173.

The results shown in Fig. 7.14 are taken from the tables of

LOWAN, A. N., 1949, *Tables of scattering functions for spherical particles*, Nat. Bur. of Standards. Appl. Math. Ser. **4.** Washington, D.C.: Gov. Printing Office.

Scattering diagrams for complex refractive indices of water drops in the infra-red spectrum have been computed by

DEIRMENDJIAN, D., CLASEN, R., and VIEZEE, W., 1961, 'Mie scattering with complex index of refraction', *J. Opt. Soc. Am.* **51,** p. 620;

STEVENS, J. J., and GERHARDT, J. R., 1961, 'Absorption cross-sections of water drops for infrared radiation', *J. Met.* **18,** p. 818;

—— 1961, 'Spectrally averaged total attenuation, scattering and absorption cross-sections for infrared radiation', ibid. **18,** p. 822; and

HERMAN, B. M., 1962, 'Infra-red absorption, scattering, and total attenuation cross-sections for water spheres', *Quart. J. R. Met. Soc.* **88,** p. 143.

8

ATMOSPHERES IN RADIATIVE EQUILIBRIUM

8.1. Introduction

THE problem of radiative equilibrium in a planetary atmosphere is the simplest problem relevant to our studies. If the radiative equilibrium state happens to be stable to all disturbances, then motions cannot develop, and radiative equilibrium would be the expected condition of the atmosphere. Comparison between observed and equilibrium states of the earth's atmosphere can therefore give important information about the genesis of atmospheric motions.

In this and the next chapter we will examine this problem making considerable use of simple heuristic models. The Eddington approximation will generally be employed; while it is not precise it omits no essential physical principles, provided that the medium is stratified. For a semi-infinite grey atmosphere, a comparison with exact computations can be made. Maximum errors occur at $\tau = 0$, where the estimated temperature is 3·5 per cent too high, e.g. of the order of 9° K for terrestrial conditions. A comparison for a finite atmosphere, corresponding more closely to a planetary atmosphere, is shown in Fig. 8.1. For $\tau^* \ll 0.1$ or $\tau^* > 10$ the Eddington approximation gives good accuracy. For $\tau^* \simeq 1$ the maximum errors are similar to those at the surface of a semi-infinite atmosphere.

8.2. Non-grey atmospheres

Formal techniques for determining approximately the radiative equilibrium in non-grey semi-infinite atmospheres (stellar conditions) have been discussed in the astrophysical literature (see Bibliography). These solutions are not generally applicable to planetary atmospheres of finite depth because an important qualitative difference exists between finite and semi-infinite atmospheres for small absorption coefficients. For a finite atmosphere, if $\tau_\nu^* \ll 1$, the presence of the atmosphere can be neglected and the flux to space is simply that from the lower boundary; for an infinite atmosphere, whose temperature will normally increase with depth, increasing transparency means that higher temperatures can communicate with outer space, and the radiation flux will increase continuously. The role of translucent spectral regions therefore differs importantly in the two cases.

In this section we will consider three formal solutions of non-grey problems, and one numerical solution based on radiation chart procedures. Comparison with calculations using grey absorption illustrates the usefulness and errors of the grey model.

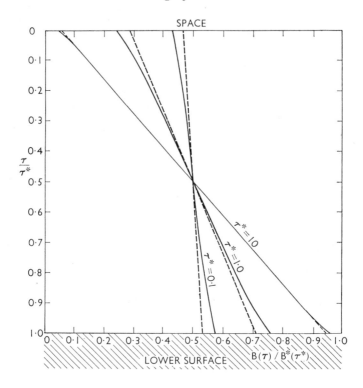

FIG. 8.1. Source functions in a grey, finite atmosphere (optical depth τ^*).

$B^*(\tau^*)$ = source function of lower boundary. - - - - - - - Eddington approximation; ——— Equation (2.143). For $\tau^* \ll 1$ both solutions give $B(\tau)/B^*(\tau^*) = 0.5$.

8.2.1. The use of Rosseland's and Chandrasekhar's means

An attempt has been made to establish the accuracy of the Rosseland and Chandrasekhar means (2.134) and (2.135) for the Elsasser band model (§ 4.4.1). The procedure is to compute the temperature distribution for a grey model whose coefficient is given by a definable average over the spectrum, and to evaluate the flux for this temperature distribution using the Elsasser transmission function in a numerical procedure equivalent to those discussed in § 6.4. If the flux computed in this way does not vary with optical depth, then the grey approximation is demonstrably good.

For a narrow band $(\overline{dB_\nu/d\theta})$ can be assumed independent of ν and taken outside the integrals in both the numerator and denominator of (2.134), which defines the Rosseland mean. Substituting the absorption coefficient for an Elsasser band with a Lorentz line shape (4.63) into the integral in the numerator, we find

$$(a\overline{k}_R)^{-1} = \frac{1}{2\pi uy} \int_{-1/2}^{+1/2} \frac{\cosh 2\pi y - \cos 2\pi x}{\sinh 2\pi y}\, dx$$

$$= \frac{1}{2\pi uy \tanh 2\pi y},$$

or
$$\bar{\tau}_R = 2\pi yu \tanh 2\pi y. \tag{8.1}$$

$\bar{\tau}_R$ is the equivalent optical depth of a pressure-independent, grey-absorbing atmosphere, which can be substituted in Yamamoto's solution for a finite atmosphere (2.143) to yield the source function as a function of optical depth (see curves marked R in Fig. 8.2).

Evaluation of the Chandrasekhar mean involves a knowledge of the flux as a function of frequency (2.135), which is not known *a priori*. Yamamoto assumes that (2.141) is valid, and that

$$\frac{F_\nu}{\pi} = \frac{4}{3} \frac{B_\nu^*}{\tau_\nu^* + 2Q(\tau_\nu^*)}, \tag{8.2}$$

where B_ν^* is the emission of the lower boundary and $Q(\tau_\nu^*)$ is a known and slowly varying function of τ_ν^* (see end of § 2.4.4). This result has only been proved for grey radiation. For the non-grey case (8.2) will not be correct since there will generally not be monochromatic radiative equilibrium in each frequency range. However, the equation gives a well-defined mean value regardless of its validity and its usefulness can be assessed by the computational procedure described above; while the resulting mean cannot strictly be called the Chandrasekhar mean, the approach is valid.

There results for the 'Chandrasekhar mean'

$$\bar{\tau}_C = \frac{2\pi yu}{\sqrt{\{(2\pi yu/Q)\coth 2\pi y + (\pi yu/Q)^2 + 1\}} - Q/2yu}. \tag{8.3}$$

Source functions based on this optical depth are also given in Fig. 8.2. Analogous expressions, based on (6.4), can be used for Lorentz lines whose width varies with the pressure in an atmosphere for which the product of the mixing ratio and line intensity is constant, and some results are illustrated in Fig. 8.2 (pressure-dependent case).

The accuracy of these approximations is illustrated by the figures in Table 8.1. The quantity computed is $100\{F(0)-F(\frac{1}{2}\tau^*)\}/F(0)$, which should be small if the approximation is good.

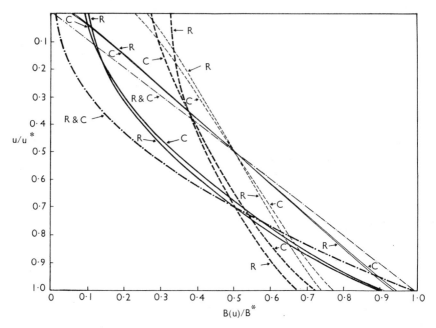

FIG. 8.2. Calculations of the source function in a finite atmosphere with a regular absorption band.

After Yamamoto (1955). R = Rosseland; C = Chandrasekhar.

$u^* =$	3·41	34·1	341
Pressure dependent	– – –	———	–·–
Pressure independent	– – –	———	–·–

In the pressure-dependent case (see text) band parameters are evaluated on the basis of the pressure at the lower surface ($y = 0·08$ for all examples).

The results in Table 8.1 suggest that, used with discretion, a grey approximation to a real atmosphere can give results of useful accuracy. No computations have been made for weak lines however ($u^* \ll 1$), and the possibility of larger errors is not eliminated. It should, moreover, be noted that serious errors in a narrow layer at the two boundaries could exist without having much influence on the computed fluxes, so that the tests used by Yamamoto do not guarantee that the solution is satisfactory in all respects. To examine this point in more detail we

TABLE 8.1

Test of constancy of flux in a non-grey model atmosphere

After Yamamoto (1955)

u^*	Mean value	$100 \times \{F(0) - F(\frac{1}{2}\tau^*)\}/F(0)$	
		Pressure dependent	Pressure independent
341	Rosseland	$+7.5$	$+7.2$
	Chandrasekhar	$+5.8$	$+2.5$
34·1	Rosseland	$+1.7$	-0.5
	Chandrasekhar	$+2.9$	-3.5
3·41	Rosseland	-6.0	-0.4
	Chandrasekhar	-6.0	-1.0

will consider in the next sections two attempts at a complete solution of the pressure-dependent case for Lorentz lines.

8.2.2. *A single Lorentz line*

We can write a flux divergence equation, such as (6.93), based on the absorption by a single, pressure-dependent line (6.5). Without serious loss of generality we may set $\xi = 1$ which is equivalent to adopting the Schwarzschild–Schuster approximation (§ 2.4.1). Equating the flux divergence to zero there results an integral equation expressing for each level a relationship between $B_i(z)$ and all other $B_i(z')$. It can be shown that, for integral $\tilde{u}$, these equations possess an exact solution:

$$B_i(x) = \frac{F_i^-(0)}{\pi} + \{F_i^+(1) - F_i^-(0)\} \sum_{r=0}^{\tilde{u}} C_r x^{t_r}, \qquad (8.4)$$

where $x = p/p_0$, p_0 being the basal pressure of the atmosphere. Substituting into the integral equations determines the constants C_r and t_r; solutions for $\tilde{u} = 1, 2$, and 3 are shown in Fig. 8.3. It may reasonably be assumed that the solutions are continuous in $\tilde{u}$ and that source functions for non-integral values can be obtained by interpolation.

The algebraic operations are tedious if $\tilde{u}$ exceeds 3, and a different method has been used to obtain the asymptotic solution for large $\tilde{u}$. From the trend of the solutions it is assumed that as $\tilde{u} \to \infty$, $B(1) \to B^*$ and $B(0) = 0$. The integral equations are then approximated by sums over discrete values of x (0·2, 0·4, 0·6, 0·7, 0·8, 0·9), and hence reduced to a set of approximate, simultaneous equations. Results are indicated in Fig. 8.3, together with similar computations for $\tilde{u} \propto x^2$. These latter computations attempt to simulate a typical water-vapour distribution; the former are more closely representative of carbon dioxide.

Comparing the profiles in Figs. 8.1 and 8.3, the most striking difference is at the upper boundary, where the line-absorbing model gives an asymptotic value $B(0) = 0$. This result can be explained in simple terms. At the upper boundary ($p = 0$) a Lorentz line is infinitely intense at the line centre, regardless of the value of $\tilde{u}$. The analogy in Fig. 8.1 is therefore the asymptotic limit $\tau^* \to \infty$, for which the discontinuity

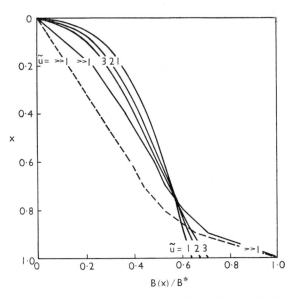

Fig. 8.3. Source functions for a finite atmosphere with a single Lorentz line. After King (1952). $x = p/p_0$. The full lines are for an atmosphere in which $\tilde{u} = \text{const}$. The broken line assumes $\tilde{u} \propto x^2$.

in the source function tends to zero at both boundaries. If we consider the situation at the lower boundary however, the lines have a finite absorption coefficient at all frequencies, and the analogy will be to grey radiation with τ^* finite; a discontinuity is therefore expected at the lower boundary.

8.2.3. *Application of the method of discrete ordinates*

The equation of radiative equilibrium (see (2.11) and (2.12)) involves integrals over both angles and frequencies. Provided that the frequency spectrum has a sufficiently regular form the latter integration can also be performed by numerical quadrature in a manner analogous to Chandrasekhar's integration over angles (§ 2.4.4).

Consider an Elsasser band model for a semi-infinite atmosphere at constant pressure. If the line profile does not change with depth, we may write

$$\tau_\nu = \frac{\bar{\tau}}{C_\nu},\tag{8.5}$$

where $\bar{\tau}$ is the arithmetic mean value of τ_ν over the band, and C_ν is a function of ν only. The equation of transfer for an isotropic source function now becomes (2.107)

$$\xi C_\nu \frac{dI(\bar{\tau}, \xi, C_\nu)}{d\bar{\tau}} = I(\bar{\tau}, \xi, C_\nu) - J(\bar{\tau}).\tag{8.6}$$

In this equation we have further assumed that the frequency range is small and that J does not therefore vary with C_ν.

The variable $x = \nu/\delta$, which varies from $-\frac{1}{2}$ to $+\frac{1}{2}$ over a repeat distance in an Elsasser band, defines C_ν uniquely. Since $\bar{\tau} \propto s$ we may write the condition for radiative equilibrium

$$-\iint \frac{dI_\nu}{ds}\, d\omega d\nu \propto \int\int_{-1/2}^{+1/2} \frac{dI(\bar{\tau}, \xi, C_x)}{d\bar{\tau}}\, d\omega dx$$

$$= \int_{-1}^{+1} 2\pi\, d\xi \int_{-1/2}^{+1/2} \frac{dx}{C_x}\{I(\bar{\tau}, \xi, C_x) - J(\bar{\tau})\}$$

$$= 0.\tag{8.7}$$

Since $\bar{\tau}$ is an arithmetic mean we have

$$\int_{-1/2}^{+1/2} \frac{dx}{C_x} = 1.\tag{8.8}$$

Eliminating $J(\bar{\tau})$ from (8.6), (8.7), and (8.8) we have

$$\xi C_x \frac{dI(\bar{\tau}, \xi, C_x)}{d\bar{\tau}} = I(\bar{\tau}, \xi, C_x) - \frac{1}{2}\int_{-1/2}^{+1/2} \frac{dx}{C_x}\int_{-1}^{+1} d\xi\, I(\bar{\tau}, \xi, C_x).\tag{8.9}$$

The double integral can be approximated by a double sum,

$$\xi_i C_j \frac{dI(\bar{\tau}, \xi_i, C_j)}{d\bar{\tau}} = I(\bar{\tau}, \xi_i, C_j) - \tfrac{1}{2}\sum_j\sum_i a_i b_j I(\bar{\tau}, \xi_i, C_j).\tag{8.10}$$

This system of linear equations is formally equivalent to those of (2.138), except that the number of equations is much greater. In the example discussed here the number of ordinates in both x- and ξ-space was limited to two.

If we employ (for example) the Gaussian method of numerical quadrature, the points of division and weights for ξ-space are given in

Table 2.2. The occurrence of the variable C_x in the denominator of the integral over x (8.9) requires a re-evaluation of divisions and weights for x-space. Details of the numerical analysis (which are extensive) for a two-ordinate Gaussian quadrature in both ξ- and x-space are to be found in papers listed in the Bibliography.

The source function as a function of optical depth is illustrated in Fig. 8.4. For this particular representation of the results the departures

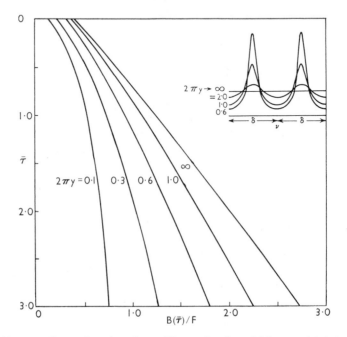

FIG. 8.4. Source functions for an Elsasser band model in a semi-infinite atmosphere.

The insert gives the absorption spectrum. After King (1956).

from grey absorption are great (the curve for $2\pi y = \infty$ is equivalent to grey absorption, see insert). This may be understood in terms of the long mean-free-path of radiation in the translucent intervals between the individual lines. At these frequencies radiation from the deep interior of the atmosphere can pass out directly to space. If therefore we impose a constraint of a constant radiative flux (as in these calculations) the temperature contrast between surface and interior must decrease continually as the opacity between lines is lowered. The same will not happen if the atmosphere is finite, since the source function near the surface in a transparent atmosphere never falls below half of the surface value (see the curves in Figs. 8.1 and 8.3). This fundamental difference between finite and semi-infinite atmospheres was forecast in the introductory remarks to this section.

8.2.4. *Numerical methods*

In two recent papers Möller and Manabe have described the computation of radiative equilibrium for conditions as close as possible to those of the earth's atmosphere. Vertical distributions of water vapour, ozone, and carbon dioxide similar to those described in § 1.3 were employed. Hence solar heating was computed as a function of altitude by the methods described in § 6.3. The temperature structure required for equilibrium with the terrestrial radiation was then computed by two different methods. The first was an iterative, marching technique, based on the repeated computation of the rate of change of temperature, and the calculation of a new temperature distribution corresponding to a time 8 hours later. The second was a matrix method, similar to the method for $\tilde{u}$ large described in § 8.2.2.

The marching technique is very flexible, and can employ any numerical method for evaluating heating rates. The method chosen by Möller and Manabe corresponds closely to that of Yamamoto and Onishi (§ 6.8.6). The reader should consult the original papers for details of the data and simplifying assumptions employed. The data are probably adequate for the purpose, and the simplifying assumptions used should not induce gross errors. In particular the authors do not have to assume that the emissivity is independent of temperature. This makes the basic flux integral (6.47) non-linear, but since it is evaluated by numerical quadrature no serious difficulties are involved.

The convergence of the marching technique is very instructive. The authors always start the integration with the same isothermal state, and stop it when the maximum rate of change of temperature becomes less than a specified value, lying between 0·03 and 0·07° K/day. These are not trivial compared with temperature changes induced by atmospheric motions, and the imbalance observed to exist in the real atmosphere (see Part 2 for measurements and computations). The authors estimate that these residual temperature changes may give rise to errors in the final equilibrium state between 3° K and 7° K. To achieve even this approximate equilibrium the integration had to be extended over 200 to 500 days at 8-hour intervals. The marching technique is therefore not an efficient method for computing radiative equilibrium. However, the information about the time to reach equilibrium is also valuable. For example, we may conclude that a process which requires more than a year to reach equilibrium will necessarily be greatly modified by diurnal and seasonal changes.

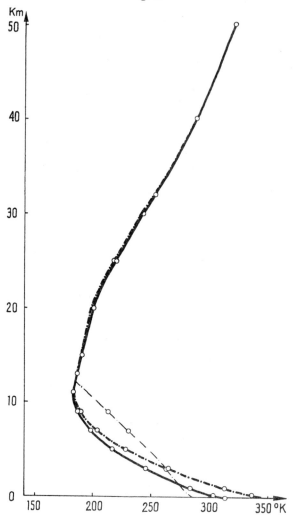

Fig. 8.5. Radiative equilibrium in the earth's atmosphere.
After Möller and Manabe (1961). (a) —○— Calculations for clear skies. (b) —·—·—·— Calculations for 6/10 cloudiness. (c) —·—○—·— Moist-adiabat with same heat content as (b). Conditions correspond to the yearly mean at latitude 40° and a mean $\xi_O = 0.5$. Results were obtained by the matrix method.

The published results from the marching technique suffer from a serious source of error in that the ground temperature was arbitrarily assumed, and not computed from a radiative balance equation. The matrix method, on the other hand, makes no such assumptions. Temperatures are computed for the ground, sixteen levels in the atmosphere, and for the outer 'skin' temperature; an 18×18 matrix has therefore to be inverted on an electronic computer. To avoid successive approximations the matrix elements must be constants, which requires that the emissivities are independent of temperature. The method therefore reduces to D. L. Brooks's technique (§ 6.8.5).

Fig. 8.5 shows one set of results from the matrix method. The discontinuity at the ground is not shown because it amounts to only $0.06°$ K. These results correspond closely below 15 km with a grey model of water vapour for $\tau^* = 4$ (see Fig. 8.7 and Table 8.2). The main difference lies in the temperature discontinuity, which is $11.3°$ K for the grey model. This discontinuity in the non-grey model has been distributed over the lowest atmospheric layers, which have a maximum lapse rate of $90°$ K km^{-1}, as opposed to $36.0°$ K km^{-1} for the grey model. This suggests that the strong rather than the average water vapour lines dominate the formation of the discontinuity (see Figs. 8.1 and 8.2).

Fig. 8.6 gives the breakdown of contributions to the equilibrium configuration. Below 15 km the balance is between heating and cooling by water vapour. Above 20 km ozone heating is of decisive importance and the temperature rises until it can be offset by carbon dioxide and water vapour emission.

8.3. The problem of the lower stratosphere

The discovery of the stratosphere by Teisserenc de Bort in the year 1900 led to a remarkable series of papers on radiative equilibrium in planetary atmospheres, for it was accepted by all investigators at that time that the characteristic feature of the lower stratosphere was an approach to radiative equilibrium. The validity of this hypothesis has been argued at length, but there is no general agreement. Winds and turbulence are observed to exist in the lower stratosphere and some heat transfer must occur by mixing. However, there is evidence from the vertical distribution of radioactive debris from nuclear explosions that there is a sharp distinction between rapid mixing in the troposphere and slow mixing in the stratosphere.

In the following sections we will pursue the ideas of Emden (1913), who inquired into the consequences of assuming radiative equilibrium, and showed that convective activity will necessarily develop in the

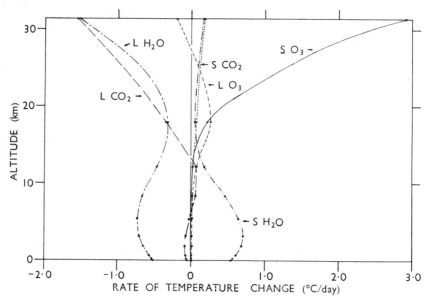

Fig. 8.6. Contributions to radiative equilibrium.

After Manabe and Möller (1961).

L CO₂	Carbon dioxide	
L H₂O	Water vapour	Terrestrial (long wave) radiation.
L O₃	Ozone	

S CO₂	Carbon dioxide	
S H₂O	Water vapour	Solar radiation.
S O₃	Ozone	

The conditions are similar to those of Fig. 8.5 but not identical. Results were obtained by the marching method, and are slightly in error near to the ground.

lower atmosphere. The power of this heuristic reasoning lies in the fact that radiative transfer provides the only fundamental heat source for driving atmospheric motions; we may therefore expect that successive mathematical approximations from an initial equilibrium state will parallel the physical process of motions developing towards a steady state from an initially static atmosphere. The achievement of a suitable approximation can be judged when agreement is obtained between theory and observation. In the next sections we will see that if a simple model of tropospheric convection is added to Emden's scheme, a remarkably accurate picture of the mean thermal structure of the lower atmosphere is already obtained.

It is an interesting reflection on the methods of a geophysical science that such a simple heuristic approach is often held to be repugnant on the ground that observational evidence demonstrates the essential complexity of atmospheric processes. Investigations such as that of Chiu and Greenfield (1959) bring out the existence of many different factors influencing stratospheric temperature change, which can be interpreted as destroying the basis for an assumption of radiative equilibrium. Such work does, indeed, demonstrate that large-scale motions, called into being by the variation of insolation with latitude, and the turbulent eddies which they create, must be included in a complete theory. Our understanding of the fluid dynamical problems involved is rapidly increasing, and it may be possible in the near future to handle the elaborate models required for a satisfactory solution. In the meanwhile, however, we may recall the essential role of a physical theory in providing numerical results which can be verified or refuted by experiment or observation. The nature of the theory is not necessarily relevant, provided that it does not without reason contradict other established theories, and its only justification need be the accuracy of its deductions. A theory is superseded if another gives more precise results, or reduces the number of hypotheses involved. On these grounds, Emden's theory, taken together with its subsequent developments, is still one of the most productive theories of the atmosphere.

A brief historical note should draw attention to the pioneer work of Schwarzschild (1906) who provided the formal basis for these studies. Humphreys (1909) and Gold (1909) approached the stratospheric problem simultaneously, but from two very different points of view. Humphreys regarded the isothermal stratosphere as a manifestation of the finite skin-temperature at the upper boundary of a medium in radiative equilibrium (see Fig. 8.1). Using the Eddington approximation (2.115), we find for the source function at $\tau = 0$:

$$B(0) = \frac{F}{2\pi} = \frac{\sigma \theta^4(0)}{\pi}. \qquad (8.11)$$

For overall balance between incoming solar and outgoing terrestrial radiation we may write

$$4\pi r^2 F = \pi r^2 (1-a)f, \qquad (8.12)$$

where a is the planetary albedo, f the solar constant, and r the radius of the earth. If $a = 0 \cdot 4$ and $f = 2$ cal cm^{-2} min^{-1}, we find $F/2\sigma = 1 \cdot 85 \times 10^9$ (°K)^4 and $\theta(0) = 207 \cdot 2°$ K, which is close to observed mid-latitude, lower-stratosphere temperatures (cf. the calculation of the mean planetary emission temperature $\theta_e = 246 \cdot 5°$ K in § 1.1; the two temperatures are related by $\theta^4(0) = \frac{1}{2}\theta_e^4$).

Gold made no attempt to explain the existence of the isothermal stratosphere but, assuming it to exist, and its temperature to equal that of the troposphere in contact with it, he computed where the base must be in order to maintain an overall radiative heat balance. He found the basal pressure to be approximately $p_0/4$, where p_0 is the ground pressure (in good agreement with observation). Milne (1922) criticized a number of aspects of Gold's work, including the assumption that a layer in radiative equilibrium is isothermal, since for a grey-absorbing atmosphere the temperature must decrease with height if F is positive (2.111).

An excellent review of these early papers was made by Pekeris (1932). They are of historical interest and throw much light on the progress of scientific thought in this area. We may note that Gold's work is not so much a theory as an examination for self-consistency. He does not explain what processes keep in existence the assumed thermal structure, but we will show that an extension of Emden's ideas leads to a criterion similar to Gold's as a necessary condition for stability.

8.4. Development of the convective troposphere

In his original discussion Emden made allowance for absorption of solar radiation in the atmosphere. This does not alter the qualitative properties of the model, and only complicates it at a juncture where we need to keep the physical meaning of the results clearly in mind. In this section, therefore, we will assume the solar radiation to be absorbed only at the earth's surface, fixing, as a boundary condition, the total amount of long-wave radiation which is to be re-emitted to space (F). We will further assume a grey absorption coefficient and the Eddington approximation, for which (2.111) and (2.115) (omitting ν-suffixes) are a complete solution. If τ^* is the total optical depth of the atmosphere, we have

$$B(\tau) = \frac{F}{2\pi}(1 + \tfrac{3}{2}\tau) = \frac{\sigma}{\pi}\theta^4(\tau),$$

$$B(0) = \frac{F}{2\pi}, \qquad\qquad\qquad (8.13)$$

$$B^*(\tau^*) = \frac{F}{2\pi}(2 + \tfrac{3}{2}\tau^*).$$

To change the independent variable from optical depth to altitude, we require a relation between the absorption coefficient and altitude. We will assume

$$\frac{k_v(z)}{k_v(0)} = \frac{\rho(z)}{\rho(0)} = e^{-z/(2\times10^5)}, \qquad\qquad (8.14)$$

i.e. an absorption coefficient scale height of 2 km, corresponding closely
to the density distribution of water vapour in the troposphere. While
water vapour is far from a grey absorber, we will attempt in this model
to simulate its physical properties as closely as possible, since it is the
most important single constituent of the lower atmosphere. The effect
of pressure on absorption, in so far as it can be taken into account by
means of a single-parameter scaling approximation (§ 6.1.2), only affects
the choice of scale height in (8.14).

We have
$$\tau = \int_z^\infty k_v \, dz = \tau^* e^{-z/(2 \times 10^5)}. \qquad (8.15)$$

Equations (8.14) and (8.15) define the temperature structure in terms
of $F/2\sigma$ and τ^*. The former quantity is known, and in Table 8.2 an
approximate global average value of $1{\cdot}82 \times 10^9$ (°K)^4 has been employed.† The temperature profiles for another value of the flux (F')

TABLE 8.2

Radiative equilibrium in a grey model atmosphere

($F/2\sigma = 1{\cdot}82 \times 10^9$ (°K)^4; absorption coefficient scale height $= 2$ km)

τ^*	$\theta(0)$	$-\Delta\theta(0)$	$-\left(\dfrac{\partial\theta}{\partial z}\right)_0$	$z_{6\cdot5}$	$\tau(s)$	$z(s)$	$\theta(s)$	$\theta'(0)$
	(°K)	(° K)	(°K km^{-1})	(km)		(km)	(°K)	(°K)
$\ll 1$	206·6	39·1	0·0	0·0	0	6·02	206·6	245·7
0·5	237·5	28·4	12·0	1·8	0·0159	6·90	207·2	252·0
1·0	259·7	22·8	19·5	3·2	0·0212	7·70	208·2	258·2
2·0	292·1	16·7	27·4	4·6	0·0239	8·87	208·3	265·9
4·0	335·9	11·3	36·0	6·0	0·0243	10·20	208·4	274·7

$\theta(0) =$ atmospheric temperature at $z = 0$ ⎫
$-\Delta\theta(0) =$ temperature discontinuity at $z = 0$ ⎬ Under radiative
$-\left(\dfrac{\partial\theta}{\partial z}\right)_0 =$ temperature lapse rate at $z = 0$ equilibrium
$z_{6\cdot5} =$ height at which lapse rate $= 6{\cdot}5°$ K km^{-1} ⎭

$\tau(s) =$ optical depth of radiative layer ⎫
$z(s) =$ height of radiative-convective transition ⎬ Under radiative-con-
$\theta(s) =$ temperature at transition vective equilibrium
$\theta'(0) =$ temperature at ground level ⎭

For $\tau^* \gg 4$ $z_s = -2 \ln \dfrac{0{\cdot}0245}{\tau^*}$ km,

$\theta(s) = 208{\cdot}5°$ K,

$\theta'(0) = 208{\cdot}5 + 6{\cdot}5z(s)$ °K.

† This differs by about 1 per cent from that for a solar constant of 2 cal cm^{-2} min^{-1}
and an albedo of 0·4. The figures 206·6° K and 245·7° K in the first line of Table 8.2
correspond to the equilibrium temperatures 207·2° K and 246·5° K mentioned in § 8.3.

can be obtained by multiplying the first three columns of Table 8.2 by the factor $(F'/F)^{1/4}$.

Data for several values of τ^* are given in Table 8.2 and three profiles are shown in Fig. 8.7. We may note the following characteristics. All profiles tend to the same asymptotic value of $\theta = 206 \cdot 6°$ K, corresponding to Humphrey's value for the stratosphere temperature. Secondly,

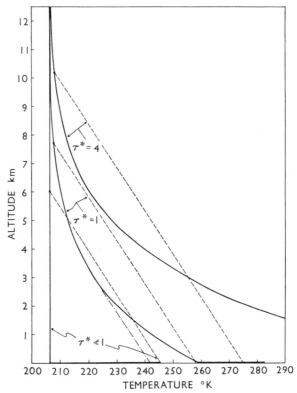

FIG. 8.7. Temperature profiles under radiative equilibrium (full lines), and as modified by tropospheric convection (broken lines).

$F/2\sigma = 1 \cdot 82 \times 10^9$ (°K)^4. Absorption coefficient scale height $= 2$ km. The chain line joining the curve for $\tau^* = 1$ at an altitude of $3 \cdot 2$ km will be explained in the text. The heavy lines on the abscissa show the magnitude of $\Delta\theta(0)$.

at the ground there is always a negative temperature discontinuity, varying from $39 \cdot 1°$ K for a transparent atmosphere to zero for a very opaque atmosphere (the step in B always equals $F/2\pi$, but this corresponds to smaller temperature steps as the temperature increases). Thirdly, the negative temperature gradient (temperature lapse) is at its maximum near the ground. The fifth column in Table 8.2 shows the height below which the lapse rate is greater than $6 \cdot 5°$ K km^{-1}.

If we consider $\tau^* = 1$ as an example, we see that the state of radiative equilibrium is hydrostatically unstable on two counts, and will break down as the result of small disturbances. In the first place, the temperature discontinuity at the ground is unstable. Secondly, a lapse rate greater than $6\cdot5°$ K km^{-1} is unstable according to direct observation.† The state of radiative equilibrium, once created, will therefore break down, giving rise to a convective layer at the surface. This, according to Emden, is the fundamental reason for the existence of the convective troposphere. Since this conclusion holds for all values of τ^*, it will not be surprising to find that it also holds in non-grey atmospheres (see § 8.5).

At this point we may consider an important criticism of Emden's work by Hergesell (1919), who pointed out that condensation sets an upper limit to the permissible water-vapour density, and it is a matter of observation that the density at any atmospheric level is in fact closely correlated with the ambient temperature. Thus a specified temperature distribution (such as one in Fig. 8.7) requires a certain value for τ^*, which should be the result of calculation, and not assumed *ab initio*. With certain reasonable assumptions about the relationship between water-vapour density and temperature Hergesell found a solution corresponding to a very transparent atmosphere, similar to that for $\tau^* \ll 1$ in Fig. 8.7. This interesting result has not received the attention that it merits, presumably because it is difficult to reconcile with the observed behaviour of the earth's atmosphere. The source of the difficulty probably lies in the assumption of grey absorption. Hergesell's conclusion depends intimately on the slope of the deduced equilibrium $\theta \sim \tau$ relation; the combined effect of the $10\,\mu$ window and the departures of band transmission from Lambert's law probably change this relation sufficiently to destroy his argument. The problem should be further investigated.

We have assumed a lapse rate of $6\cdot5°$ K km^{-1} to be the limit of convective stability; the first stage in the breakdown of the radiative state will therefore be the development of a convective layer up to $z_{6\cdot5}$. This will not change the temperatures above $z_{6\cdot5}$ for, from (8.13), these are governed by the imposed flux and the optical depth only. Consequently,

---

† This figure is less than the dry adiabatic lapse rate of $9\cdot8°$ K km^{-1}, but larger on the average than the moist adiabatic lapse rate (see, e.g., Fig. 8.5). Atmospheric convection is partly moist and partly dry and, if there were no other factors to be considered, some mean value would presumably be observed. However, radiative transfer cannot be neglected even in the presence of active convection, and will have some influence on the tropospheric temperature distribution. To keep the complexity within reasonable bounds we will accept $6\cdot5°$ K km^{-1} as the tropospheric stability limit on observational grounds.

the state of the atmosphere below will not affect the radiative equilibrium at a given point, provided only that the lower atmosphere delivers the necessary upward flux component to yield the required net flux. The ability to provide this upward flux is the upper boundary condition imposed upon the convective layer, which can now be completely specified if we assume that active convection gives rise to a neutral lapse rate (6·5° K km^{-1}), and irons out any temperature discontinuities at the lower boundary.

One important consequence of these assumptions is illustrated by the chain line attached to the profile $\tau^* = 1$ in Fig. 8.7. This follows the conventions established for the structure of the convective layer with, in addition, continuity of temperature at $z_{6·5}$ (3·2 km, in this particular example). We shall show that this profile cannot deliver the required upward flux, and is therefore not the profile which we are seeking. In the context of the assumptions we have made about the convective layer, the only possible adjustment is to move the chain curve bodily sideways, destroying the continuity of temperature at $z_{6·5}$. This is not a surprising result, for we have already had many examples to show that there is, in general, no requirement for temperature continuity at the boundary of a region in radiative equilibrium.

We may justify the statement made in the last paragraph by comparing full and chain lines below 3·2 km on the $\tau^* = 1$ curve in Fig. 8.7. The full line indicates higher temperatures at all levels, including the earth's surface. The upward flux of radiation at 3·2 km is therefore greater from the radiative than from the convective atmosphere. Since, as a matter of definition, the radiative atmosphere gives the required upward flux at 3·2 km, it follows that this particular convective curve gives less than the required flux. To give the correct flux the chain curve must be moved bodily to the right, i.e. to higher temperatures. The discontinuity at $z_{6·5}$ will therefore be negative.

A negative discontinuity is unstable; this circumstance therefore demands that the convective layer must rise above $z_{6·5}$. The logical consequence of this argument is that the convective layer must rise (in this particular example) to 7·70 km, where it can join to the radiative layer without an unstable discontinuity. The lapse rate in the radiative layer is now much smaller than 6·5° K km^{-1} and the temperature configuration both at the boundary and in the radiative region is stable to small disturbances.

Data for the height of the transition $z(s)$, the temperature at the point of transition $\theta(s)$, and the temperature at $z = 0$ ($\theta'(0)$, equal by assumption to the ground temperature) for several values of τ^*, are given in Table 8.2. The calculation of the upward flux at the top of the convective layer used an integral equation consistent with the Eddington approximation. This is equivalent to the use of a diffusivity factor $r = 1\cdot5$ (see § 6.2). The lowest possible transition level for the assumed mean flux is for a very transparent atmosphere, and lies at $6\cdot04$ km.

If we compare the model with the observed behaviour of the tropopause and lower atmosphere (§ 1.1), we see that the general nature and altitude of the sudden change taking place at the tropopause are explained. The observed change of phase of the seasonal temperature wave is also to be expected from a model which envisages a change of heat transfer mechanism at the surface of transition. These are remarkable achievements for such a simple model. In two major respects, however, it is deficient. Firstly, it predicts only negative temperature gradients in the lower stratosphere (an invariable feature of a grey model with positive radiative flux), while positive gradients are observed at low latitudes. Secondly, it predicts the wrong latitude variation of lower-stratosphere temperature. This temperature never differs greatly from the skin temperature at $z = \infty$, and according to (8.11) should vary as $F^{1/4}$. Since the solar flux is greatest in the tropics, so, too, on this model should be the stratosphere temperature, whereas the opposite is observed. This conclusion can be modified by allowing for some solar absorption in the lower stratosphere, or by postulating that the terrestrial flux differs greatly from the local solar flux because of heat transfer between latitudes by atmospheric motions; neither suggestion, however, accounts fully for the discrepancy. We will see in the following sections that the difficulty does not occur for a non-grey atmosphere. At the same time, the problem of the temperature gradient in the lower stratosphere also resolves itself, for this, too, is a consequence of strict adherence to a grey model without any direct solar heating.

8.5. Extension to a non-grey atmosphere

According to the results of Möller and Manabe (1961), the coexistence of water vapour, carbon dioxide, and ozone, each with separate absorption bands, and the inclusion of direct solar absorption do not alter the picture of an unstable boundary layer created by radiative processes. Nor, according to Goody (1949), do they change the equilibrium criterion of temperature continuity at the tropopause. An analytic solution is no longer possible, but the existence and stability of the equilibrium configuration in the absence of solar absorption have been demonstrated by numerical methods; their plausibility has already been established because the qualitative results of §§ 8.3 and 8.4 do not depend upon the optical depth of the atmosphere.

Calculations of the height and temperature of the transition in the absence of direct solar absorption are given in Table 8.3. In place of

TABLE 8.3

The temperature and height of the tropopause in a cloud-free atmosphere, without direct solar absorption

After Goody (1949)

Ground temperature (°K)	Tropopause temperature (°K)		Tropopause height (km)	
	Calculated	*Observed*	*Calculated*	*Observed*
300	205	193	14·6	16·5
290	208	208	12·6	12·6
280	210	214	10·8	10·2
270	213	217	8·8	8·2
260	214	219	7·1	6·3

specifying the outgoing flux, these results have been computed on the basis of an assumed ground temperature. To simplify the computations the stratosphere was assumed to be isothermal, and cloud and haze were not considered. The absorption data used are not as good as now exist and direct solar absorption was neglected. The agreement with observation is probably better than it deserves to be under these circumstances. It is particularly interesting to note the out-of-phase relation between ground and stratosphere temperatures.

This unusual relation is not a simple consequence of non-grey absorption. The varying absorption coefficient permits direct radiation to space from all atmospheric levels. This destroys the coupling between skin temperature and outgoing flux which, in the case of grey absorption, gives rise to the upper boundary condition in (8.13). To give a simple example: if the spectrum has a transparent window, we can create a large flux to space with a high ground temperature even though the atmosphere itself is cool; if the atmosphere is uniformly grey, however, a large flux demands a high temperature close to the level of unit optical depth. Without this boundary restraint,† the temperature of

---

† The removal of this boundary restraint destroys the convenient result, used in connexion with Fig. 8.7, that the structure of the radiative layer is independent of events in the convective layer. There is now a certain degree of coupling, and the thermal state of the radiative layer must be computed by successive approximations.

the lower stratosphere depends instead on the complex interaction between the three gases present and the factors which control their densities. If a single important feature has to be isolated, it is the effect of temperature upon water-vapour density through the phenomenon of saturation which is responsible for the figures in Table 8.3.

A somewhat similar result with respect to water vapour is given by the radiative equilibrium calculations of Manabe and Möller (1961). These authors identify the minimum temperature near 10 km (see Fig. 8.5) with the tropopause, and there is indeed a superficial resemblance. However, this is not a satisfactory interpretation of the results for it accepts a highly unstable troposphere and equilibrium time constants up to 500 days. Both theory and observation indicate a convective troposphere, and the mechanism described in the previous section will result. The interplay of radiation and convection will then determine the position of the tropopause. Nevertheless, increased cooling and decreased heating will certainly cool the tropopause and raise its level, whatever the mechanism. Manabe and Möller show that in the tropics solar heating by ozone is decreased and cooling by water vapour is increased. Thus the inclusion of direct solar heating should exaggerate the out-of-phase relation between ground and tropopause temperature shown in Table 8.3.

Solar heating is also particularly important as regards the phase change of the seasonal temperature wave as it passes from troposphere to stratosphere. If computations are made without solar heating a wave of the observed amplitude is obtained, but it has not the correct phase, and agreement cannot be obtained without including solar heating.

As illustrated by Fig. 8.6, solar heating by ozone is very important when we come to consider the temperature gradient in the lower stratosphere. However, there are also important formal differences between grey and non-grey atmospheres which can lead to positive gradients under radiative equilibrium, even in the absence of direct solar heating.

To illustrate this point, consider a gas mixture in which each gas has a different vertical distribution of absorption coefficient. Without important loss of generality, we may assume

$$\frac{1}{k_{v,\nu}(z)} \frac{dk_{v,\nu}(z)}{dz} = \alpha_i, \qquad (8.16)$$

where α_i varies from band to band, as specified by the subscript (i).

We will use Curtis's formulation of the heating equation (6.57) and write

$$H_i(z) = \int\limits_0^1 \{B_i(z') - B_i(z)\}\, dK_i^{(1)} + \int\limits_0^1 \{B_i(z') - B_i(z)\}\, dK_i^{(2)}. \quad (8.17)$$

The condition for radiative equlibrium is

$$h(z) = \sum_i h_i(z) = \sum_i 2\pi H_i(z)\bar{k}_{v,i}(z)\,\Delta v = 0, \quad (8.18)$$

where

$$\bar{k}_{v,i}(z) = \frac{\int\limits_i k_{v,i}(z)\,dv}{\int\limits_i dv}, \quad (8.19)$$

is the arithmetic mean absorption coefficient for the ith spectral range. For convenience we will introduce a new function

$$M_i^{(1,2)}(z',z) = \frac{\int k_{v,v}^2(z)E_1\{\tau_v(z',\infty) - \tau_v(z,\infty)\}\,dv}{(\bar{k}_{v,i})^2\,\Delta v_i}, \quad (8.20)$$

where the indices (1) and (2) denote $z' < z$ and $z' > z$ respectively. By virtue of (8.16) we can demonstrate that

$$\frac{\partial K_i(z',z)}{\partial z} = \pm \bar{k}_{v,i}(z)M_i(z',z), \quad (8.21)$$

the positive sign referring to $z' < z$, and the negative to $z' > z$.

Equation (8.18) holds identically for all z in the radiative region, and we may therefore differentiate with respect to z and set the result equal to zero. After rearranging terms we obtain the expression

$$\sum_i 4\pi\bar{k}_{v,i}\,\Delta v_i \frac{\partial B_i(z)}{\partial z} = \sum_i \alpha_i h_i(z) - \sum_i 2\pi(\bar{k}_{v,i})^2\,\Delta v_i \times$$
$$\times \left\{ \int\limits_0^\infty B_i(z')\, dM_i^{(1)}(z',z) - \int\limits_0^\infty B_i(z')\, dM_i^{(2)}(z',z) \right\}. \quad (8.22)$$

For a small range of temperature the l.h.s. is proportional to the temperature gradient.

To compare this result with those of § 8.4 we may consider what happens when we reintroduce the assumption of grey absorption. Since all bands now overlap, there is no physical distinction between the gases and hence only one value of α_i can be used. Since $\sum_i h_i = 0$, it follows that the first term on the r.h.s. of (8.22) is also identically zero. In the second term M can be replaced by E_1 and, if we employ the Eddington approximation, E_1 differs from E_3 by a numerical factor only. Thus, from (2.96), this term is proportional to the net flux, and (8.22) reduces, as expected, to (2.111)

$$\frac{dB}{dz} = -\frac{3k(z)}{4\pi} F. \tag{8.23}$$

Thus, the second term in (8.22) which, by analogy, can be called the *flux term*, demands a negative temperature gradient for a positive flux, and the qualitative results of § 8.4 are preserved. The first term on the r.h.s. of (8.22) has entirely different properties, however, which cannot even be visualized using a grey model. Since this term is non-zero if the α_i are not all equal, it may be called the *distribution term*.

The distribution term is positive if α_i and h_i are positively correlated, i.e. if the most rapidly attenuated gases give the greatest cooling. From Fig. 8.6 we see that water vapour is the most important cooling gas in the lower stratosphere; it is also the most rapidly attenuated with height. In the earth's atmosphere, therefore, the distribution term tends to produce a positive gradient of temperature, and numerical estimates indicate that it may be of the order of magnitude of 1°K km^{-1}.

Two factors are responsible for the positive correlation between α_i and h_i; one is accidental, in the sense that it might not exist for a different gas mixture; the second is a necessary consequence of the coexistence of more than one gas, and can be expected in other planetary atmospheres. To take the second first: other things being equal, a gas which cools strongly at the tropopause will have a high density at this level and a small superincumbent mass in the stratosphere to interfere with the escape of thermal radiation to space; this condition is met if the gas is rapidly attenuated in the lower stratosphere.

The first factor depends upon the relative positions of active bands in the infra-red spectrum. It can be demonstrated from a simple generalization of (8.13) to the case of a single band following Lambert's law, that the skin temperature decreases as the wavelength of the band increases, if all other factors are held constant. As it happens, the three most important atmospheric bands ($9 \cdot 6\,\mu$ ozone, $15\,\mu$ carbon dioxide, rotation band of water vapour) fall in the order of λ increasing as α_i decreases, giving a positive correlation between α_i and h_i for terrestrial radiation, and a further positive contribution to the distribution term.

BIBLIOGRAPHY

8.1. Introduction

8.2. Non-grey atmospheres

Astrophysical work is usually concerned with situations differing from those appropriate to a planetary atmosphere, and most of the elegant work on non-grey stellar atmospheres is inapplicable to our problems. The effect of a discontinuous change in the absorption coefficient, as at the Balmer limit, for example, is discussed by

CARRIER, G. F., and AVRETT, E. H., 1961, 'A non-gray radiative-transfer problem', *Astrophys. J.* **134**, p. 469; and

STONE, P. H., and GAUSTAD, J. E., 1961, 'The application of a moment method to the solution of non-gray radiative-transfer problems', ibid., p. 456.

The treatment of small departures from grey absorption is discussed by CHANDRASEKHAR (1950) (see § 2.1).

§ 8.2.1 follows

YAMAMOTO, G., 1955, 'Radiative equilibrium of the earth's atmosphere. II. The use of Rosseland's and Chandrasekhar's means in the line absorbing case', *Science Rept., Tohoku Univ., Series 5, Geophysics*, **6**, No. 3, p. 127.

The treatment of § 8.2.2 is given by

KING, J. I. F., 1952, 'Line absorption and radiative equilibrium', *J. Met.* **9**, p. 311.

§ 8.2.3 follows

KING, J. I. F., 1955a, 'The Gaussian quadrature for an Elsasser band-absorption model', *Astrophys. J.* **121**, p. 425;

—— 1955b, 'Radiative equilibrium of a line-absorbing atmosphere. I', ibid., p. 711; and

—— 1956, 'Radiative equilibrium of a line-absorbing atmosphere. II', ibid. **124**, p. 272.

§ 8.2.4 follows

MANABE, S., and MÖLLER, F., 1961, 'On radiative equilibrium and heat balance of the atmosphere', *Mon. Weath. Rev.* **89**, p. 503; and

MÖLLER, F., and MANABE, S., 1961, 'Über das Strahlungsgleichgewicht der Atmosphäre', *Z. für Met.* **15**, p. 3.

8.3. The problem of the lower stratosphere

The early work in this field is reviewed by

PEKERIS, C. L., 1932, *The development and the present status of the theory of the heat balance in the atmosphere*, M.I.T. Professional Notes, No. 5.

Some of the most important contributions are by:

EMDEN, R., 1913, 'Über Strahlungsgleichgewicht und atmosphärische Strahlung', *Sitz. d. Bayerische Akad. d. Wiss.*, Math.-Phys. Klasse, p. 55;

SCHWARZSCHILD, K., 1906, 'Über das Gleichgewicht der Sonnenatmosphäre', *Nach. d. k. Gesell. d. Wiss. zu Göttingen*, Math.-Phys. Klasse, Heft 1;

HUMPHREYS, W. J., 1909, 'Vertical temperature gradient of the atmosphere in the region of the upper inversion', *Astrophys. J.* **29**, p. 14;

GOLD, E., 1909, 'The isothermal layer of the atmosphere and atmospheric radiation', *Proc. Roy. Soc.* A, **82**, p. 43; and

MILNE, E. A., 1922, 'Radiative equilibrium, the insolation of the atmosphere', *Phil. Mag.* **144**, p. 872.

An empirical analysis of the heat sources in the lower stratosphere is by

CHIU, W. C., and GREENFIELD, R. S., 1959, 'The relative importance of different heat-exchange processes in the lower stratosphere', *J. Met.* **16**, p. 271.

8.4. The development of the convective troposphere

The difficulties associated with a condensation mechanism influencing gaseous density in a grey atmosphere were pointed out by

HERGESELL, H., 1919, 'Die Strahlung der Atmosphäre unter Zugrundelegung von Lindenbergen Temperatur- und Feuchtigkeitsmessungen', *Arb. preuss. aero. Obs.* **13**.

An attempt to avoid this difficulty has been described by

PEKERIS, C. L., 1930, 'Radiation equilibrium and humidity distribution in a semigrey atmosphere', *Beitr. Geophys.* **28**, p. 337.

Contradictions inherent in a grey model of the atmosphere were first emphasized quantitatively in the classical paper by

SIMPSON, G. C., 1928, *Further studies in terrestrial radiation*, Mem. R. Met. Soc. **3**, No. 21, p. 1.

MILNE (1922) (see § 8.3) introduced direct solar absorption to explain the latitude variation of stratosphere temperature.

YAMAMOTO, G., 1953, 'Radiative equilibrium of the earth's atmosphere. I. The grey case', *Science Rept., Tohoku Univ., Series 5, Geophysics*, **5**, No. 2, p. 45, shows that by suitably varying τ^* the latitude variation of ground and stratosphere temperatures are consistent, but he does not explain how polar radiation fluxes are made to exceed those in the tropics, as a grey model demands.

8.5. Extension to a non-grey atmosphere

The treatment in this section follows MÖLLER and MANABE (1961) (§ 8.2);

GOODY, R. M., 1949, 'The thermal equilibrium at the tropopause and the temperature of the lower stratosphere', *Proc. Roy. Soc.* A, **197**, p. 487; and
—— 1949, *Radiative heat exchange in the lower stratosphere*, Ph.D. Thesis, Cambridge University.

Some of the fundamental ideas are to be found in a paper by

DOBSON, G. M. B., BREWER, A. W., and CWILONG, B. M., 1946, 'Meteorology of the lower stratosphere', *Proc. Roy. Soc.* A, **185**, p. 144.

A discussion of the distribution term has also been given by

KAPLAN, L. D., 1954, 'The infra-red spectrum of the lower stratosphere and its importance in the heat balance', *Sci. Proc. I.A.M. Xth Gen. Assembly,* Rome, 1954, p. 583.

STRONG and PLASS (1950) (see § 6.1) have also attempted to explain the positive temperature gradients in the lower stratosphere, but according to the ideas expressed here this is more properly considered in § 6.1.

RADIATIVE TRANSFER AND FLUID MOTIONS

9.1. Dissipation time for a radiative process

9.1.1. *A general solution*

W E will continue the line of inquiry of the previous chapter, and seek some insight into the nature of the motions created in the convective troposphere, again using simple heuristic models.

A different approach is possible where radiation computations have been cast into a form suitable for combining with numerical solutions of the equations of fluid motion (the methods of numerical weather forecasting). The work of Möller and Manabe (§ 8.2.4) was developed for just this purpose, but it has yet to be applied to a dynamical problem. Even when such computational techniques are available, however, there will still be a need to consider simple models because of the insight which they give into the physical processes.

The heating rate in a fluid system is determined by

$$h = -\nabla \cdot \mathbf{F} = -\nabla \cdot (\mathbf{F}_r + \mathbf{F}_a + \mathbf{F}_t + \mathbf{F}_m), \tag{9.1}$$

where
$$\mathbf{F} = \text{total heat flux,}$$
$$\mathbf{F}_r = \text{radiative heat flux,}$$
$$\mathbf{F}_a = \text{advective heat flux,}$$
$$\mathbf{F}_t = \text{turbulent heat flux,}$$
$$\mathbf{F}_m = \text{molecular heat flux.}$$

$\mathbf{F}_a$ and $\mathbf{F}_t$ are distinguished by the fact that the former is attributable to motions with a non-zero time average, and the latter to motions with a zero time average. Under some circumstances there is a limited analogy between $\mathbf{F}_t$ and $\mathbf{F}_m$; then it will be convenient to combine them into a single diffusive term $\mathbf{F}_d$.

When we come to assess the relative importance of these different processes, the rate at which each returns a disturbed system to its steady state will be a critical parameter. If the equations can be linearized any disturbance can be represented by a Fourier sum of simple-harmonic disturbances; consequently our first step will be to inquire into the rate of dissipation of a sinusoidal disturbance of small amplitude in an unlimited medium. Spiegel (1957) has shown that, to first order, the effect of variation over the fluid of absorption coefficient with temperature can be neglected. We will therefore confine our attention to a medium with constant k_ν.

We take (2.83) as a starting-point, considering absorption only, and this independent of position, so that τ_ν can be replaced by $k_{\nu,\nu}s$. Planck's function is used as the source function (the distinction between $-\mathbf{s}$ and $\mathbf{s}$ in (2.83) thus disappears), and this is taken to be time-dependent. There results

$$\bar{I}_\nu(P,t) = \int \frac{k_{\nu,\nu} B_\nu(\mathbf{s},t)e^{-k_{\nu,\nu}s}}{4\pi s^2} \, dV(\mathbf{s}). \tag{9.2}$$

Consider small departures from an equilibrium state, for which $\bar{I}$, B, and θ have the values $\bar{I}^{(0)}$, $B^{(0)}$, and $\theta^{(0)}$. Let

$$\bar{I}_\nu = \bar{I}_\nu^{(0)} + \bar{I}_\nu^{(1)},$$
$$B_\nu = B_\nu^{(0)} + B_\nu^{(1)}, \tag{9.3}$$
$$\theta = \theta^{(0)} + \theta^{(1)},$$

where $I_\nu^{(1)}$, $B_\nu^{(1)}$, and $\theta^{(1)}$ are small quantities.

The increase in thermal energy of the fluid can be written, with the help of (2.18),

$$h = \rho c \frac{\partial \theta(P,t)}{\partial t} = \int 4\pi k_{\nu,\nu}\{\bar{I}_\nu(P,t) - B_\nu(P,t)\} \, d\nu, \tag{9.4}$$

where ρ is the density and c the specific heat of the fluid (we can leave unspecified the precise nature of the specific heat, although for numerical examples in the atmosphere the specific heat at constant pressure is correct). $(\bar{I}, B, \theta)$ and $(\bar{I}^{(0)}, B^{(0)}, \theta^{(0)})$ are both solutions of (9.4). Since it is linear in these quantities, $(\bar{I}^{(1)}, B^{(1)}, \theta^{(1)})$ must also be a solution; similarly $\bar{I}^{(1)}$ and $B^{(1)}$ are related by (9.2). Since we assume $B^{(1)}$ to be small, we may linearize the equations in the same manner as for (2.133),

$$\rho c \frac{\partial \theta^{(1)}(P,t)}{\partial t} = \int_0^\infty d\nu \, 4\pi k_{\nu,\nu} \overline{\frac{dB_\nu}{d\theta}}\left(\int \frac{k_{\nu,\nu}\theta^{(1)}(\mathbf{s},t)e^{-k_{\nu,\nu}s}}{4\pi s^2} \, dV(\mathbf{s}) - \theta^{(1)}(P,t)\right). \tag{9.5}$$

Let us now suppose the disturbance to be periodic, with wave number vector $\mathbf{n}$ and with a time-dependent amplitude, $\Phi(\mathbf{n},t)$. We have

$$\theta^{(1)}(\mathbf{n},\mathbf{x},t) = \Phi(\mathbf{n},t)e^{i\mathbf{n}\cdot\mathbf{x}}. \tag{9.6}$$

Since the equations governing $\theta^{(1)}$ are linear, we can construct an arbitrary disturbance by adding solutions of the type (9.6).

Now introduce a new origin of coordinates, such that the point P is defined by a position vector $\mathbf{d}$. A point which was previously defined by the position vector $\mathbf{s}$ now has the position vector $(\mathbf{s}+\mathbf{d})$. Thus we write $\theta^{(1)}(P,t) = \theta^{(1)}(\mathbf{d},t)$ and $\theta^{(1)}(\mathbf{s},t) = \theta^{(1)}(\mathbf{d}+\mathbf{s},t)$. Substituting (9.6) in (9.5), making this transformation, and removing a factor $e^{i\mathbf{n}\cdot\mathbf{d}}$ from both sides, we find

$$\rho c \frac{\partial \Phi(\mathbf{n},t)}{\partial t} = \Phi(\mathbf{n},t) \int_0^\infty d\nu 4\pi k_{v,\nu} \frac{d\overline{B}_\nu}{d\theta} \left(\int \frac{k_{v,\nu} e^{i\mathbf{n}\cdot\mathbf{s}} e^{-k_{v,\nu}s}\, dV(\mathbf{s})}{4\pi s^2} - 1 \right). \quad (9.7)$$

Equation (9.7) can be written

$$\frac{\partial \Phi(\mathbf{n},t)}{\partial t} = -N(n)\Phi(\mathbf{n},t), \quad (9.8)$$

where

$$N(n) = \frac{4\pi}{\rho c} \int_0^\infty d\nu k_{v,\nu} \frac{d\overline{B}_\nu}{d\theta}\left(1 - \int \frac{k_{v,\nu} e^{i\mathbf{n}\cdot\mathbf{s}} e^{-k_{v,\nu}s}\, dV(\mathbf{s})}{4\pi s^2}\right), \quad (9.9)$$

is a reciprocal time.

Equation (9.8) has the solution

$$\Phi(\mathbf{n},t) = \Phi(\mathbf{n},0)e^{-N(n)t}, \quad (9.10)$$

where $\Phi(\mathbf{n},0)$ is the imposed initial condition at $t = 0$.

The generalization to an arbitrary disturbance can be made by the usual methods of Fourier analysis. The amplitude component can be found from the integral

$$\Phi(\mathbf{n},t) = \frac{1}{(2\pi)^{3/2}} \int \theta^{(1)}(\mathbf{s},t)e^{-i\mathbf{n}\cdot\mathbf{s}}\, dV(\mathbf{s}). \quad (9.11)$$

Given $\theta^{(1)}(\mathbf{s},0)$, (9.11) determines the initial conditions for (9.10). $\Phi(\mathbf{n},t)$ is therefore known and $\theta^{(1)}(\mathbf{s},t)$ can be found by inverting (9.11)

$$\theta^{(1)}(\mathbf{s},t) = \frac{1}{(2\pi)^{3/2}} \int e^{-i\mathbf{n}\cdot\mathbf{s}}\Phi(\mathbf{n},t)\, dV(\mathbf{n}). \quad (9.12)$$

This completes the formal solution, and it remains to determine $N(n)$. Let $\xi = \cos(\mathbf{n},\mathbf{s})$. From (2.82) and (2.91)

$$dV(\mathbf{s}) = -2\pi\, d\xi\,.s^2\, ds, \quad (9.13)$$

and

$$N(n) = \frac{4\pi}{\rho c} \int_0^\infty d\nu\, k_{v,\nu} \frac{d\overline{B}_\nu}{d\theta}\left\{1 - \frac{1}{2} \int_{-1}^{+1} d\xi \int_0^\infty ds\, k_{v,\nu} e^{ins\xi} e^{-ik_{v,\nu}s}\right\}. \quad (9.14)$$

We may first perform the integral with respect to ξ, and subsequently integrate with respect to either s or ν, depending upon the nature of the problem. In the former case we have

$$N(n) = \frac{4\pi}{\rho c} \int_0^\infty d\nu \, k_{v,\nu} \frac{\overline{dB_\nu}}{d\theta} \left\{ 1 - \frac{k_{v,\nu}}{n} \tan^{-1}\left(\frac{n}{k_{v,\nu}}\right) \right\}, \tag{9.15}$$

and in the latter (after one partial integration with respect to s)

$$N(n) = \frac{4\pi \overline{(dB/d\theta)}}{\rho c} \int_0^\infty \frac{ds}{s} \frac{\partial \epsilon_c^*(s)}{\partial s} \left\{ \frac{\sin ns}{ns} - \cos ns \right\}, \tag{9.16}$$

where

$$\epsilon_c^*(s) = \frac{\int_0^\infty d\nu \, (\overline{dB_\nu/d\theta})(1 - e^{-k_{v,\nu}s})}{\int_0^\infty d\nu \, (\overline{dB_\nu/d\theta})}, \tag{9.17}$$

is the modified emissivity introduced in (6.46).

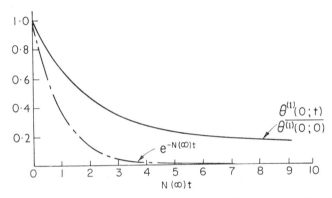

FIG. 9.1. Decay of central temperature in a spherically-symmetric disturbance.
After Spiegel (1958). The initial disturbance had the form
$\theta^{(1)}(s, 0)/\theta^{(1)}(0, 0) = \exp(-k_v s)$.

From (9.15) we have for grey absorption

$$N(n) = \frac{4\pi k_v \overline{(dB/d\theta)}}{\rho c} \left\{ 1 - \frac{k_v}{n} \tan^{-1}\left(\frac{n}{k_v}\right) \right\}$$

$$= N(\infty) \left\{ 1 - \frac{k_v}{n} \tan^{-1}\left(\frac{n}{k_v}\right) \right\}, \tag{9.18}$$

where $N(\infty)$ is the reciprocal decay time for a disturbance small compared to the mean-free-path of the radiation.

A solution to (9.12) for a particular initial disturbance, using $N(n)$ from (9.18), is shown in Fig. 9.1. The difference from a simple exponential time decay is striking. As time progresses, and the disturbance grows, the dominant wave number tends to zero; $N(n)$ also tends to zero, and the dissipation becomes progressively slower.

9.1.2. *The effective absorption coefficient*

We can write
$$N(n) = \frac{4\pi(\overline{dB/d\theta})}{\rho c} \tilde{k}_v(n),$$

where, from (9.15) and (9.16),

$$\tilde{k}_v(n) = \int_0^\infty \frac{ds}{s} \frac{\partial \epsilon_c^*(s)}{\partial s} \left(\frac{\sin ns}{ns} - \cos ns \right) \tag{9.19}$$

$$= \frac{\int_0^\infty dv\, k_{v,\nu}(\overline{dB_\nu/d\theta})\{1 - (k_{v,\nu}/n)\tan^{-1}(n/k_{v,\nu})\}}{\overline{dB/d\theta}}, \tag{9.20}$$

is an *effective absorption coefficient* for wave number n.

As $n \to \infty$
$$\tilde{k}_v(n) \to \check{k}_{v,\mathrm{P}}, \tag{9.21}$$

where $\check{k}_{v,\mathrm{P}}$ is the Planck mean (see (2.133)). As $n \to 0$, we have

$$\tilde{k}_v(n) \to \frac{1}{3} \frac{n^2}{\check{k}_{v,\mathrm{R}}}, \tag{9.22}$$

where $\check{k}_{v,\mathrm{R}}$ is the Rosseland mean (see (2.134)).

The main contributions to the Planck and Rosseland means come from independent spectral regions; the former is dominated by the strong line centres, and the latter by the weak continua between bands. Fig. 9.2 illustrates this point for water vapour in dilute mixture with air. A word of explanation about the data used for the Planck mean is required. These make use of a tabulation by Cowling (1950) of the amount of water $(a_{0,i})$ required to give a mean transmission of $\frac{1}{2}$ in the ith spectral interval. From (5.5)

$$\frac{1 \cdot 97}{a_{0,i}} = \left(\frac{\sigma}{\delta} \right)_i = \frac{\int_i k_\nu\, dv}{\int_i dv}. \tag{9.23}$$

We may therefore write (2.133) in the approximate form

$$\check{k}_{\mathrm{P}} \simeq \frac{\sum_i (1 \cdot 97/a_{0,i})(dB/d\theta)_i}{\sum_i (dB/d\theta)_i}. \tag{9.24}$$

The data of Cowling, used in Fig. 9.2, are not the same as the data discussed in § 5.4.4 and for Cowling's data a numerical factor 2·5 should replace 1·97 in (9.24). Note that (9.23) is independent of pressure;

and so too must be $\bar{k}_P$. $\bar{k}_P$ depends slightly upon temperature, although this can be neglected in the present context. The data yield for 290° K

$$\bar{k}_{v,P}(H_2O) = 203\rho_{H_2O} \text{ cm}^{-1}. \qquad (9.25)$$

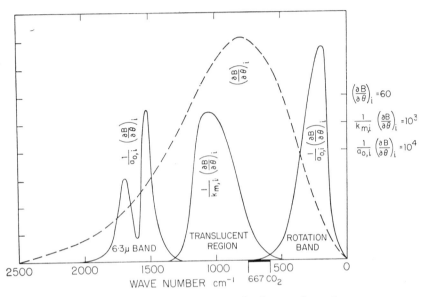

FIG. 9.2. Estimation of Planck and Rosseland means for water vapour.

The water vapour is in dilute mixture with air at 290° K and a total pressure of 1 atmosphere. The curves marked $(dB/d\theta)_i/a_{0,i}$ show contributions to the Planck mean. $a_{0,i}$ is the amount of water vapour (g cm^{-2}) required to give $\bar{T}_i = \frac{1}{2}$ in the ith spectral range (Cowling, 1950). The curve marked $(dB/d\theta)_i/k_{m,i}$ relates to the Rosseland mean (see text); the data follow Saiedy (1960). The extent of the 15 μ CO$_2$ band is marked on the frequency axis.

Carbon dioxide and ozone also contribute to the Planck mean, mainly through their strong bands at $15\,\mu$ and $9\cdot6\,\mu$ respectively. Over these bands $dB_\nu/d\theta$ is approximately constant and (2.133) becomes

$$\bar{k}_P \simeq \frac{(dB/d\theta)_i S_{\text{band}}}{dB/d\theta}. \qquad (9.26)$$

Using the values for the band intensity of CO$_2$ (00^{0}0–01^{1}0) from Table 5.16, and O$_3(\nu_3)$ from § 5.6.3, we find, at 293° K,

$$\bar{k}_{v,P}(CO_2) = 96\rho_{CO_2} \text{ cm}^{-1}, \qquad (9.27)$$

$$\bar{k}_{v,P}(O_3) = 136\rho_{O_3} \text{ cm}^{-1}. \qquad (9.28)$$

The Planck mean volume absorption coefficient for a mixture of these gases is the sum of (9.25), (9.27), and (9.28).

It is a matter of observation (see § 5.4.7) that the most transparent spectral regions of the lower atmosphere near $10\,\mu$ are dominated by a water-vapour continuum, and it can readily be demonstrated that these regions alone contribute to the Rosseland mean. Data for this continuum have been used to construct the curve marked $(dB/d\theta)_i/k_{m,i}$ in Fig. 9.2, and integration of this curve leads to

$$\bar{k}_{v,\mathrm{R}}(\mathrm{H_2O}) = 0{\cdot}21\rho_{\mathrm{H_2O}}\!\left(\frac{p}{p_0}\right)\,\mathrm{cm^{-1}}. \tag{9.29}$$

The pressure factor in (9.29) is not established experimentally, but it is a reasonable inference from our interpretation of the continuum as formed by the far wings of pressure-broadened lines (§ 5.4.7).

The strong ν_2 band of CO_2 influences the Rosseland mean by reducing the integrand in the numerator of (2.134) to zero in the frequency range covered by the band. Its extent is indicated in Fig. 9.2, showing that its effect will be small and can be neglected for most atmospheric applications.

Between the two asymptotic forms involving the Planck and Rosseland means (9.19) can be integrated numerically if the necessary data are available. In Fig. 9.3, the data of Elsasser and Culbertson (1960) have been used to construct section II of the curves.

9.2. Radiative-diffusive boundary layers

9.2.1. *The stationary state*

The Eddington approximation with grey absorption for a homogeneous stratified fluid in thermodynamic equilibrium gives for the radiative flux (2.110)

$$\frac{\partial^2 F_{\mathrm{r}}}{\partial \tau^2} - 3F_{\mathrm{r}} = -4\pi\frac{\partial B}{\partial \tau}, \tag{9.30}$$

where partial differentials have been introduced to permit time-dependent solutions.†

We will include a diffusive heat flux, defined in terms of a conductivity coefficient, K,

$$F_{\mathrm{d}} = -K\frac{\partial \theta}{\partial z}. \tag{9.31}$$

This may be the molecular flux, F_{m}, or, if the mixing-length theory of turbulence should be valid, it can also be the turbulent flux, F_{t}. The magnitude of K differs greatly in the two cases, but the flux equation

---

† This simple method of introducing temporal variability is possible only because it is implicitly assumed that the velocity of light is infinitely large compared with the velocities of fluid motions.

is the same. τ and ξ are measured in the direction of the stratification and fluxes are also in this direction.

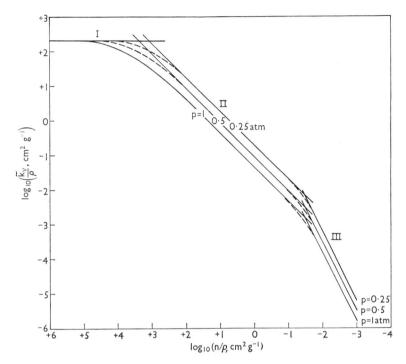

FIG. 9.3. Effective absorption coefficient for water vapour broadened by air. The curves fall into three distinct sections. I follows (9.21) and corresponds to a *weak-line* approximation. III follows (9.22) and is dominated by absorption in the $10\,\mu$ window. Section II has unit slope, corresponding to $d\epsilon^*/d\log a = $ const, a relation which is approximately obeyed over a wide range of a, and corresponds to a *strong-line* condition. The broken lines are hypothetical.

Equation (9.30) is linearized by writing

$$\delta B = \left(\overline{\frac{\partial B}{\partial \theta}}\right)\delta\theta, \tag{9.32}$$

and treating $\overline{(\partial B/\partial\theta)}$ as a constant. The equations can be closed with a heat equation,

$$\rho c\frac{\partial\theta}{\partial t} = -\frac{\partial F}{\partial z}. \tag{9.33}$$

Throughout this section, we assume that there are no mean motions and therefore no advected heat flux.

As regards boundary conditions, the diffusive heat transfer will ensure that there are no temperature discontinuities. Consequently, we must employ (2.118) and (2.119):

$$\left. \begin{array}{l} \left(\dfrac{\partial F_\mathrm{r}}{\partial \tau}\right)_{\tau=\tau^*} = -2rF_\mathrm{r}(\tau^*) \\[3mm] \left(\dfrac{\partial F}{\partial \tau}\right)_{\tau=0} = +2rF_\mathrm{r}(0) \end{array} \right\}, \tag{9.34}$$

where

$$r = \frac{1+\alpha}{1-\alpha},$$

and

$$\alpha = \text{reflectivity of the boundary.}$$

Eliminate F_r and F_d from (9.30), (9.31), and (9.33) and write

$$\left. \begin{array}{l} d\tau = -k_v\,dz \\[3mm] \chi = \dfrac{4\pi\,\overline{d\,B/d\theta}}{3k_v\,K} \\[3mm] N(\infty) = \dfrac{4\pi k_v\,\overline{d\,B/d\theta}}{\rho c} \end{array} \right\}. \tag{9.35}$$

We find

$$\frac{3\chi}{N(\infty)}\left(\frac{\partial^2}{\partial \tau^2}-3\right)\frac{\partial\theta}{\partial t} = \left\{\left(\frac{\partial^2}{\partial \tau^2}-3\right)-3\chi\right\}\frac{\partial^2\theta}{\partial \tau^2}. \tag{9.36}$$

The parameter χ is the ratio of an effective radiative conductivity $\{4\pi(\overline{d\,B/d\theta})/3k_v\}$ to the thermal conductivity (K). Some early meteorological literature is based on the presumption that this parameter alone defines the relative importance of radiation and diffusion as alternative means of heat transfer. Such a point of view rests on an assertion that the radiative flux can be described by means of a diffusion equation, like (9.31). This is so if the first term on the l.h.s. in (9.30) can be neglected, implying that F_r varies only slowly with τ. Far from the influence of boundaries or other discontinuities, it is to be expected that this condition will be obeyed, e.g. in the deep interior of a star. One of the characteristic features of the earth's atmosphere, however, is its transparency in certain spectral regions through which there is direct heat exchange with the boundaries. In such a boundary region no restrictions on the rate of change of F_r can be foreseen, and the first term on the l.h.s. of (9.30) cannot be neglected.

We will now consider a stationary problem $(\partial/\partial t = 0)$ with boundaries at $\tau = 0$ and τ^*, at which the temperatures are $\theta(0)$ and $\theta(\tau^*)$. The l.h.s. of (9.36) is now zero, and eliminating F_r and its derivatives from

(9.30), (9.31), and (9.34) with the help of the relation

$$\frac{\partial F}{\partial \tau} = \frac{\partial (F_r + F_d)}{\partial \tau} = 0,$$ (9.37)

we find, as boundary conditions

$$\left.\begin{array}{l}\dfrac{1}{3}\left(\dfrac{\partial^3 \theta}{\partial \tau^3}\right)_{\tau=\tau^*} + \dfrac{1}{2r}\left(\dfrac{\partial^2 \theta}{\partial \tau^2}\right)_{\tau=\tau^*} - \chi\left(\dfrac{\partial \theta}{\partial \tau}\right)_{\tau=\tau^*} = 0 \\[3mm] \dfrac{1}{3}\left(\dfrac{\partial^3 \theta}{\partial \tau^3}\right)_{\tau=0} - \dfrac{1}{2r}\left(\dfrac{\partial^2 \theta}{\partial \tau^2}\right)_{\tau=0} - \chi\left(\dfrac{\partial \theta}{\partial \tau}\right)_{\tau=0} = 0 \end{array}\right\}.$$ (9.38)

The solution to (9.36) follows immediately, and is conveniently expressed in terms of a ratio of the temperature gradient to its mean value across the fluid

$$\frac{\partial \theta}{\partial \tau}\bigg/\frac{\overline{\partial \theta}}{\partial \tau} = \frac{\chi \cosh g(\tau - \tau^*/2) + \cosh g\tau^*/2 + (g/2r)\sinh g\tau^*/2}{(2\chi/g\tau^*)\sinh g\tau^*/2 + \cosh g\tau^*/2 + (g/2r)\sinh g\tau^*/2},$$ (9.39)

where $$g = \sqrt{\{3(1+\chi)\}}.$$

If the heat transfer were governed by a diffusion equation only, the temperature gradient would be constant and

$$\frac{d\theta}{d\tau}\bigg/\frac{\overline{d\theta}}{d\tau} = 1.$$ (9.40)

This limit can be reached in two ways. If $\chi \ll 1$ the first terms in both numerator and denominator of (9.39) can be neglected compared to the other terms. If $g\tau^* \ll 1$, the first terms in the numerator and denominator are equal. In either case, we have the result (9.40); in the first case the medium is opaque, and in the second it is transparent.

If τ^* and χ are both large, however, an exponential boundary layer develops. Let us examine the boundary layer near $\tau = \tau^*$ on the assumption that $\tau^* \gg 1$. It can be shown that (9.39) becomes

$$\frac{\partial \theta}{\partial z} \simeq -\frac{F}{K}\left\{\frac{1}{1+\chi} + \frac{2\chi}{1+\chi}\frac{\exp g(\tau - \tau^*)}{2 + (g/r)}\right\}.$$ (9.41)

In an optically thick atmosphere the predicted conditions near to a surface are therefore a constant lapse rate equal to $(F/K)\{1/(1+\chi)\}$, with a superposed exponential term, whose slope at the surface is (for large χ) approximately $\sqrt{\chi}$ greater than the constant term. This exponential term can be identified with the temperature discontinuity found for strict radiative equilibrium, smoothed out by thermal diffusion.

In a translucent medium the upper and lower boundary layers merge, and give rise to an S-shaped temperature profile. Fig. 9.4 shows some calculated profiles, which attempt to predict the behaviour of a 2 cm layer of gaseous ammonia in a laboratory experiment. There are three different boundary conditions; of these the asymmetric condition with one mirror and one black body requires a simple extension of the analysis presented in this section. It is interesting to note how much the profile is changed by destroying the symmetry.

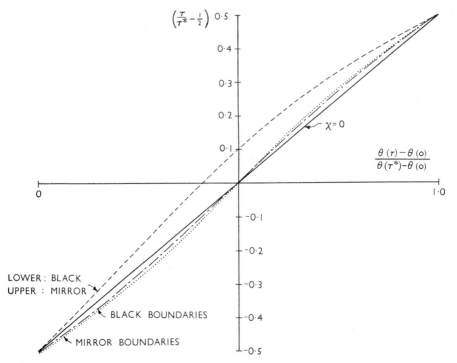

FIG. 9.4. Temperature profiles for a 2 cm layer of NH_3 in diffusive-radiative equilibrium under different boundary conditions.

$$\chi = 5 \cdot 46.$$

It has been suggested that the temperature profile observed in the atmosphere over a surface on a very hot, wind-free day is that given by (9.41). On occasion the observed temperatures fit an exponential curve well, but, nevertheless, it is doubtful whether this is a correct interpretation of the phenomenon. Consider the numerical magnitude of the scale height in the exponent of (9.41),

$$H = \frac{z}{g(\tau - \tau^*)} = \frac{1}{k_v \sqrt{\{3(1+\chi)\}}}. \tag{9.42}$$

The *minimum* value of H is that for which k_v is a *maximum* and K is a *minimum*. These extrema can be determined with some precision.

The mean value of k_v cannot exceed the Planck mean, which for a water-vapour pressure of 10 mb (a pressure which is rarely exceeded) is $1\cdot4\times10^{-3}$ cm^{-1}. K cannot under any circumstances be less than the thermal conductivity of air, which is $2\cdot38\times10^3$ erg cm^{-1} s^{-1}. These figures give

$$H \geqslant 9 \text{ cm.}$$

It is an interesting coincidence that $H \simeq 8$ cm has been observed over a surface in the tropics, but it can scarcely be more than a coincidence since this particular boundary layer was observed to be in active convection and the effective conduction coefficient should far exceed the molecular conductivity. Moreover, under similar circumstances in the laboratory, scale heights of a few millimetres have been observed, and these are much too small to be attributed to the mechanism under discussion.

9.2.2. Transmission of an harmonic wave

The diurnal temperature wave is propagated through the lower atmosphere by radiative transfer, turbulent diffusion, and advection. Close to the earth's surface the first two modes of heat transfer predominate, and (9.36) governs the temperature changes which take place, provided that turbulence can be described by the diffusion equation (9.31).

Consider an optically thick atmosphere, whose temperature is

$$\theta = \theta^{(0)} + \theta^{(1)}, \tag{9.43}$$

where $\theta^{(1)}$ has a zero time-average, and has the simple-harmonic form

$$\theta^{(1)} = \theta_s^{(1)} e^{2\pi i v t} e^{-(a+ib)\zeta}, \tag{9.44}$$

where $\theta_s^{(1)}$ = amplitude of $\theta^{(1)}$ at the surface,

and $\zeta = \sqrt{3}(\tau^* - \tau).$

The constants a and b govern respectively the change of amplitude and the change of phase with height.

Equation (9.41) is the solution for $\theta^{(0)}$; here we will only consider the time-dependent term $\theta^{(1)}$. The solution will be compared with that for diffusion alone, for which circumstance we have from (9.31) and (9.33)

$$\rho c \frac{\partial \theta^{(1)}}{\partial t} = K \frac{\partial^2 \theta^{(1)}}{\partial z^2}. \tag{9.45}$$

Substituting (9.44) in (9.45), we find the familiar result

$$\zeta a = \zeta b = z \sqrt{\left(\frac{\pi \rho c v}{K}\right)}. \tag{9.46}$$

Observed diurnal changes in the atmosphere do not follow this simple form. However, they are often described in the literature as if they did but with a and b both varying with height. It is possible to define two coefficients $K(\text{amp})$ and $K(\text{phase})$ by

$$\left.\begin{aligned}\frac{d\zeta a}{dz} &= \sqrt{\Big/\!\left(\frac{\pi\rho c\nu}{K(\text{amp})}\right)}\\\frac{d\zeta b}{dz} &= \sqrt{\Big/\!\left(\frac{\pi\rho c\nu}{K(\text{phase})}\right)}\end{aligned}\right\}, \tag{9.47}$$

but this is simply a descriptive device, with no physical significance unless K happens to be a constant.

We will now briefly describe the general form of the solution to (9.36). A new parameter enters as a result of substituting

$$\frac{\partial}{\partial t} \equiv 2\pi i\nu. \tag{9.48}$$

This is the dimensionless period

$$\xi = \frac{N(\infty)}{2\pi\nu}. \tag{9.49}$$

If ξ is large or small, the oscillations are *slow* or *fast* respectively. Using (9.48), the boundary conditions equivalent to (9.38) can be stated, involving two new, imaginary terms. These need only be applied to the lower boundary; the upper boundary conditions being satisfied if the disturbance decreases exponentially with height.

The equations permit two solutions which will be designated (1) and (2) respectively. They involve the parameters ξ and χ, but the discussion may be restricted to $\chi \gg 1$, for if χ is small, the solution is (9.46), and hence radiative transfer plays no part in the phenomenon.

The most interesting properties of this dual wave system are that $a(1)$ always exceeds $a(2)$, and that $\theta_s(1) > \theta_s(2)$ for fast oscillations, while $\theta_s(1) < \theta_s(2)$ for slow oscillations. Thus if the oscillations are sufficiently fast, solution (1) dominates near the surface, but is more rapidly attenuated than solution (2), which therefore overtakes and dominates above some value of $\zeta = \zeta_c$ (or $z = z_c$). For slow oscillations solution (2) dominates at all levels. The qualitative nature of the solution therefore depends greatly upon the value of the parameter ξ.

The following figures are appropriate to a diurnal wave at ground level in the earth's atmosphere. We choose $\tilde{k}_v = 4{\cdot}5\times10^{-5}$ cm^{-1}, corresponding to a disturbance of about 14 m characteristic length (Fig. 9.3). A representative value of $K/\rho c$ for turbulent diffusion in the earth's boundary layer is of the order of 10^2 cm^2 s^{-1}, although it is a highly variable quantity. With these data, there results

$$\chi = 120, \qquad \xi = 1\cdot0, \qquad z_c = 10 \text{ m.}$$

The character of the wave therefore changes at 10 m, for this particular example. With the definitions of (9.47) we find $K(\text{amp})/\rho c \simeq 40$ cm^2 s^{-1} for $z < z_c$ and $K(\text{amp})/\rho c \simeq 10^4$ cm^2 s^{-1} for $z > z_c$. $K(\text{phase})$ is about six times larger than $K(\text{amp})$.

Some of the features of this solution are observed in the earth's boundary layer. They are usually described in terms of modifications to (9.45), for which fluid-dynamical explanations are sought. The present solution suggests that it is dangerous to seek a purely dynamical solution, and that the complex behaviour of the diurnal wave in the boundary layer may be partly ascribed to radiative transfer.

9.2.3. *A numerical solution*

The numerical results in Table 9.1 concern the rate of cooling at night of an initially isothermal atmosphere under the influence of both turbulence and radiation. A representative turbulent conductivity was used and allowed to vary with height according to observation. Radiative cooling was computed from a modified radiation chart and the temperature change was obtained by a marching technique.

TABLE 9.1

Temperature drop in 4 hours for an initially isothermal atmosphere cooling to space

After Gaevskaya, Kondratiev, and Yakushevskaya (1961)

	Temperature drop in 4 hours ($^{\circ}K$)		
Altitude	*Turbulence alone*	*Radiation alone*	*Turbulence + radiation*
(cm)			
0	4·29	1·8 (approx.) at	3·76
2·0	3·11	all levels	2·78
4·8	2·82		2·54
24·0	2·26		2·06
48·0	2·05		1·89
200·0	1·52		1·45
480·0	1·22		1·20
2400·0	0·71		0·81
24000·0	0·19		0·43

From these numerical values alone it is difficult to gain insight into the physical processes, but they appear to confirm the tentative conclusions of the last section. The non-additivity of turbulent and radiative contributions emphasizes the fundamental difference between the two processes; it also illustrates the pitfall of attempting to subtract computed radiative changes when discussing observed turbulent transfer in the boundary layer. In this particular example it would be better to ignore the radiative term rather than to subtract it, but since both transfer mechanisms are demonstrably important this would be a dangerous conclusion to apply generally.

At low levels there is the curious result that the two mechanisms cool less together than turbulence alone. Somewhere between 4·8 m and 24 m there is a change and the net cooling is greater than the turbulent but less than the radiative contribution. It may well be that this change is to be associated with the change of wave régime at 10 m, which was discussed in the last section.

9.3. The onset of cellular convection

9.3.1. *Dimensional argument*

A fluid heated from below will start to convect if the buoyancy forces overcome the stabilizing viscous forces. Suppose that a disturbance has a characteristic dimension l, and that its density differs from the surroundings by $\rho^{(1)}$. The buoyancy force will be of the order of $g\rho^{(1)}l^3$. If ν is the kinematic viscosity and $\partial w/\partial x$ a typical velocity gradient, the viscous force will be of the order of $\nu\rho l^2(\partial w/\partial x)$. The criterion for onset of convection can therefore be written

$$g\rho^{(1)}l^3 \geqslant \text{const } \nu\rho l^2\frac{\partial w}{\partial x}. \qquad (9.50)$$

If α is the coefficient of thermal expansion

$$\rho^{(1)} = -\alpha\rho\theta^{(1)}, \qquad (9.51)$$

and since the buoyancy is associated with motions of dimension l

$$\theta^{(1)} \simeq \beta l, \qquad (9.52)$$

where β is the temperature gradient. Finally, if we regard the velocity w as being created during the lifetime, t, of the disturbance, we can write

$$\frac{\partial w}{\partial x} \simeq t^{-1}. \qquad (9.53)$$

The criterion for onset of convection now becomes

$$-\frac{g\alpha\beta l^2}{\nu}t \geqslant \text{const.} \qquad (9.54)$$

The constant can be evaluated by referring to a special case for which a complete solution is available, namely parallel plates with spacing h, with diffusive transfer alone. Referring to (9.45), we may expect that as regards order of magnitude

$$t_{\text{diff}} \simeq \frac{\rho c}{K} l^2. \tag{9.55}$$

It is therefore natural to use the Rayleigh number

$$Ra = -\frac{g\alpha\beta h^4 \rho c}{K\nu}, \tag{9.56}$$

to describe the stability for this particular problem. According to (9.54) the fluid first becomes unstable when the Rayleigh number exceeds some critical value, which we will designate Ra_c.

For diffusion alone Ra_c is known to be 657 for mechanically free surfaces, and 1708 for rigid surfaces. These two numbers will be represented by the symbol $Ra_c(\text{diff})$. From (9.54) it follows that, in a more general case

$$Ra_c = Ra_c(\text{diff})\frac{t(\text{diff})}{t}. \tag{9.57}$$

If two competitive processes occur, such as radiation and diffusion, we may expect

$$\frac{1}{t} \simeq \frac{1}{t(\text{diff})} + \frac{1}{t(\text{rad})}. \tag{9.58}$$

From (9.58), (9.55), and (9.18), identifying the wave number with l^{-1}, we have

$$Ra_c = Ra_c(\text{diff})\left\{1 + \frac{\lambda^2}{a^2}\left(1 - \frac{\lambda}{a\sqrt{(3\chi)}}\tan^{-1}\frac{a\sqrt{(3\chi)}}{\lambda}\right)\right\}, \tag{9.59}$$

where $a = h/l$ defines the size of disturbance in terms of the plate separation, and λ^2/a^2 is the ratio of $t(\text{diff})$ to $t(\text{rad})$ for very large wave numbers:

$$\lambda^2 = t(\text{diff})N(\infty)a^2$$
$$= \frac{\rho c l^2}{K}\frac{4\pi(\overline{dB/d\theta})k_v}{\rho c}\frac{h^2}{l^2}$$
$$= 3\chi\tau^{*2}, \tag{9.60}$$

where $\tau^* = k_v h = k_v al.$

We cannot determine the value of a from these elementary considerations. Clearly it is of order of magnitude unity, and when its correct value is known we will find (9.59) to give results close to those from more elaborate theory. An important feature of this simple theory is that, being stated only in terms of dissipation times, it can be applied wherever these are known, e.g. to non-grey radiation, using the time constants in § 9.1.2.

9.3.2. *Variational methods*

Consider an initial static state, with a vertical temperature gradient β but no horizontal variation, on which an infinitesimally small disturbance, with vertical velocity w develops. The partial derivative on the l.h.s. of the heat equation (9.33) must now be replaced by a substantive derivative, and the three-dimensional character of the flux must be recognized. The equation becomes

$$-\nabla.\mathbf{F} = \rho c \frac{\partial \theta}{\partial t} + \rho c w \beta. \tag{9.61}$$

Here, and throughout the following analysis, the fluid is assumed incompressible. This restriction is not as serious as it may seem at first sight, for under certain conditions the effect of compressibility is taken into account by adding to β the adiabatic lapse rate. This does not modify the analysis in any important way.

If density variations through the fluid are small, they only enter the equations of motion through buoyancy terms, i.e. in the form of products with gravity. With this restriction (the Boussinesq approximation), the perturbation equations for an incompressible, non-rotating fluid reduce to

$$\frac{\partial}{\partial t} \nabla^2 w = \nu \nabla^4 w + g\alpha \nabla_1^2 \theta^{(1)}, \tag{9.62}$$

where
$$\nabla^2 = \frac{\partial^2}{\partial x^2} + \frac{\partial^2}{\partial y^2} + \frac{\partial^2}{\partial z^2} = \nabla_1^2 + \frac{\partial^2}{\partial z^2}.$$

The time derivatives in (9.61) and (9.62) must be treated with care. Suppose that we can write
$$w \propto e^{\nu t},$$

or
$$\frac{\partial w}{\partial t} = \nu w. \tag{9.63}$$

If ν is real, then $\nu > 0$ describes a growing disturbance, while $\nu < 0$ describes one which is decaying. For marginal stability, when the disturbance is on the point of growing, $\nu = 0$ and time-dependent terms can be omitted. However, ν may be complex, in which case there are oscillatory solutions. For diffusion alone this possibility has been ruled out for both free and rigid boundaries. For mixed radiative-diffusive transfer we may reasonably expect the same result, although it has only been demonstrated in the case of β constant and for rigid boundaries. We will therefore set $\partial/\partial t = 0$ in (9.61) and (9.62).

$\nabla.\mathbf{F}$ can be written in the form

$$\nabla.\mathbf{F} = \nabla.\mathbf{F}^{(0)} + \nabla.\mathbf{F}^{(1)}. \tag{9.64}$$

$\mathbf{F}^{(0)}$ is the flux in the initial static state and, by definition,

$$\nabla.\mathbf{F}^{(0)} = -\rho c \frac{\partial \theta^{(0)}}{\partial t} = 0. \tag{9.65}$$

From (9.31) and (9.5) both radiative and diffusive heating rates can be expressed by equations linear in θ,

$$\nabla.\mathbf{F} = -\mathscr{L}(\theta), \tag{9.66}$$

where $\mathscr{L}$ is a linear operator. It follows that

$$\nabla.\mathbf{F}^{(1)} = -\mathscr{L}(\theta^{(1)}), \tag{9.67}$$

and, eliminating $\theta^{(1)}$ between (9.61), (9.62) and (9.67), we have

$$\frac{-\rho c \beta g \alpha}{\nu} \nabla_1^2 w = \mathscr{L}(\nabla^4 w). \tag{9.68}$$

Equation (9.68) together with the appropriate boundary conditions for free or rigid surfaces is sufficient to specify the problem. The Rayleigh number is an eigenvalue of the equation and its minimum value (with respect to all free parameters) is the critical Rayleigh number. Variational techniques can yield reasonably accurate eigenvalues without having to make a complete solution of (9.68). The first step is to separate the space variables. If h is the extent of the fluid in the z-direction, we write

$$w = e^{ia_\xi \xi + ia_\eta \eta} W(\zeta), \tag{9.69}$$

where
$$\xi, \eta, \zeta = \frac{x}{h}, \frac{y}{h}, \left(\frac{z}{h} - \frac{1}{2}\right),$$

and a_ξ, a_η define the horizontal wave number of the disturbance. An arbitrary horizontal structure can be synthesized by Fourier summation, if required, but the eigenvalues of (9.68) depend only on the parameter

$$a^2 = a_\xi^2 + a_\eta^2, \tag{9.70}$$

which is treated as a free parameter for minimizing the Rayleigh number. Writing
$$D = \partial/\partial\zeta,$$

$$Ra = -\frac{g\alpha\bar\beta h^4 \rho c}{K\nu},$$

$$\bar\beta = \text{mean value of } \beta,$$

we find

$$Ra\,W(\zeta)e^{ia_\xi\xi+ia_\eta\eta}\frac{\beta}{\bar{\beta}} = \frac{h^2}{K}\mathscr{L}\left(e^{ia_\xi\xi+ia_\eta\eta}\frac{(D^2-a^2)^2}{a^2}\,W(\zeta)\right). \qquad (9.71)$$

Let M be an operator, which gives a non-zero result when applied to (9.71), then

$$Ra = \frac{h^2 M \cdot \mathscr{L}\left[e^{ia_\xi\xi+ia_\eta\eta}\{(D^2-a^2)^2/a^2\}W(\zeta)\right]}{KM\cdot(\beta/\bar{\beta})\,W(\zeta)e^{ia_\xi\xi+ia_\eta\eta}}. \qquad (9.72)$$

The aim of the variational method is to find a form for M which makes (9.72) stationary with respect to small variations of $W(\zeta)$ from the correct solution. The form of M will depend upon the boundary conditions, but if a suitable choice can be made it is generally sufficiently accurate to use in (9.72) any physically reasonable trial function $W^{(0)}(\zeta)$ which obeys the boundary conditions.

9.3.3. Numerical results

Two numerical solutions of (9.72) are available, for slightly different conditions. Spiegel (1960) has treated the case of radiative transfer alone, for rigid boundaries, under the assumption that β is constant. This last assumption is not consistent with the results of § 9.2.1, although it is appropriate to the particular circumstance of the deep interior of a star, which Spiegel had in mind.

An operator with the required minimal properties was shown to be

$$M = \int_{-\frac{1}{2}}^{+\frac{1}{2}} (D^2-a^2)^2 W(\zeta)\,d\zeta = \int_{-\frac{1}{2}}^{+\frac{1}{2}} L(\zeta)\,d\zeta, \qquad (9.73)$$

and, using the expression (9.5) for the heating rate, there results

$$Ra_c = \frac{3\tau^{*2}\chi}{a^2}\,\frac{\displaystyle\int_{-\frac{1}{2}}^{+\frac{1}{2}} L^2(\zeta)\,d\zeta - \int_{-\frac{1}{2}}^{+\frac{1}{2}} d\zeta \int_{-\frac{1}{2}}^{+\frac{1}{2}} d\zeta'\,K(\tau^*,\zeta-\zeta')L(\zeta)L(\zeta')}{\displaystyle\int_{-\frac{1}{2}}^{+\frac{1}{2}} L(\zeta)W(\zeta)\,d\zeta}, \qquad (9.74)$$

where

$$K(\tau^*,\zeta-\zeta') = \int_{-\infty}^{+\infty} d\xi' \int_{-\infty}^{+\infty} d\eta'\,\tau^*\,\frac{e^{-\tau^*\sqrt{((\xi-\xi')^2+(\eta-\eta')^2+(\zeta-\zeta')^2)}}e^{ia_\xi(\xi'-\xi)+ia_\eta(\eta'-\eta)}}{4\pi\{(\xi-\xi')^2+(\eta-\eta')^2+(\zeta-\zeta')^2\}}. \qquad (9.75)$$

Equation (9.74) can be solved by numerical quadrature. By inspection, it can be seen that Ra_c/χ can be expressed as a function of τ^* only. Two trial functions were employed,

$$L_1^{(0)}(\zeta) = \cos\pi\zeta, \qquad (9.76)$$

$$L_2^{(0)}(\zeta) = 1+\cos 2\pi\zeta, \qquad (9.77)$$

which give similar numerical results. The smallest computed values of Ra_c, after minimization with respect to a (minimum at a_c), are shown in Table 9.2. Also shown, in the final column, are the results from the approximate equation (9.59), neglecting the first term in the brackets (the diffusion term, because Spiegel did not include diffusion), and using a_c from Table 9.2. The agreement is remarkably good considering the simple nature of the dimensional discussion.

TABLE 9.2

Critical Rayleigh numbers for rigid boundaries with radiative transfer alone

After Spiegel (1960)

τ^*	a_c	Ra_c/χ	Ra_c/χ [from (9.59)]
0	4·77	0	0
0·1	4·75	2·04	2·2
0·5	4·55	45·3	52
1	4·32	155	189
π	3·56	681	1030
5	3·27	1080	1377
10	3·14	1440	1610
1000	3·11	1716	1708

Further numerical results use free-surface boundary conditions including a diffusion term, and permit β to vary according to (9.39) which is the correct initial state for the problem. However, a complete solution of the integral equation for radiative heating was not attempted; instead two asymptotic solutions were obtained, one for the *opaque* and one for the *transparent* approximation ((2.123) and (2.130)). In these limits, the minimal condition required for (9.72) can be obtained with the operator

$$M = \int_{-\frac{1}{2}}^{+\frac{1}{2}} W(\zeta)\, d\zeta. \tag{9.78}$$

Expressions for the Rayleigh number are

$\tau^* \gg a$ (opaque):

$$Ra = \frac{\int_{-\frac{1}{2}}^{+\frac{1}{2}} W(\zeta)(1+\chi)\{(D^2-a^2)^3/a^2\}W(\zeta)\, d\zeta}{\int_{-\frac{1}{2}}^{+\frac{1}{2}} W(\zeta)(\beta/\bar{\beta})W(\zeta)\, d\zeta}, \tag{9.79}$$

$\tau^* \ll a$ (transparent):

$$Ra = \frac{\int\limits_{-\frac{1}{2}}^{+\frac{1}{2}} W(\zeta)\{(D^2-a^2)^2/a^2\}\{(D^2-a^2)-\lambda^2\}W(\zeta)\,d\zeta}{\int\limits_{-\frac{1}{2}}^{+\frac{1}{2}} W(\zeta)(\beta/\bar{\beta})W(\zeta)\,d\zeta}. \qquad (9.80)$$

Two trial functions were employed. The first corresponds to (9.76) and represents a motion through the whole of the fluid (*body motion*). The second took account of the boundary-layer character of (9.39) for large τ^* and χ, and envisaged the possibility of a motion concentrated near the boundary (*boundary-layer motion*). A suitable trial function for boundary-layer motions for the half space $\zeta > 0$ is

$$W^{(0)}(\zeta) = \tfrac{1}{2}e^{b\zeta}\left\{\left(\zeta - \frac{1}{2} - \frac{1}{b}\right)^3 + \frac{1}{b^3}\right\}, \qquad (9.81)$$

where b is a free parameter, with respect to which Ra_c is a minimum. Solutions exist for this trial function, but only in the transparent approximation, indicating that the boundary layer is always narrower than the mean-free-path of the radiation. However, since Ra_c for these motions is generally greater than Ra_c for body motions (see Fig. 9.5), the boundary-layer motions are unlikely to be observed.

The solutions shown in Fig. 9.5 leave virtually no doubt as to their behaviour in the transition region between opaque and transparent approximations. They agree well with the results in Table 9.2 provided that $\lambda \gg 1$ and if allowance is made for the different boundary conditions used in the two problems. The main difference is near $\lambda = 10^2$, when the curves in Fig. 9.5 show a kink, whose presence is attributable to the variable β. This gives an extra stabilization by decreasing the negative temperature gradient over the central region of the fluid.

If λ is not large compared to unity, the results of Table 9.2 and Fig. 9.5 differ greatly, owing to the inclusion of diffusive effects in the latter. However, the results of Fig. 9.5 agree reasonably well with the approximate expression (9.59), which appears to possess all the essential features of the complete solution.

9.3.4. *Application to the atmosphere*

To clarify the relationship between the theoretical work of this section and possible atmospheric processes, the data in Table 9.3 have been computed for a wide size range of disturbances in an ideal atmosphere imitative of the lower troposphere. Effective absorption coefficients have been taken from Fig. 9.3 and therefore take account of the non-grey character of the water-vapour absorption spectrum.

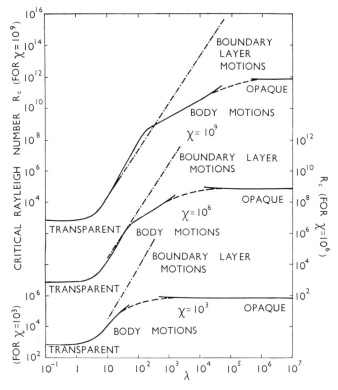

FIG. 9.5. Approximate critical Rayleigh numbers.

After Goody (1956). The full lines (body motions) are computed using the trial function (9.76) in both opaque and transparent approximations. The broken line represents an attempt at interpolation between the two asymptotic solutions. The chain line (boundary-layer motions) comes from the use of (9.81) in the transparent approximation.

The tabulated data contain many points of interest. At first sight, the characteristic times of column 7 may appear to be too long to have physical significance. However, observations on the atmosphere suggest that $2°$ K day^{-1} is an important rate of change of temperature for planetary-scale motions, while $1°$ K hour^{-1} is typical of weather phenomena. In terms of characteristic times $\{(1/\theta)\partial\theta/\partial t\}^{-1}$ these are 10^7 s and 10^6 s respectively, which are comparable to the characteristic times in Table 9.3 for motions of similar scale.

A curious feature is that τ^* is always small, and varies little for scales between 10 m and 100 km. This is because those spectral regions are important for which the transparency is low enough to permit a long mean-free-path, but for which the absorption coefficient is high enough to permit strong interaction between atmospheric layers. For path lengths less than 10 m the line centres are no longer capable of providing sufficient absorption. For path lengths much greater than 100 km the minimum continuum absorption coefficient between the ν_2 and rotation bands is no longer translucent. For all other scales there is a spectral region for which $\tau^* \simeq 0.1$, in which interaction by radiative heat transfer will principally take place.

TABLE 9.3

Values of parameters for a homogeneous atmosphere at s.t.p. containing 3 mb of water vapour ($\rho_{H_2O} = 2.4 \times 10^{-6}$ g cm^{-3})

Numerical values are chosen as follows:

$$a_c = 4.7,$$

$$\pi \, \overline{dB/d\theta} = 5.6 \times 10^3 \text{ erg cm}^{-2} \text{ s}^{-1} \text{ deg}^{-1},$$

$$K = 2 \times 10^3 \text{ erg cm}^{-1} \text{ s}^{-1} \text{ deg}^{-1},$$

$$\rho c = 1.36 \times 10^4 \text{ erg cm}^{-3} \text{ deg}^{-1}.$$

$\tilde{k}$ is taken from Fig. 9.3; χ is from (9.35); $\tau^* = l\tilde{k}_v a_c$ (9.60); $\lambda^2 = 3\chi\tau^{*2}$ (9.60); $Ra_c/Ra_c(\text{diff})$ is from (9.59); N is from (9.19).

1 l	2 $\tilde{k}_v$	3 χ	4 τ^*	5 λ	6 $Ra_c/Ra_c(\text{diff})$	7 $(N)^{-1}$
	(cm^{-1})					(s)
1 cm	4.9×10^{-4}	8.2×10^3	2.3×10^{-3}	3.2×10^{-1}	1	1.2×10^3
10 cm	4.4×10^{-4}	8.2×10^3	2.1×10^{-2}	3.2	1.7	1.2×10^3
1 m	1.9×10^{-4}	2.1×10^4	8.9×10^{-2}	2.1×10	2.4×10	3.1×10^3
10 m	3.1×10^{-5}	1.3×10^5	1.5×10^{-1}	8.3×10	3.7×10^2	1.9×10^4
100 m	4.3×10^{-6}	9.3×10^5	2.0×10^{-1}	3.1×10^2	5.2×10^2	1.4×10^5
1 km	4.3×10^{-7}	9.3×10^6	2.0×10^{-1}	9.7×10^2	5.2×10^4	1.4×10^6
10 km	4.3×10^{-8}	9.3×10^7	2.0×10^{-1}	3.1×10^3	5.2×10^5	1.4×10^7
100 km	4.3×10^{-9}	9.3×10^8	2.0×10^{-1}	9.7×10^3	5.2×10^6	1.4×10^8

The role of molecular viscosity is illustrated by the figures in column 6. For path lengths of 10 m or greater no modification would result if it were neglected. However, if the table had been computed on the basis of turbulent diffusion, for which the effective conductivity can be larger than molecular by a factor of 10^3 or more, diffusion would still be important for motions on the scale of kilometres.

Columns 3 and 6 represent, respectively, the upper limit to and the actual stabilizing influence of radiation. The figures are large, and radiation can therefore be expected to have a profound influence on convective systems. However, the influence must be somewhat indirect, for Rayleigh numbers in the atmosphere are commonly very large indeed, and the problem of the onset of the first convective mode has little direct relevance. For example, a $1°$ K temperature difference over a 10 km path represents a Rayleigh number of about 10^{20}; even a stabilizing factor as large as 5×10^5 is trivial in this context.

This does not mean that the foregoing analysis is irrelevant to atmospheric problems, for it bears directly upon current theories of turbulence. Malkus (1954) bases a theory of turbulent convection on an analysis of the fundamental convective modes, and an important parameter is the smallest eddy size which can grow on the mean temperature field; this can be profoundly affected by the radiative transfer. In the next section the problem of turbulence will be investigated from an entirely different viewpoint, but with the same general conclusion as to the importance of radiative transfer in turbulent systems.

9.4. The interaction of radiation with turbulent motions

Before considering the interaction of turbulence and radiative transfer, we will briefly recapitulate the classical discussion of Richardson on the conditions required to maintain a turbulent field in the presence of a stable temperature gradient. The adiabatic displacement of a small parcel of air is considered. In a stable atmosphere restoring buoyancy forces are generated, and work has to be done on the parcel to displace it. On the other hand, the eddy stresses acting upon the parcel normally do work and accelerate the displacement. If the rate of working required to overcome buoyancy exceeds the rate of working by eddy stresses the intensity of turbulence will decrease, and vice versa. This criterion can be expressed in terms of the Richardson number

$$Ri = \frac{(g/\theta)(\partial\theta/\partial z + g/c_p)}{(\partial u/\partial z)^2}, \qquad (9.82)$$

in which it is assumed that the wind is horizontal (velocity u), and that only derivatives with respect to z are important. The numerator of (9.82) is proportional to the rate of working against buoyancy forces and the denominator to the rate of working by eddy stresses; turbulence can therefore be expected to *increase* if the Richardson number is *below* some number of order unity.

Townsend (1958) has discussed the mechanism of the maintenance of turbulence in detail and his conclusions can also be expressed in terms of a critical Richardson number (Ri_c) which marks the transition

between growth and decay of turbulence. His considerations are sufficiently general to permit the inclusion of heat transfer mechanisms
other than eddy diffusion.

Let us consider briefly what we may expect if radiative transfer is
permitted in the context of Richardson's argument. Radiation destroys
temperature anomalies at a rate approximately equal to $N(n)$ (9.9),
and the buoyancy of a disturbed parcel of air will be decreased. A greater
initial buoyancy will therefore be required to achieve a neutral balance,
and the value of Ri_c will be increased. Since $\partial u/\partial z$ has the dimensions
of a reciprocal time, we may expect on dimensional grounds that Ri_c is
an increasing function of $N(n)/(\partial u/\partial z)$. The results obtained from
Townsend's detailed analysis are

$$\left. \begin{array}{ll} Ri_c \simeq \dfrac{1}{12}\left(1 + \dfrac{6N(n)}{\partial u/\partial z}\right)^2 & \text{for } \dfrac{N(n)}{\partial u/\partial z} < \dfrac{1}{6} \\[4mm] Ri_c \simeq \dfrac{2N(n)}{\partial u/\partial z} & \text{for } \dfrac{N(n)}{\partial u/\partial z} > \dfrac{1}{6} \end{array} \right\}. \tag{9.83}$$

Note that, according to (9.83), the inclusion of radiative transfer
decreases the stability of the fluid, while the results of § 9.3 for cellular
convection indicated an *increase* in stability. Radiation, by adding
a new mode of heat transfer, decreases the influence of temperature
stratification, whether it be stable or unstable, bringing the fluid closer
to a neutral condition. In the previous section the thermal stratification
was unstable; in this section it is stable, and hence the apparently
paradoxical results.

The importance of radiative transfer in the maintenance of atmospheric turbulence can be illustrated by three examples:

The upper atmosphere

Between 30 and 50 km and also between 80 and 100 km the atmosphere has a positive temperature gradient (see § 1.2). Typically
$\partial\theta/\partial z = +2 \times 10^{-5}$ °K cm^{-1}, while the wind shear is about $2 \cdot 5 \times 10^{-3}$
s^{-1}; with $\theta = 240°$ K, (9.82) gives $Ri = 79$. According to (9.83),
$Ri_c = 1/12$ without radiative transfer, so that $Ri \simeq 10^3 Ri_c$ and turbulent fluctuations should decay rapidly if radiative transfer plays no
role. The observational evidence indicates turbulence, however, particularly in the upper layer, and it is puzzling to know how this can be
in the face of the large Richardson number. Townsend suggests that
radiative transfer may account for the discrepancy. For Ri_c to exceed
79, (9.83) requires that $N(n)$ should be larger than $0 \cdot 1$ s^{-1}. At high
altitudes the atmosphere is transparent, and we may use (9.25), (9.27),

and (9.28) to determine the effective absorption coefficient. These equations indicate that 10 per cent of the atmosphere by mass must consist of polyatomic gases if N is to be greater than $0 \cdot 1$ s^{-1}. As far as is known, carbon dioxide is the densest polyatomic gas at these levels, with a mixing ratio of about $3 \cdot 8 \times 10^{-4}$ or less, which fails by a substantial margin to explain the observed phenomenon. The solution to this problem must lie elsewhere than in radiative transfer.

The whole troposphere

We may take $\partial u / \partial z = 2 \times 10^{-3}$ s^{-1} as a typical wind shear. According to (9.83), if $N(n) \simeq 1 \cdot 4 \times 10^{-4}$ s^{-1}, the critical Richardson number will be doubled through the influence of radiative transfer. From Table 9.3 $N(n)$ will exceed this value for eddies smaller than a few metres in diameter. For eddies larger than 100 m the influence of radiative transfer can be neglected.

The boundary layer

Close to the earth's surface water-vapour pressure and temperature can be higher than is assumed in Table 9.3. As a reasonable upper limit N^{-1} can be 10^2 for eddies of size less than 1 m, which are the most important eddies in the lowest metre or two of the atmosphere. Large shears are observed in this layer. With $\partial u / \partial z = 10^{-1}$ s^{-1} we find that radiative effects increase the critical Richardson number by a factor 2.6. A change of this magnitude could be measured with relative ease. The difference in potential temperature between top and bottom of a 1 m layer would be $0 \cdot 08°$ K if the turbulence were maintained without radiative transfer; with radiative transfer the comparable figure is $0 \cdot 21°$ K. An observational test of this prediction should not be difficult to devise.

BIBLIOGRAPHY

9.1. Dissipation time for a radiative process

§ 9.1.1 follows

Spiegel, E. A., 1957, 'The smoothing of temperature fluctuations by radiative transfer', *Astrophys. J.* **126**, p. 202.

References to the data used in § 9.1.2 are to be found in Chapter 5.

9.2. Radiative-diffusive boundary layers

References to the properties of the earth's boundary layer are given in the Bibliography to § 1.2.

The theoretical development of this section follows

Goody, R. M., 1956, 'The influence of radiative transfer on cellular convection', *J. Fl. Mech.* **1**, p. 424; and

—— 1960, 'The influence of radiative transfer on the propagation of a temperature wave in a stratified diffusing medium', ibid. **9**, p. 445.

For earlier work on radiative-diffusive heat transfer, see

Brunt, D., 1944, *Physical and dynamical meteorology*, 2nd ed., p. 129. Cambridge University Press;

Ramdas, L. A., and Malurkar, S. L., 1932, 'Surface convection and variation of temperature near a hot surface', *Ind. J. Phys.* **7**, p. 1; and

Malurkar, S. L., and Ramdas, L. A., 1932, 'Theory of extremely high lapse-rates of temperatures very near the ground', ibid. **6**, p. 495.

The latter paper discusses measurements over the hot ground in the tropics.

Papers on the temperature profile over a surface in the laboratory are by

Thomas, D. B., and Townsend, A. A., 1957, 'Turbulent convection over a heated horizontal surface', *J. Fl. Mech.* **2**, p. 473; and

Croft, J. F., 1958, 'The convective regime and temperature distribution above a horizontal heated surface', *Quart. J. R. Met. Soc.* **84**, p. 418.

§ 9.2.3 follows

Gaevskaya, G. N., Kondratiev, K. Y., and Yakushevskaya, K. E., 1961, *The heat radiative flux divergence and the heat regime in the near-ground layer of the atmosphere*, the reports at the Symposium on Radiation in Vienna. Leningrad University Press. This paper gives a valuable review of investigations of radiative transfer close to the ground.

9.3. The onset of cellular convection

This section is based upon Goody (1956) (§ 9.2); and

Spiegel, E. A., 1960, 'The convective instability of a radiating fluid layer', *Astrophys. J.* **132**, p. 716.

Also referred to is a paper by

Malkus, W. V. R., 1954, 'The heat transport and spectrum of thermal turbulence', *Proc. Roy. Soc.* A, **225**, p. 196.

9.4. The interaction of radiation with turbulent motions

The work described in the section follows

Townsend, A. A., 1958, 'The effects of radiative transfer on turbulent flow of a stratified fluid', *J. Fl. Mech.* **4**, p. 361.

APPENDIX 1

Definitions of symbols

Symbol	Defined or first used	Definition
a a_ν	§ 1.1	Albedo or reflectivity / Albedo at frequency ν
a a_0 $\tilde{a}_{w,c}$	§ 2.1.2	Amount of matter / Amount of matter for $\bar{T} = \frac{1}{2}$ / Scaled amount of matter for water vapour or carbon dioxide
a	§ 2.2.3	An Einstein coefficient
$a_{i,j}$	§ 2.4.4	Weights for numerical integration
a	§ 3.2	Antisymmetric
$a(l)$	§ 3.4	Function of l
a	§ 3.4	Term symbol
a	§ 3.6.4.1	Amplitude
a	§ 6.9	Fitting constant
a	§ 7.2	Particle radius
a_n	§ 7.6	Coefficient in Mie's expansion
a	§ 9.2.2	Attenuation coefficient of thermal wave
a a_ξ a_η a_ζ	§ 9.3.1	h/l / ξ-component / η-component / ζ-component
a	A.14.4	Reduction constant
A	§ 2.2.2	Non-dimensional quantity
A	§ 2.2.3	An Einstein coefficient
A	§ 3.2	Symmetry class
A	§ 3.2	Rotational constant
A	§ 3.4	Term symbol
A A_ν $\bar{A}$ $A*$	§ 4.1	Absorption / Absorption at frequency ν / Mean absorption / Mean absorption
$A_i^{(r,s)}$	§ 6.7.2	Matrix for heating computations
A	§ 7.3	Function of polarizability
A	§ 7.4	Altitude function
$\mathbf{A}$	A.2	Avagadro's number
Å	A.3	Angstrom unit
b	§ 2.2.3	An Einstein coefficient
b	§ 3.4, § 6.9	Fitting constant
b_k	§ 6.7.4	Power series coefficients

Symbol	*Defined or first used*	*Definition*
b_n	§ 7.6	Coefficients in Mie's theory
b_i	§ 8.2.3	Weights for numerical integration
b	§ 9.2.2	Phase lag coefficient
b	§ 9.3.3	Free parameter
B		Planck function
B_ν		Planck function at frequency ν
B_λ		Planck function at wavelength λ
B_s		Planck function at level s
B_i	§ 2.2.1	Planck function at level i
B^*		Planck function of boundary
B_ν^*		Planck function of boundary at frequency ν
$B^{(0)}$		Equilibrium value of Planck function
$B^{(1)}$		Perturbation of Planck function
B	§ 2.2.3	An Einstein coefficient
B		Rotational constant
B_e	§ 3.2	Rotational constant without vibrational energy
B_v		Rotational constant with v vibrational quanta
B	§ 7.3	Function of polarizability
$\mathbf{c}$		Velocity of light *in vacuo*
$\mathbf{c'}$	§ 2.1.1	Velocity of light in a medium
c_α	§ 2.4.4	Numerical coefficient
c	§ 3.4	Term symbol
c	§ 5.4.4	Coefficient
c	§ 5.5.3	Empirical constant
c	§ 6.1.1	Mixing ratio
c		Specific heat
c_p	§ 9.1.1	Specific heat at constant pressure
°C	§ 1.2	Degrees Celsius
$C_{1,2}$	§ 2.2.1	Radiation constants
C	§ 3.2	Rotational constant
C	§ 3.4	Term symbol
C	§ 5.4.4	Coefficient
$C_k^{(r,s)}$	§ 6.7.4	Heating function matrix
C_r	§ 8.2.2	Coefficient
C_ν	§ 8.2.3	Spectral variable
d	§ 2.1.1	Displacement
$\mathbf{d}$	§ 2.1.1	Displacement vector
d	§ 3.4	Term symbol
d_i	§ 3.5	Degeneracies
d	§ 3.6	Doppler–Lorentz profile parameter
d	§ 5.4.4	Coefficient
D	§ 3.2	Second rotational constant
D	§ 3.4	Term symbol
D	§ 5.4.4	Coefficient
D	§ 7.4	Angular factor
D	§ 9.3.2	$\partial/\partial\zeta$

Symbol	Defined or first used	Definition
e		Extinction coefficient
e_ν		Extinction coefficient at frequency ν
e_λ		Extinction coefficient at wavelength λ
e_v	§ 2.1.1	Extinction coefficient per unit volume
e_m		Extinction coefficient per unit mass
e_n		Extinction coefficient per molecule
e_s		Extinction coefficient per cm s.t.p.
e'		Decadic extinction coefficient
e	§ 6.9	Water vapour partial pressure
E		Energy
E_ν		Radiative energy at frequency ν
E_λ		Radiative energy at wavelength λ
E_r	§ 2.1.1	Rotational energy
E_e		Electronic energy
E_v		Vibrational energy
E_t		Translational energy
E_n	§ 2.3.3	Exponential integral of degree n
E	§ 3.2	Symmetry class
E	§ 4.4.1	Elsasser function
$\mathbf{E}$		Electric vector
$E^{(l)}$		Component of electric vector
$E^{(r)}$	§ 7.1	Component of electric vector
$E_0^{(l)}$		Electric vector of incident beam
$E_0^{(r)}$		Electric vector of incident beam
E	A.14.1	Coronal line emission
f	§ 1.1	The solar constant
f_ν	§ 2.1.1	Irradiation at frequency ν
f_λ		Irradiation at wavelength λ
$f(J)$	§ 2.2.2	Fraction in J-state
f	§ 3.4	Term symbol
f	§ 3.6.1	Line shape factor
f	§ 4.2.1	Averaging factor
f	§ 4.7.2	Adjustable constant
$f(\phi)$	§ 5.6.3	Function of ϕ
f	§ 6.6	Relative humidity
$f(Z)$	§ 7.5	Airy integral
F		Heat flux
F_ν		Radiative heat flux at frequency ν
F_λ		Radiative heat flux at wavelength λ
F^+		Upward radiative heat flux
F^-		Downward radiative heat flux
$F_{\nu,d}$	§ 2.1.1	d-component of radiative heat flux
F_r		Radiative heat flux
F_a		Advected heat flux
F_t		Turbulent heat flux
F_m		Molecular heat flux
$F^{(0)}$		Equilibrium heat flux
$F^{(1)}$		Perturbation of heat flux

Symbol	Defined or first used	Definition
F	§ 2.2.1	A function
F	§ 3.2	Symmetry class
$F(J)$	§ 3.2	Measure of rotational energy
F	§ 3.4	Term symbol
$F(\mu, \epsilon)$	§ 3.6.4.3	Correcting factor
$F(u)$	§ 4.4.4	Function of u
F_{ij}	§ 7.2	Transformation matrix
F	§ 7.4	Correcting factor
F	A.14.1	Emission from F-corona
g	§ 2.2.2	Statistical weight
g	§ 3.4	Symmetry symbol
g	§ 4.4.3	$\log_{10}\left[\dfrac{2\pi y u}{\sinh 2\pi y}\right]$
g	§ 6.1.1	Acceleration due to gravity
$\left.\begin{array}{l}g^{(r)}\\ g^{(l)}\end{array}\right\}$	§ 7.5	Scattering efficiency factor Scattering efficiency factor
g	§ 9.2.1	$\sqrt{\{3(1+\chi)\}}$
g	A.14.4	Number of spot groups
G	§ 7.4	Geometric cross-section
$\left.\begin{array}{l}h\\ h_\nu\\ h_\lambda\\ h_i\end{array}\right\}$	§ 2.1.1	Heating rate Heating rate for frequency ν Heating rate for wavelength λ Heating rate for spectral range i
$\mathbf{h}$	§ 2.2.1	Planck's constant
h	§ 4.4.3	$\log_{10}(u_1/u_2)$
h	§ 7.5	A constant
h	§ 9.3.1	Separation of parallel plates
$\left.\begin{array}{l}H_i\\ H_i^{(0)}\\ H_i^{(1)}\end{array}\right\}$	§ 2.1.2	Heating function Zero approximation to heating function First approximation to heating function
H	§ 2.2.2	Scale height
H	§ 2.4.4	H-function
$\mathbf{H}$	§ 7.1	Magnetic vector
H	§ 7.6	Bessel function
i	§ 2.1.1	Integral variable
i_1	§ 7.2	$\lvert S_1 \rvert^2$
i_2	§ 7.2	$\lvert S_2 \rvert^2$
$\left.\begin{array}{l}I\\ I_\nu\\ I_\lambda\\ \bar{I}\\ I^+\\ I^-\\ \bar{I}^{(0)}\\ \bar{I}^{(1)}\end{array}\right\}$	§ 2.1.1	Intensity Intensity at frequency ν Intensity at wavelength λ Intensity averaged over angle Upward intensity Downward intensity Equilibrium mean intensity Perturbation of mean intensity

Symbol	*Defined or first used*	*Definition*
$I_\nu^{(i)}$		Stokes parameter
$I_\nu^{(l)}$	§ 2.1.3	Stokes parameter
$I_\nu^{(r)}$		Stokes parameter
I	§ 3.2	Nuclear spin
I		Moment of inertia
I_A		A-component of moment of inertia
I_B	§ 3.2	B-component of moment of inertia
I_C		C-component of moment of inertia
I_1		Bessel function with imaginary argument
I_0	§ 4.2	Bessel function with imaginary argument
$\mathscr{I}$	§ 7.2	Imaginary part of
j	§ 2.4.4	Integral variable
J		Source function
J_ν		Source function at frequency ν
J_λ		Source function at wavelength λ
$J+$	§ 2.1.2	Source function in upward direction
$J-$		Source function in downward direction
J_i		Source function in ith spectral range
$\bar{J}$		Angular average of source function
J		Rotational quantum number
J'	§ 2.2.3	Rotational quantum number of upper state
J''		Rotational quantum number of lower state
$\mathbf{J}$	§ 3.2	Total angular momentum
J	§ 7.4	Bessel function
$\mathbf{J}$	A.2	Joule equivalent
k		Absorption coefficient
k_λ		Absorption coefficient at wavelength λ
k_ν		Absorption coefficient at frequency ν
k_v		Absorption coefficient per unit volume
k_m		Absorption coefficient per unit mass
k_s	§ 2.1.2	Absorption coefficient per cm s.t.p.
k_n		Absorption coefficient per molecule
$\tilde{k}$		Effective absorption coefficient
k_P		Planck mean absorption coefficient
k_R		Rosseland mean absorption coefficient
k_C		Chandrasekhar mean absorption coefficient
$\bar{k}$		Arithmetic mean absorption coefficient
$\mathbf{k}$	§ 2.2.1	Boltzmann's constant
k	§ 5.4.4	Coefficient
$°\mathrm{K}$	§ 1.1	Degrees Kelvin
K	§ 3.2	A quantum number
$\mathbf{K}$	§ 3.2	Component of angular momentum
K	§ 3.4	Electron shell
K	§ 4.2.2	A constant
K	§ 5.4.4	Coefficient
$K_{\mathrm{L},i}$	§ 6.7.1	Function of Lorentz shape

Symbol	Defined or first used	Definition
$K_{D,i}$	§ 6.7.1	Function of Doppler shape
K	§ 7.4	Scattering function
K		Conductivity
$K(\text{amp})$	§ 9.2.1	Empirical conductivity
$K(\text{phase})$		Empirical conductivity
$K(\zeta)$	§ 9.3.3	Function of ζ
K	A.14.1	Emission from K-corona
$\mathbf{l}$	§ 2.1.3	Parallel vector
l	§ 2.2.2	Molecular free path
$\bar{l}$		Mean free path
l	§ 2.4.2	Scale length
$\mathbf{l}$	§ 3.3	Angular momentum
l	§ 3.3	Quantum number
l	§ 9.3.1	Characteristic length
L_{α}	§ 2.4.4	Constant of integration
$L_{-\alpha}$		Constant of integration
L	§ 3.4	Electron shell
$\mathbf{L}$	§ 3.4	Electronic angular momentum
L	§ 3.4	Quantum number
L	§ 4.1	Ladenburg and Reiche function
$L(\zeta)$	§ 9.3.3	Function of ζ
$\mathscr{L}$	§ 6.1.1	Linear operator
m	§ 2.4.4	Integral variable
m_l	§ 3.4	Magnetic quantum number
m_s		Magnetic quantum number
m	§ 3.5	Sequential numbering of rotation lines
m	§ 3.6.1	Mass of a molecule
m	§ 7.1	Refractive index
$\tilde{m}$		Refractive index
$\mathbf{M}$		Magnetic moment
M_A	§ 3.1	A-component of magnetic moment
M_B		B-component of magnetic moment
M_C		C-component of magnetic moment
M	§ 3.4	Electron shell
M_L	§ 3.4	Quantum number
M_S		Quantum number
M_i	§ 8.5	Function of optical depth
M	§ 9.3.2	Operator
M	A.2	Atomic or molecular weight
n	§ 2.1.2	Number of molecules per cm^3
n_s	§ 2.1.2	Loschmidt's number
$n(E)$	§ 2.2.2	Number of molecules with energy E
n	§ 3.4	Principal quantum number
n	§ 4.6.1	Integral variable
n	§ 5.5.3	Empirical constant

Symbol	Defined or first used	Definition
n_s	§ 5.5.3	Strong line variable
n_w	§ 5.5.3	Weak line variable
n	§ 6.1.1	Pressure-scaling factor
n		Refractive index (real part)
$\tilde{n}$	§ 7.1	Refractive index (real part)
n'		Refractive index (imaginary part)
$\tilde{n}'$		Refractive index (imaginary part)
$\mathbf{n}$	§ 9.1.1	Wave-number vector
N	§ 3.3	Number of atoms in a molecule
N	§ 3.4	Electron shell
N	§ 4.2.2	Number of absorption lines
N	§ 7.1	Magnitude of Poynting vector
$N^{(i)}$	§ 7.1	Stokes parameters
N	§ 7.2	Number density of scattering centres
$N(n)$	§ 9.1.1	Radiative rate of dissipation
p	§ 3.4	Term symbol
p	§ 3.6.1	Probability
p		Pressure
p_0	§ 3.6.4.2	Standard pressure
$\tilde{p}$		Effective pressure
p	§ 7.5	Ordinal numbers
P	§ 2.1.1	Point in space
P_{ij}	§ 2.1.3	Phase matrix
P'_{ij}		Phase matrix
P	§ 2.1.3	Phase function
P'		Phase function
P_m	§ 2.4.4	Legendre polynomial of degree m
P	§ 3.2	Branch of molecular spectrum
P	§ 3.4	Term symbol
$P(p)$	§ 6.1.1	Pressure factor in k
P	§ 7.5	Degree of polarization
q	§ 2.4.4	Hopf function
q	§ 5.4.3	Quadrupole moment
q	§ 6.7.3	Line parameter
q	§ 6.9	Cloud correction factor
q_r	§ 7.5	Fresnel reflection coefficient
Q_ν	§ 2.1.3	A Stokes parameter
$Q(\tau^*)$	§ 2.4.4	Modified Hopf function
Q	§ 2.4.4	Constant of integration
Q	§ 3.2	Partition function
Q	§ 3.2	Branch of molecular spectrum
Q	§ 4.6.2	Generating function
Q_e		Efficiency factor for extinction
Q_a	§ 7.2	Efficiency factor for absorption
Q_s		Efficiency factor for scattering

Symbol	*Defined or first used*	*Definition*
r	§ 1.1	Earth's radius
$\mathbf{r}$	§ 2.1.3	Perpendicular vector
r_e	§ 3.5	Nuclear equilibrium separation
r	§ 3.6.4.1	Inter-molecular separation
r	§ 4.4.3	Δ/δ
r	§ 6.2	Diffusivity factor
r	§ 6.7	Integral variable
r	§ 6.9	Correlation coefficient
r	§ 9.2.1	$\dfrac{1+\alpha}{1-\alpha}$
r	A.14.1	Distance in units of the solar radius
R_{ij}	§ 3.1	Matrix element of dipole moment
R	§ 3.2	Branch of molecular spectra
R	§ 3.4	Rydberg's constant
R	§ 6.8.4	Auxiliary function for radiation charts
Ra		Rayleigh number
Ra_c	§ 9.3.1	Critical Rayleigh number
$Ra_c(\text{diff})$		Critical Rayleigh number for diffusion
Ri	§ 9.4	Richardson number
Ri_c		Critical Richardson number
$R_\odot$	A.2	Solar radius
$\mathbf{R}$		Gas constant—universal
R_a	A.2	Gas constant for air
R_w		Gas constant for water vapour
R	A.14.4	Sunspot number
$\mathscr{R}$	§ 3.6.4.1	Real part of
s		Displacement
$\mathbf{s}$	§ 2.1.1	Displacement vector
$s_\odot$		Displacement in sun's direction
s		Scattering coefficient
s_λ		Scattering coefficient at wavelength λ
s_ν		Scattering coefficient at frequency ν
s_v	§ 2.1.2	Scattering coefficient per unit volume
s_m		Scattering coefficient per unit mass
s_n		Scattering coefficient per molecule
s_s		Scattering coefficient per cm s.t.p.
s	§ 3.2	Symmetric
$\mathbf{s}$	§ 3.4	Spin vector
s	§ 3.4	Quantum number
s	§ 3.4	Term symbol
s	§ 6.7.2	Integral variable
s	A.14.4	Number of spots
S		Band or line intensity
S_v		Band or line intensity per unit volume
S_m	§ 2.2.4	Band or line intensity per unit mass
S_n		Band or line intensity per molecule
S_s		Band or line intensity per cm s.t.p.

Symbol	Defined or first used	Definition
S_{ij}	§ 2.4.5	Reflection matrix
$S^{(m)}$	§ 2.4.5	Term in expansion
S	§ 3.4	Term symbol
$\mathbf{S}$	§ 3.4	Total electronic spin
S	§ 3.4	Quantum number
$\mathbf{S}$	§ 7.2	Amplitude scattering matrix
t		Time
$t(\text{diff})$	§ 2.1.1	Diffusion time
$t(\text{rad})$		Radiation time
t	§ 2.3.3	$\tau_\nu(z', \infty)$, dummy variable
t	§ 4.3.1	Dummy variable
t_r	§ 8.2.2	rth index
T_{ij}	§ 2.4.5	Transmission function
$T^{(m)}$		Element of transmission function
T		Transmission
T_ν		Transmission at frequency ν
T'_ν	§ 4.1	Transmission at frequency ν
$\bar{T}$		Mean transmission
$\bar{T}_0$		Mean transmission
$\bar{\bar{T}}$		Mean transmission
u		Energy density
u_λ	§ 2.1.1	Energy density at wavelength λ
u_ν		Energy density at frequency ν
u	§ 3.4	Symmetry symbol
u	§ 3.6.1	Velocity component
$\bar{u}$	§ 3.6.4.3	Mean velocity
u	§ 4.2.1	$Sa/2\pi\alpha_L$
$\tilde{u}$	§ 6.1.1	$\dfrac{S}{2\pi}\dfrac{da}{d\alpha_L}$
u	§ 7.5	Rectangular coordinate
u	§ 9.4	Horizontal velocity
U	§ 2.1.3	Stokes parameter
U_ν		Stokes parameter at frequency ν
v	§ 2.1.2	Volume
v	§ 2.2.3	Vibrational quantum number
v	§ 7.5	Rectangular coordinate
V	§ 2.1.3	Stokes parameter
V_ν		Stokes parameter at frequency ν
w	§ 2.3.3	Dummy variable
w		Nuclear weight factor
w_a	§ 3.2	Nuclear weight factor
w_s		Nuclear weight factor
w_J		Nuclear weight factor
w	§ 4.2.3	$Sa/\alpha_D\sqrt{\pi}$
w_j	§ 6.7.3	$\int S_j\, da/\alpha_{D,j}\sqrt{\pi}$

Symbol	Defined or first used	Definition
w	§ 9.3.1	Fluid velocity
W_{ij}	§ 2.1.3	Intensity transformation matrix
W		Equivalent width
W'	§ 4.1	Equivalent width
W_L		Equivalent width
$\overline{W}$		Mean value of equivalent width
W	§ 6.8.6	Scaling function for radiation diagrams
$W(\zeta)$	§ 9.3.2	ζ factor of w
$W^0(\zeta)$	§ 9.3.3	Trial function
x	§ 2.1.1	Rectangular coordinate (horizontal)
$\mathbf{x}$	§ 2.1.1	Unit vector
x	§ 2.1.1	θ/ν
x	§ 4.2.1	ν/δ
x	§ 4.2.3	ν/α_D
x	§ 6.2	Dummy variable
x	§ 7.2	$2\pi a/\lambda_0$
x	§ 8.2.2	p/p_0
X	§ 2.4.5	X-function
X	§ 3.4	Term symbol
$X_{ij}^{+,-}$	§ 4.3.3	Overlap function
X	§ 6.7.3	Dummy variable
X	§ 6.8.1	Auxiliary function for radiation chart
y	§ 2.1.1	Rectangular coordinate (horizontal)
$\mathbf{y}$	§ 2.1.1	Unit vector
y	§ 4.2.1	α_L/δ or $K\alpha_L/\delta$
Y	§ 2.4.5	Y-function
Y	§ 6.8.1	Auxiliary function for radiation chart
z	§ 2.1.1	Rectangular coordinate (vertical)
$\mathbf{z}$	§ 2.1.1	Unit vector
z_c	§ 9.2.2	Critical height
Z	§ 2.4.4	Tabulated function
$\mathbf{Z}$	§ 3.4	Atomic number
Z	§ 6.2	Diffuse transmission function
Z	§ 7.5	Argument of Airy function
α_ν	§ 2.4.1	Spectral reflectivity
α_i	§ 3.5	Coefficients
α		Width of spectral line
α_D		Doppler width of spectral line
α_L	§ 3.6.1	Lorentz width of spectral line
α_N		Natural width of spectral line
$\tilde{\alpha}$		Effective width of spectral line
$\bar{\alpha}$		Mean width of spectral line
$\boldsymbol{\alpha}$	§ 7.2	Polarizability matrix
α_i	§ 8.5	Logarithmic derivative of $k_{v,\nu}$
α	§ 9.3.1	Coefficient of expansion

Symbol	Defined or first used	Definition
β	§ 2.1.3	Axial ratio of polarization ellipse
β	§ 3.6.4.1	A constant
$\beta_i^{(s)}$	§ 6.7.4	Normalized band intensity
β	§ 7.4	$\tan^{-1}\dfrac{\tilde{n}'}{\tilde{n}-1}$
$\left.\begin{array}{l}\beta \\ \bar{\beta}\end{array}\right\}$	§ 9.3.1	Temperature gradient Mean temperature gradient
γ	§ 6.7.3	Auxiliary function
γ	§ 6.8.6	A constant
$\left.\begin{array}{l}\gamma \\ \gamma^{(r)} \\ \gamma^{(l)}\end{array}\right\}$	§ 7.2	An angle An angle An angle
γ	A.2	Ratio of specific heats
γ	A.8	Euler's constant
Γ	§ 6.1.1	Γ-function
δ	§ 2.2.2	$8\pi\nu^2/\mathbf{c}^3$
δ	§ 2.4.1	δ-function
δ	§ 4.2.1	Mean line spacing
δ	§ 7.1	Phase difference
Δ	§ 3.4	Term symbol
Δ	§ 4.3.2	Frequency interval
Δ	§ 7.3	Depolarization factor
ϵ	§ 3.6.4.3	Line shape parameter
ϵ_c	§ 5.4.8	Column emissivity
ϵ_s	§ 5.4.8	Slab emissivity
$\epsilon_{c.s}^*$	§ 6.4	Function related to emissivity
ϵ	§ 7.4	A small quantity
ϵ	§ 7.5	A function of the Fresnel coefficients
$\left.\begin{array}{l}\zeta \\ \zeta_\odot\end{array}\right\}$	§ 2.3.3	Zenith angle Solar zenith angle
ζ	§ 5.5.3	Adjustable constant
$\zeta(a)$	§ 5.6.3	Function of a
ζ_n	§ 7.6	Mie function
ζ	§ 9.2.2	$\sqrt{3}(\tau^*-\tau)$
ζ	§ 9.3.2	$z/h-\frac{1}{2}$
η	§ 2.2.2	Relaxation time
η	§ 3.6.4.1	Phase shift
$\eta(\phi)$	§ 5.6.3	Function of ϕ
η	§ 9.3.2	y/h
η	A.2	Viscosity

Symbol	Defined or first used	Definition
θ		Temperature
θ_s		Standard temperature
θ_0		Ice point temperature
θ_e		Emission temperature
θ^*		Boundary temperature
$\tilde{\theta}$	§ 1.1	Effective temperature
θ_f		Frost-point temperature
θ_s		Surface temperature
θ_t		Cloud top temperature
θ_b		Cloud base temperature
$\theta^{(0)}$		Equilibrium temperature
$\theta^{(1)}$		Perturbation of temperature
θ	§ 2.4.5	Scattering angle
Θ	§ 6.8.6	Auxiliary function for radiation diagram
κ_l	§ 2.4.5	Coefficient
κ	§ 3.2	Function of moments of inertia
λ	§ 2.1.1	Wavelength
λ_0		Wavelength *in vacuo*
λ	§ 6.9	Cloud correction factor
λ	§ 9.3.1	$3\chi\tau^{*2}$
Λ	§ 3.4	A quantum number
μ	§ 1.1	Micron
μ	§ 3.6.4.3	Line shape parameter
ν		Frequency
ν_0	§ 2.1.1	Frequency
ν'		Frequency
ν_p	§ 3.6.4.3	Critical frequency
ν	§ 9.3.1	Kinematic viscosity
ν	§ 9.3.2	Inverse time
ξ		$\cos\zeta$
$\xi_\odot$	§ 2.3.3	$\cos\zeta_\odot$
ξ_z		$\cos\zeta$ at altitude z
ξ	§ 4.2.1	Dummy variable
ξ	§ 9.1.1	$\cos(\mathbf{n}, \mathbf{s})$
ξ	§ 9.2.2	$N(\infty)/2\pi\nu$
ξ	§ 9.3.2	x/h
π		Area
π_s	§ 2.1.1	Area
π_d		Area
$\pi(\Delta)$	§ 4.5.1	Probability
π_n	§ 7.3	Mie function
Π	§ 3.4	Term symbol

Symbol	Defined or first used	Definition
ρ		Density
ρ_s		Standard density
ρ_0		Density at s.t.p.
ρ_a		Air density
ρ_w	§ 2.1.2	Water vapour density
ρ_c		Carbon dioxide density
$\tilde{\rho}_\mathrm{w}$		Scaled water vapour density
$\tilde{\rho}_\mathrm{c}$		Scaled carbon dioxide density
$\rho^{(1)}$		Perturbation of density
ρ	§ 5.4.5	Water vapour band
ρ	§ 7.4	$2x(\tilde{m}-1)$
σ	§ 1.1	Stefan–Boltzmann constant
σ_i	§ 3.6.4.2	Collision diameter
σ		Mean line intensity
σ_s	§ 4.2.2	Mean line intensity for strong lines
σ_w		Mean line intensity for weak lines
σ	§ 3.4	Water-vapour band
Σ	§ 3.4	Term symbol
τ		Optical depth or path
τ_ν		Optical depth or path for frequency ν
$\bar{\tau}_\mathrm{C}$	§ 2.1.2	Mean value of optical depth or path
$\bar{\tau}_\mathrm{R}$		Mean value of optical depth or path
τ_ν^*		Optical depth or path for finite atmosphere
τ^*		Optical depth or path for finite atmosphere
τ	§ 3.2	Numbering parameter
τ	§ 3.4.6.2	Time between collisions
$\bar{\tau}$		Mean time between collisions
τ	§ 5.4.5	Water-vapour band
$\tau_{0,j}$	§ 6.7.3	Auxiliary optical function
$\tau_{1,j}$		Auxiliary optical function
τ	§ 7.4	Angular variable
τ_n	§ 7.6	Mie function
ϕ	§ 2.2.3	Natural lifetime of an excited state
ϕ	§ 2.3.4	Azimuth angle
$\phi_\odot$		Azimuth angle of sun
ϕ_l^m	§ 2.4.5	Auxiliary function
ϕ	§ 5.6.3	Function of a and p
ϕ	§ 6.1.1	Factor in absorption coefficient
$\phi_{\mathrm{L},i}$	§ 6.7.1	Auxiliary function
$\phi_{\mathrm{D},i}$		Auxiliary function
Φ	§ 3.4	Term symbol
Φ	§ 5.4.5	Water-vapour band
Φ	§ 9.1.1	Fourier transform of temperature
χ	§ 2.1.3	An angle
χ	§ 5.4.4	Scaling function for radiation diagram

Symbol	Defined or first used	Definition
χ	§ 6.8.2	Water-vapour band
χ	§ 9.2.1	$\dfrac{4\pi\, \overline{dB/d\theta}}{3k_v K}$
ψ	§ 2.4.5	Auxiliary function
ψ	§ 3.1	Wave function
ψ	§ 5.4.4	Water-vapour band
$\psi(\nu)$	§ 6.1.1	Factor of absorption coefficient
ψ_n	§ 7.6	Mie function
Ψ	§ 2.4.5	Characteristic function
ω		Solid angle
ω_s	§ 2.1.1	Solid angle
ω_d		Solid angle
$\omega_\odot$		Solid angle of sun
ω	A.8	Dummy variable
Ω	§ 3.4	Quantum number
Ω	§ 5.4.5	Water-vapour band

APPENDIX 2

Physical constants

General constants (on the chemical mass scale)

Stefan–Boltzmann constant, $\boldsymbol{\sigma}$ $= (5\cdot6698\pm0\cdot0011)\times10^{-5}\ \text{erg cm}^{-2}\ \text{deg}^{-4}\ \text{s}^{-1}$
$= (8\cdot1278)\times10^{-11}\ \text{cal cm}^{-2}\ \text{deg}^{-4}\ \text{min}^{-1}\dagger$

Velocity of light *in vacuo*, **c** $= (2\cdot99791\pm0\cdot00001)\times10^{10}\ \text{cm s}^{-1}$

Boltzmann constant, **k** $= (1\cdot38024\pm0\cdot00007)\times10^{-16}\ \text{erg deg}^{-1}$

Planck constant, **h** $= (6\cdot6237\pm0\cdot0002)\times10^{-27}\ \text{erg s}$

Joule equivalent, **J** $= (4\cdot1855\pm0\cdot0004)\times10^{7}\ \text{erg cal}^{-1}$

Mass of unit atomic weight,
m/M $= (1\cdot66019\pm0\cdot00006)\times10^{-24}\ \text{g}$

Loschmidt number, n_s $= (2\cdot6875\pm0\cdot0001)\times10^{19}\ \text{cm}^{-3}$

Avogadro's number, **A** $= (6\cdot0238\pm0\cdot0002)\times10^{23}\ \text{mole}^{-1}$

Gas constant, **R** $= (8\cdot3144\pm0\cdot0004)\times10^{7}\ \text{erg deg}^{-1}\ \text{mole}^{-1}$

Standard surface gravity, g $= 980\cdot665\ \text{cm s}^{-2}$

Ice point, θ_0 $= 273\cdot155\pm0\cdot015°\ \text{K}$

Standard pressure, p_0 $= 760\ \text{mm Hg}$
$= (1\cdot013246\pm0\cdot000004)\times10^{6}\ \text{dyn cm}^{-2}$

Astrophysical data

Earth's equatorial radius, r $= 6378\cdot24\pm0\cdot12\ \text{km}$

Earth–Sun mean distance, $R_\odot$ $= 1\cdot4960\times10^{13}\ \text{cm}$

Mean solar angular diameter,
$d\omega_\odot$ $= 31'\cdot988$

Effective solar emission temperature, θ_e $= 5785°\ \text{K}$

Data on dry air

Effective molecular weight, M $= 28\cdot970$

Density at s.t.p., ρ_0 $= 1\cdot2928\times10^{-3}\ \text{g cm}^{-3}$

Gas constant, R_a $= 2\cdot870\times10^{6}\ \text{erg deg}^{-1}\ \text{g}^{-1}$

Specific heats at s.t.p., c_p $= 0\cdot2403\ \text{cal g}^{-1}\ \text{deg}^{-1}$
$= 1\cdot006\times10^{7}\ \text{erg g}^{-1}\ \text{deg}^{-1}$

c_v $= 0\cdot1715\ \text{cal g}^{-1}\ \text{deg}^{-1}$
$= 7\cdot18\times10^{6}\ \text{erg g}^{-1}\ \text{deg}^{-1}$

$\gamma = c_p/c_v$ $= 1\cdot401$

Viscosity at s.t.p., η $= 1\cdot72\times10^{-4}\ \text{gm cm}^{-1}\ \text{s}^{-1}$

Kinematic viscosity, $\nu = \eta/\rho_0$ $= 0\cdot133\ \text{cm}^2\ \text{s}^{-1}$

Thermal conductivity, K $= 5\cdot6\times10^{-5}\ \text{cal cm}^{-1}\ \text{s}^{-1}\ \text{deg}^{-1}$
$= 2\cdot34\times10^{3}\ \text{erg cm}^{-1}\ \text{s}^{-1}\ \text{deg}^{-1}$

For data on water vapour, see Appendix 6.

BIBLIOGRAPHY

ALLEN, C. W., 1955, *Astrophysical quantities*. London: Athlone Press.

† Conversion factors for energy units are to be found in Appendix 3.

APPENDIX 3

Spectroscopic units

Wavelength (λ)

TABLE A.3.1

Unit	Symbol	Size	Spectral region
Ångström unit .	Å	10^{-8} cm	Soft X-ray, ultra-violet, and visible
Micron . . .	μ	10^{-4} cm	Visible and infra-red
Millimetre . .	mm	10^{-1} cm	
Centimetre . .	cm	1 cm	Micro- and radio-waves
Metre . . .	m	10 cm	

The use of wavelength to specify the quality of electromagnetic radiation suffers from the defect that it is inversely proportional to the refractive index of the medium in which it is measured. For this reason, wavelengths are commonly corrected to their values *in vacuo*. The difference between wavelengths in air and *in vacuo* is not large (see Appendix 4 for the refractive index of dry air), and is of little interest in atmospheric studies excepting the problem of line identification in the rich solar spectrum.

Wave number and frequency

According to Planck's relation the energy of a quantum is proportional to its frequency. Analysis of spectra is therefore conveniently performed in terms of this unit. Since frequency is not a function of the physical properties of the medium in which it is measured, its use gives rise to no confusion as to whether or not vacuum corrections have been applied. The symbol ν is universally accepted, and the units are c.p.s. (s^{-1}), kilocycle p.s. (10^3 c.p.s.), and megacycle p.s. (10^6 c.p.s.).

The kilo- and megacycle p.s. are in general use by radio engineers. Spectroscopists, however, have shown a strange aversion to the use of frequency units, preferring the *wave number* or reciprocal wavelength, equal to ν/c'. This practice is too widely adopted to attempt reform, but two difficulties should be noted. First, the refractive index of the medium has reappeared through its influence on the velocity of light. Secondly, there is no general agreement on the name or unit for this quantity. Some authors use ν for both wave number and frequency, and refer to both as 'frequencies'. However, quantities with different physical dimensions deserve distinct symbols, and there have been a number of attempts to introduce the symbol k and the unit *kayser*. It would have been desirable to support this convention in these volumes but for the fact that in quantitative spectroscopy k is generally pre-empted to denote the *absorption coefficient*. Consequently, we will use an older form, and adopt the symbol ν/c together with the unit cm^{-1}.

Energy

One of the great achievements of physical science has been to demonstrate the unity of different forms of energy. For all of them one unit should suffice.

Quite irrationally, the trend of physics has been to adopt a new unit of energy for each field of study; nowhere is this profusion of units richer or more confusing than in spectroscopy. Of the five units in the following table only the erg is defined in a straightforward manner. The unit cm^{-1} denotes the energy ($h\nu$) associated with a quantum for which $\nu/c = 1$; the unit A cal gives the energy per molecule for 1 cal per mole; the electron-volt (eV) is the potential energy of an electron with respect to a 1 V potential drop; k deg is the thermal energy $k\theta$ associated with $\theta = 1$ deg.

TABLE A.3.2

Conversion factors for energy units

Unit	cm^{-1}	erg	A cal	eV	k deg
1 cm^{-1}	1	$1\cdot98570 \times 10^{-16}$	$2\cdot8576$	$1\cdot2396 \times 10^{-4}$	$1\cdot43867$
1 erg	$5\cdot0360 \times 10^{15}$	1	$1\cdot43291 \times 10^{16}$	$6\cdot2428 \times 10^{11}$	$7\cdot2454 \times 10^{15}$
1 A cal	$3\cdot4994 \times 10^{-1}$	$6\cdot9788 \times 10^{-17}$	1	$4\cdot3378 \times 10^{-5}$	$5\cdot0344$
1 eV	$8\cdot0671 \times 10^{3}$	$1\cdot60184 \times 10^{-12}$	$2\cdot3053 \times 10^{4}$	1	$1\cdot16059 \times 10^{4}$
1 k deg	$6\cdot9509 \times 10^{-1}$	$1\cdot38019 \times 10^{-16}$	$1\cdot9863$	$0\cdot86163 \times 10^{-4}$	1

Extinction coefficient

As discussed in Chapter 2, the *extinction coefficient* is defined in four different ways according to the measure of the amount of matter in the optical path. The following table gives the dimensions of these four quantities and the conversion factors between them.

TABLE A.3.3

Dimensions and conversion factors for extinction coefficients

Symbol Name Dimensions	e_v Volume e.c. cm^{-1}	e_m Mass e.c. $g^{-1} cm^2$	e_n Molecular e.c. cm^2	e_s E.c. per cm s.t.p. cm^{-1}
$e_v = 1$	1	ρ^{-1}	n^{-1}	$n_s\, n^{-1}$
$e_m = 1$	ρ	1	m	ρ_s
$e_n = 1$	n	m^{-1}	1	n_s
$e_s = 1$	nn_s^{-1}	ρ_s^{-1}	n_s^{-1}	1

ρ = density of absorbing gas (g cm^{-3}).
n = molecular number density (cm^{-3}).
n_s = molecular number density at s.t.p. (cm^{-3}).
m = mass of molecule (g).

The volume coefficient e_v is the quantity entering atmospheric computations, where distance is the independent variable. Laboratory data cannot, however, be presented in this unit without also stating the gaseous density, and it is better to use one of the other three coefficients. Of these, e_s is the popular choice of many infra-red spectroscopists. It is, however, an unfortunate unit since it contains in its definition a reference to a standard temperature when the measurements may be made at some very different temperature. Thus, it is possible to find such ambiguous statements as 'the extinction coefficient per cm at s.t.p., at room temperature is . . .'. A different name would resolve the difficulty, for we see that e_s differs from the completely satisfactory unit e_n only by the universal constant n_s (Loschmidt's number). The units of e_n (cm^2), and the alternative title *molecular cross-section* convey a valuable picture of the extinction process in terms of an impact area between photon and particle.

Some authors prefer to use *decadic* coefficients rather than the *napierian* coefficients used in this volume,

$$e' = \frac{e}{2 \cdot 3026}.$$

Related to the decadic coefficient is the power loss in db km^{-1}, sometimes used by microwave spectroscopists

$$e \text{ (db km}^{-1}) = 10^6 e'_e \text{ (cm}^{-1}).$$

Band and line intensities

The integrated absorption coefficient over a band or line has the dimensions of an extinction coefficient times a wavelength (usually in microns), wave number, or frequency. Since e_m, e_n, and e_s are equally suitable for the extinction, a band or line intensity can have nine different units. Fortunately these can be distinguished since no two have the same dimensions (see the table below). Most commonly band intensities are quoted in units of cm^{-2} or cm.

TABLE A.3.4

Physical dimensions of intensities

		e_m g^{-1} cm^2	e_n cm^2	e_s cm^{-1}
Wavelength	cm	g^{-1} cm^3	cm^3	1
Wave number	cm^{-1}	g^{-1} cm	cm	cm^{-2}
Frequency	s^{-1}	g^{-1} cm^2 s^{-1}	cm^2 s^{-1}	cm^{-1} s^{-1}

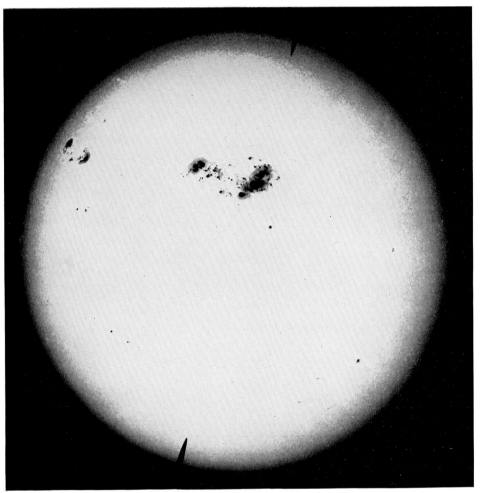

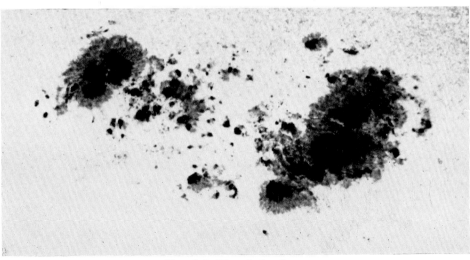

FIG. A.3. The photosphere.
Showing a large sunspot group in white light on 7 April 1947.

Courtesy of the Mt. Wilson and Palomar Observatories

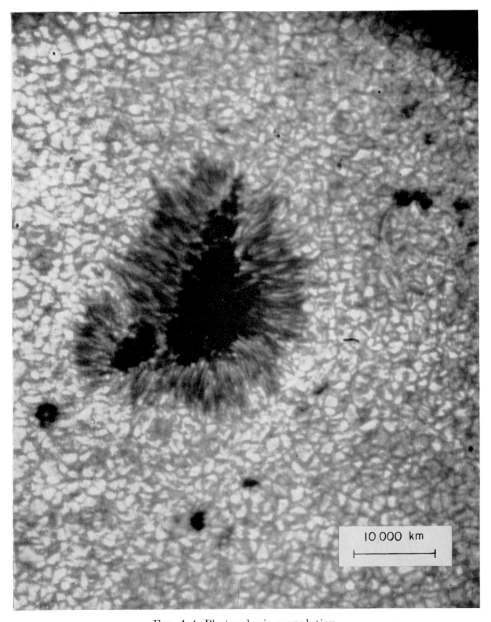

10 000 km

Fig. A.4. Photospheric granulation.
Photographed from a balloon flying in the stratosphere.

By courtesy of Professor Martin Schwarzschild and Project Stratoscope of Princeton University

APPENDIX 4

Refractive index of air

At $p = 760$ mm Hg and $\theta = 15°$ C, Edlén's formula gives

$$\{n(15,760)-1\} \times 10^6 = 64 \cdot 328 + 29\,498 \cdot 1(146 - \lambda^{-2})^{-1} + 255 \cdot 4(41 - \lambda^{-2})^{-1},$$

where the wavelength λ is in microns.

At pressure p (mm Hg) and temperature θ (°C)

$$\{n(\theta, p)-1\} = \{n(15,760)-1\} \frac{p\{1 + (1 \cdot 049 - 0 \cdot 0157\theta) \times 10^{-6}p\}}{720 \cdot 883(1 + 0 \cdot 003\,661\theta)}.$$

For a water-vapour pressure, e (mm Hg), the quantity $(n-1) \times 10^{-6}$ for dry air must be *reduced* by

$$\Delta(n-1) \times 10^{-6} = \frac{0 \cdot 0624 - 0 \cdot 000\,680\lambda^{-2}}{1 + 0 \cdot 003\,661\theta}\, e.$$

Penndorf points out that the water-vapour correction can be neglected unless pressure and temperature are known with great precision.

TABLE A.4.1

Refractive index of dry air $(n-1) \times 10^8$

Wavelength	$-30°$ C	$0°$ C	$+30°$ C
(μ)			
0·2	38 406	34 187	30 802
0·3	34 552	30 756	27 711
0·4	33 509	29 828	26 875
0·5	33 060	29 428	26 514
0·6	32 824	29 218	26 325
0·7	32 684	29 093	26 213
0·8	32 594	29 013	26 140
0·9	32 533	28 959	26 091
1·0	32 489	28 920	26 056
2·0	32 351	28 797	25 946
3·0	32 326	28 775	25 925
4·0	32 317	28 767	25 918
5·0	32 314	28 763	25 915
6·0	32 311	28 761	25 913
7·0	32 309	28 760	25 912
8·0	32 309	28 759	25 912
9·0	32 308	28 759	25 911
10·0	32 308	28 758	25 911
12·0	32 307	28 758	25 910
14·0	32 307	28 757	25 910
16·0	32 306	28 757	25 910
18·0	32 306	28 757	25 910
20·0	32 306	28 757	25 910
∞	32 305·7	28 756·5	25 909·2

BIBLIOGRAPHY

PENNDORF, R., 1957, 'Tables of the refractive index for standard air and the Rayleigh scattering coefficient for the spectral region between 0·2 and 20·0 μ and their application to atmospheric optics', *J. Opt. Soc. Am.* **47,** p. 176.

ALLEN, C. W., 1955, *Astrophysical quantities.* London : Univ. of London Press.

APPENDIX 5

Model atmospheres

TABLE A.5.1

The ARDC model atmosphere 0 to 30 km

Altitude	Temperature	Pressure	Density	Molecular number density	Scale height
(km)	(°K)	(mb)	(g cm^{-3})	(cm^{-3})	(km)
0	288·16	$1·01325 \times 10^3$	$1·2250 \times 10^{-3}$	$2·5476 \times 10^{19}$	8·4344
1	281·66	$8·9876 \times 10^2$	1·1117	2·3118	8·2468
2	275·16	7·9501	1·0066	2·0933	8·0591
3	268·67	7·0121	$9·0926 \times 10^{-4}$	1·8909	7·8713
4	262·18	6·1660	8·1935	1·7039	7·6835
5	255·69	5·4048	7·3643	1·5315	7·4957
6	249·20	4·7217	6·6011	1·3728	7·3077
7	242·71	4·1105	5·9002	1·2270	7·1198
8	236·23	3·5651	5·2578	1·0934	6·9317
9	229·74	3·0800	4·6706	$9·7130 \times 10^{18}$	6·7436
10	223·26	2·6500	4·1351	8·5993	6·5554
11	216·78	2·2700	3·6480	7·5864	6·3672
12	216·66	1·9399	3·1194	6·4870	6·3656
13	216·66	1·6579	2·6659	5·5441	6·3676
14	216·66	1·4170	2·2785	4·7385	6·3696
15	216·66	1·2112	1·9475	4·0501	6·3716
16	216·66	1·0353	1·6647	3·4619	6·3736
17	216·66	$8·8496 \times 10$	1·4230	2·9593	6·3756
18	216·66	7·5652	1·2165	2·5298	6·3776
19	216·66	6·4674	1·0399	2·1627	6·3796
20	216·66	5·5293	$8·8909 \times 10^{-5}$	1·8490	6·3816
21	216·66	4·7274	7·6015	1·5808	6·3836
22	216·66	4·0420	6·4995	1·3516	6·3856
23	216·66	3·4562	5·5575	1·1557	6·3876
24	216·66	2·9554	4·7522	$9·8828 \times 10^{17}$	6·3896
25	216·66	2·5273	4·0639	8·4513	6·3916
26	219·34	2·1632	3·4359	7·1453	6·4728
28	225·29	1·5949	2·4663	5·1290	6·6525
30	231·24	1·1855	1·7861	3·7144	6·8323

In Fig. A.1, the ARDC model atmosphere is compared with three other models more suitable to certain extreme conditions.

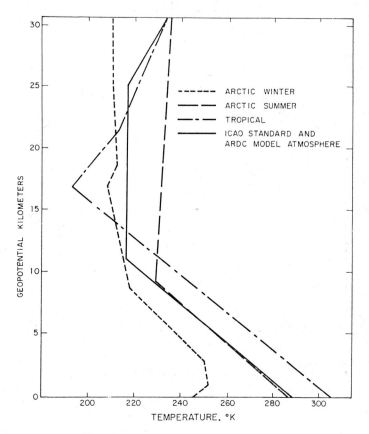

FIG. A.1. Model atmospheres, 0 to 30 km.

TABLE A.5.2

Density from 25 to 100 km $\{-\log_{10}\rho(g\ cm^{-2})\}$

Summer

Pressure level	Latitude								
	0°	10°	20°	30°	40°	50°	60°	70°	80°
(mb)									
20	4·50	4·50	4·50	4·51	4·51	4·52	4·52	4·53	4·53
16	4·60	4·60	4·61	4·61	4·61	4·62	4·62	4·63	4·63
12	4·74	4·74	4·74	4·74	4·75	4·75	4·75	4·76	4·76
10	4·82	4·82	4·82	4·83	4·83	4·83	4·84	4·84	4·84
8	4·92	4·92	4·93	4·94	4·94	4·94	4·94	4·95	4·94
6	5·06	5·06	5·07	5·07	5·08	5·08	5·08	5·08	5·07
4	5·26	5·26	5·26	5·27	5·27	5·27	5·27	5·27	5·26
3	5·40	5·40	5·40	5·41	5·41	5·41	5·40	5·40	5·39
2·5	5·48	5·48	5·49	5·49	5·50	5·49	5·49	5·48	5·48
2·0	5·59	5·59	5·59	5·60	5·60	5·60	5·59	5·59	5·58
1·6	5·69	5·69	5·70	5·70	5·70	5·70	5·70	5·69	5·69
1·2	5·82	5·82	5·82	5·82	5·83	5·84	5·84	5·83	5·82
1·0	5·90	5·90	5·91	5·90	5·91	5·93	5·93	5·92	5·91
0·8	6·00	6·00	6·00	6·00	6·01	6·02	6·02	6·02	6·01
0·6	6·13	6·13	6·12	6·12	6·12	6·13	6·14	6·14	6·14
0·4	—	6·29	6·29	6·28	6·28	6·29	6·30	6·30	6·30
0·3	—	—	6·40	6·39	6·39	6·40	6·41	6·41	6·41
0·2	—	—	—	6·55	6·54	6·55	6·56	—	—
0·15	—	—	—	6·67	6·66	6·66	6·67	—	—
0·10	—	—	—	6·83	6·81	6·81	6·82	—	—
0·08	—	—	—	6·91	6·90	6·90	6·91	—	—
0·04	—	—	—	7·19	7·17	7·17	7·16	—	—
0·02	—	—	—	7·46	7·45	7·43	7·43	—	—
0·01	—	—	—	7·76	7·75	7·73	7·71	—	—
0·004	—	—	—	8·17	8·16	8·14	8·12	—	—
0·002	—	—	—	8·50	8·48	8·47	8·45	—	—
0·001	—	—	—	8·82	8·81	8·80	8·78	—	—
0·0005	—	—	—	9·14	9·14	9·14	9·13	—	—

TABLE A.5.2 (*continued*)

Density from 25 to 100 km $\{-\log_{10}\rho(g\ cm^{-2})\}$

Winter

Pressure level	Latitude								
	0°	10°	20°	30°	40°	50°	60°	70°	80°
(mb)									
20	4·49	4·49	4·49	4·49	4·49	4·49	4·48	4·47	4·46
16	4·59	4·59	4·59	4·59	4·59	4·58	4·57	4·56	4·56
12	4·73	4·72	4·72	4·72	4·71	4·71	4·70	4·69	4·68
10	4·81	4·81	4·81	4·80	4·80	4·79	4·78	4·77	4·76
8	4·92	4·92	4·91	4·91	4·90	4·89	4·88	4·88	4·85
6	5·05	5·05	5·05	5·04	5·03	5·02	5·01	5·00	4·99
4	5·25	5·25	5·24	5·23	5·23	5·21	5·20	5·18	5·17
3	5·38	5·38	5·38	5·37	5·36	5·35	5·33	5·32	5·31
2·5	5·47	5·47	5·47	5·46	5·45	5·43	5·42	5·40	5·40
2·0	5·58	5·58	5·57	5·57	5·55	5·54	5·52	5·51	5·50
1·6	5·68	5·68	5·68	5·67	5·66	5·64	5·63	5·61	5·61
1·2	5·82	5·82	5·81	5·80	5·79	5·77	5·76	5·74	5·74
1·0	5·90	5·90	5·89	5·89	5·87	5·86	5·84	5·83	5·82
0·8	6·00	6·00	6·00	5·99	5·97	5·96	5·94	5·93	5·92
0·6	6·13	6·13	6·12	6·12	6·10	6·09	6·07	6·05	6·05
0·4	6·30	6·30	6·30	6·29	6·28	6·27	6·25	6·24	6·23
0·3	—	6·41	6·41	6·41	6·41	6·40	6·39	6·37	6·37
0·2	—	6·57	6·57	6·57	6·57	6·57	6·57	6·56	6·56
0·15	—	—	—	6·68	6·69	6·70	6·70	—	—
0·10	—	—	—	6·83	6·84	6·86	6·88	—	—
0·08	—	—	—	6·92	6·93	6·95	6·97	—	—
0·04	—	—	—	7·20	7·20	7·21	7·22	—	—
0·02	—	—	—	7·47	7·48	7·48	7·49	—	—
0·01	—	—	—	7·76	7·77	7·77	7·79	—	—
0·004	—	—	—	8·17	8·18	8·19	8·21	—	—
0·002	—	—	—	8·49	8·50	8·52	8·53	—	—
0·001	—	—	—	8·81	8·83	8·84	8·85	—	—
0·0005	—	—	—	9·13	9·15	9·16	9·16	—	—

TABLE A.5.3

Heights of pressure contours (geodynamic km) 25 to 100 km

Summer

Pressure level	Latitude								
	0°	10°	20°	30°	40°	50°	60°	70°	80°
(mb)									
20	26·5	26·5	26·6	26·9	27·1	27·3	27·4	27·5	27·6
16	27·9	28·0	28·1	28·4	28·6	28·8	28·9	29·0	29·1
12	29·8	29·9	30·0	30·3	30·6	30·8	30·9	31·0	31·1
10	31·1	31·1	31·2	31·6	31·8	32·1	32·2	32·3	32·4
8	32·6	32·6	32·8	33·1	33·4	33·7	33·8	33·9	34·0
6	34·6	34·6	34·8	35·2	35·5	35·7	35·9	36·0	36·1
4	37·5	37·5	37·8	38·2	38·5	38·8	38·9	39·0	39·1
3	39·7	39·7	40·0	40·4	40·7	41·0	41·1	41·2	41·2
2·5	41·1	41·1	41·4	41·8	42·2	42·5	42·5	42·6	42·6
2·0	42·8	42·9	43·2	43·6	44·0	44·2	44·2	44·4	44·3
1·6	44·6	44·6	44·9	45·4	45·8	46·1	46·1	46·1	46·0
1·2	47·0	47·0	47·3	47·7	48·2	48·5	48·5	48·5	48·4
1·0	48·4	48·5	48·8	49·2	49·7	50·1	50·1	50·0	49·9
0·8	50·3	50·3	50·6	51·1	51·6	52·0	52·0	51·9	51·8
0·6	52·7	52·7	53·0	53·4	54·0	54·4	54·5	54·4	54·2
0·4	—	56·0	56·2	56·6	57·2	57·7	57·8	57·8	57·6
0·3	—	—	58·4	58·8	59·4	60·0	60·1	60·1	59·9
0·2	—	—	—	61·8	62·3	63·0	63·2	—	—
0·15	—	—	—	63·9	64·4	65·1	65·4	—	—
0·10	—	—	—	66·7	67·1	67·8	68·2	—	—
0·08	—	—	—	68·3	68·6	69·3	70·0	—	—
0·04	—	—	—	72·7	73·0	73·6	74·0	—	—
0·02	—	—	—	77·0	77·1	77·6	78·0	—	—
0·01	—	—	—	81·1	81·1	81·4	81·8	—	—
0·004	—	—	—	86·6	86·4	86·5	86·7	—	—
0·002	—	—	—	90·9	90·6	90·6	90·5	—	—
0·001	—	—	—	95·4	95·1	94·9	94·7	—	—
0·0005	—	—	—	100·2	99·8	99·5	99·2	—	—

TABLE A.5.3 (*cont.*)

Heights of pressure contours (geodynamic km) 25 to 100 km

Winter

Pressure level	Latitude								
	0°	10°	20°	30°	40°	50°	60°	70°	80°
(mb)									
20	26·1	26·1	26·2	26·2	26·2	26·1	25·7	25·2	25·0
16	27·5	27·5	27·6	27·6	27·6	27·5	27·1	26·6	26·3
12	29·4	29·4	29·5	29·5	29·4	29·3	28·8	28·3	28·0
10	30·6	30·6	30·7	30·6	30·6	30·4	30·0	29·4	29·0
8	32·1	32·1	32·2	32·1	32·0	31·8	31·3	30·7	30·3
6	34·1	34·0	34·1	34·0	33·9	33·7	33·1	32·4	32·0
4	37·0	37·0	37·0	36·8	36·6	36·3	35·7	34·9	34·5
3	39·1	39·0	39·0	38·9	38·7	38·2	37·6	36·7	36·2
2·5	40·4	40·4	40·4	40·2	40·0	39·5	38·8	37·9	37·6
2·0	42·1	42·1	42·1	41·9	41·6	41·1	40·3	39·3	38·8
1·6	43·9	43·8	43·8	43·6	43·2	42·6	41·8	40·8	40·3
1·2	46·2	46·1	46·1	45·8	45·4	44·7	43·8	42·8	42·2
1·0	47·6	47·6	47·5	47·2	46·8	46·1	45·1	44·0	43·5
0·8	49·5	49·4	49·3	49·0	48·5	47·7	46·7	45·6	45·0
0·6	51·8	51·8	51·6	51·3	50·7	49·8	48·7	47·6	47·0
0·4	55·2	55·1	55·0	54·5	53·9	52·9	51·7	50·4	49·7
0·3	—	57·4	57·2	56·8	56·1	55·1	53·8	52·5	51·8
0·2	—	60·5	60·4	60·0	59·3	58·3	56·9	55·5	54·8
0·15	—	—	—	62·1	61·4	60·4	59·1	—	—
0·10	—	—	—	65·0	64·4	63·6	62·2	—	—
0·08	—	—	—	66·5	66·0	65·2	63·9	—	—
0·04	—	—	—	71·1	70·6	70·0	68·9	—	—
0·02	—	—	—	75·4	75·0	74·4	73·5	—	—
0·01	—	—	—	79·6	79·2	78·6	77·8	—	—
0·004	—	—	—	85·0	84·7	84·3	83·7	—	—
0·002	—	—	—	89·3	89·1	88·9	88·4	—	—
0·001	—	—	—	93·7	93·7	93·6	93·3	—	—
0·0005	—	—	—	98·4	98·6	98·6	98·3	—	—

TABLE A.5.4

Atmospheric temperature (°K), 25 to 100 km

Summer

Pressure level	Latitude								
	0°	10°	20°	30°	40°	50°	60°	70°	80°
(mb)									
20	220	220	222	224	226	230	233	235	236
16	223	223	225	227	229	232	234	236	237
12	227	227	229	231	233	236	237	239	239
10	230	230	232	234	237	239	240	242	242
8	234	234	237	239	242	244	244	245	244
6	240	240	243	246	249	251	250	250	248
4	251	251	254	257	261	261	259	257	254
3	260	260	263	266	269	267	264	261	257
2·5	265	265	267	270	273	271	267	264	260
2·0	271	271	272	274	277	276	273	268	265
1·6	274	274	276	277	281	282	279	274	271
1·2	278	278	279	279	285	290	289	284	279
1·0	279	279	280	279	285	293	293	289	283
0·8	280	280	280	278	283	291	294	292	288
0·6	280	279	277	274	277	285	290	291	290
0·4	—	274	270	265	265	272	276	278	280
0·3	—	—	262	259	254	261	267	269	270
0·2	—	—	—	248	244	247	254	—	—
0·15	—	—	—	242	237	238	245	—	—
0·10	—	—	—	233	227	226	232	—	—
0·08	—	—	—	228	222	220	226	—	—
0·04	—	—	—	214	208	205	203	—	—
0·02	—	—	—	202	197	189	186	—	—
0·01	—	—	—	200	194	186	180	—	—
0·004	—	—	—	208	202	194	185	—	—
0·002	—	—	—	218	212	204	195	—	—
0·001	—	—	—	229	226	217	210	—	—
0·0005	—	—	—	241	243	238	234	—	—

TABLE A.5.4 (cont.)

Atmospheric temperature (°K), 25 to 100 km

Winter

Pressure level	Latitude								
	0°	10°	20°	30°	40°	50°	60°	70°	80°
(mb)									
20	214	213	213	214	214	214	209	204	201
16	218	217	216	216	214	214	208	203	200
12	222	221	220	218	216	213	208	203	199
10	226	225	223	220	217	213	208	203	199
8	230	229	228	224	220	214	209	203	199
6	237	236	234	230	225	218	212	206	202
4	246	245	243	239	234	226	219	212	208
3	253	253	251	247	240	232	224	217	213
2·5	257	257	256	252	245	236	228	220	217
2·0	263	262	260	256	249	241	233	225	221
1·6	268	267	264	260	253	244	236	229	225
1·2	274	273	269	265	257	248	238	232	229
1·0	277	276	272	267	260	250	239	233	231
0·8	279	279	274	270	262	252	241	235	232
0·6	281	280	277	273	265	255	245	237	234
0·4	276	276	275	274	268	259	250	243	239
0·3	—	270	269	269	267	262	255	247	244
0·2	—	257	256	257	259	263	259	254	251
0·15	—	—	—	249	253	261	262	—	—
0·10	—	—	—	236	243	254	262	—	—
0·08	—	—	—	232	237	247	258	—	—
0·04	—	—	—	218	220	224	229	—	—
0·02	—	—	—	207	209	211	216	—	—
0·01	—	—	—	200	204	207	214	—	—
0·004	—	—	—	205	210	218	224	—	—
0·002	—	—	—	214	220	228	236	—	—
0·001	—	—	—	224	233	239	244	—	—
0·0005	—	—	—	236	245	250	250	—	—

TABLE A.5.5

Atmospheric temperature, pressure, density, 100 *to* 1000 *km*

Altitude (km)	Temperature (°K)		Pressure (dyne cm⁻²)		Density (gm cm⁻³)	
	Day	Night	Day	Night	Day	Night
100	210	210	$4 \cdot 0 \times 10^{-1}$	$4 \cdot 0 \times 10^{-1}$	$6 \cdot 6 \times 10^{-10}$	$6 \cdot 6 \times 10^{-10}$
120	235	235	$4 \cdot 3 \times 10^{-2}$	$4 \cdot 3 \times 10^{-2}$	$3 \cdot 6 \times 10^{-11}$	$3 \cdot 6 \times 10^{-11}$
140	500	500	$1 \cdot 2 \times 10^{-2}$	$1 \cdot 2 \times 10^{-2}$	$6 \cdot 8 \times 10^{-12}$	$6 \cdot 8 \times 10^{-12}$
160	779	649	$5 \cdot 0 \times 10^{-3}$	$4 \cdot 6 \times 10^{-3}$	$1 \cdot 8 \times 10^{-12}$	$2 \cdot 0 \times 10^{-12}$
180	967	759	$2 \cdot 9 \times 10^{-3}$	$2 \cdot 3 \times 10^{-3}$	$8 \cdot 0 \times 10^{-13}$	$8 \cdot 0 \times 10^{-13}$
200	1100	836	$1 \cdot 9 \times 10^{-3}$	$1 \cdot 3 \times 10^{-3}$	$4 \cdot 3 \times 10^{-13}$	$4 \cdot 0 \times 10^{-13}$
300	1385	980	$3 \cdot 6 \times 10^{-4}$	$1 \cdot 5 \times 10^{-4}$	$5 \cdot 8 \times 10^{-14}$	$3 \cdot 3 \times 10^{-14}$
400	1457	1005	$9 \cdot 9 \times 10^{-5}$	$2 \cdot 7 \times 10^{-5}$	$1 \cdot 4 \times 10^{-14}$	$5 \cdot 3 \times 10^{-15}$
500	1480	1011	$3 \cdot 2 \times 10^{-5}$	$5 \cdot 4 \times 10^{-6}$	$4 \cdot 3 \times 10^{-15}$	$1 \cdot 0 \times 10^{-15}$
600	1489	1013	$1 \cdot 1 \times 10^{-5}$	$1 \cdot 2 \times 10^{-6}$	$1 \cdot 4 \times 10^{-15}$	$2 \cdot 2 \times 10^{-16}$
800	1494	1014	$1 \cdot 5 \times 10^{-6}$	$6 \cdot 4 \times 10^{-8}$	$1 \cdot 9 \times 10^{-16}$	$1 \cdot 2 \times 10^{-17}$
1000	1496	1015	$2 \cdot 3 \times 10^{-7}$	$4 \cdot 8 \times 10^{-9}$	$2 \cdot 9 \times 10^{-17}$	$7 \cdot 7 \times 10^{-19}$

BIBLIOGRAPHY

Details of standard atmospheres below 30 km are given by

WARES, G. W., 1960, *Handbook of geophysics*, New York: Macmillan, p. 1.

The data between 30 and 100 km follow

MURGATROYD, R. J., 1957, 'Winds and temperatures between 20 km and 100 km—a review', *Quart. J. R. Met. Soc.* **83**, p. 417.

Between 100 and 1000 km we have followed

JASTROW, R., and KYLE, L., 1961, 'The upper atmosphere' in *Handbook of astronautical engineering*. New York: McGraw-Hill.

APPENDIX 6

Properties of water vapour

Molecular weight, $M = 18$
Density at s.t.p., $\rho_0 = 8 \cdot 031 \times 10^{-4}$ g cm^{-3}
Gas constant, $R_\mathrm{w} = 4 \cdot 618 \times 10^6$ erg deg^{-1} g^{-1}

TABLE A.6.1

Saturated vapour pressure (mb) over pure liquid water

°C	−30	−20	−10	0	+10	+20	+30	+40
0	0·5088	1·2540	2·8627	6·1078	12·272	23·373	42·430	73·777
+1	0·5589	1·3664	3·0971	6·5662	13·119	24·861	44·927	77·802
+2	0·6134	1·4877	3·3484	7·0547	14·017	26·430	47·554	82·015
+3	0·6727	1·6186	3·6177	7·5753	14·969	28·086	50·307	86·423
+4	0·7371	1·7597	3·9061	8·1294	15·977	29·831	53·200	91·034
+5	0·8070	1·9118	4·2148	8·7192	17·044	31·671	56·236	95·855
+6	0·8827	2·0755	4·5451	9·3465	18·173	33·608	59·422	100·89
+7	0·9649	2·2515	4·8981	10·013	19·367	35·649	62·762	106·16
+8	1·0538	2·4409	5·2753	10·722	20·630	37·796	66·264	111·66
+9	1·1500	2·6443	5·6780	11·474	21·964	40·055	69·934	117·40

TABLE A.6.1 (cont.)

Saturation vapour pressure (mb) over pure ice

°C	−100	−90	−80	−70	−60	−50	−40	−30	−20	−10	0
0	1.403×10^{-5}	9.672×10^{-5}	5.472×10^{-4}	2.618×10^{-3}	1.080×10^{-2}	3.935×10^{-2}	1.283×10^{-1}	3.798×10^{-1}	1.032	2.597	6.107
+1	1.719×10^{-5}	1.160×10^{-4}	6.444×10^{-4}	3.032×10^{-3}	1.236×10^{-2}	4.449×10^{-2}	1.436×10^{-1}	4.213×10^{-1}	1.135	2.837	
+2	2.101×10^{-5}	1.388×10^{-4}	7.577×10^{-4}	3.511×10^{-3}	1.413×10^{-2}	5.026×10^{-2}	1.606×10^{-1}	4.669×10^{-1}	1.248	3.097	
+3	2.561×10^{-5}	1.658×10^{-4}	8.894×10^{-4}	4.060×10^{-3}	1.612×10^{-2}	5.671×10^{-2}	1.794×10^{-1}	5.170×10^{-1}	1.371	3.379	
+4	3.117×10^{-5}	1.977×10^{-4}	1.042×10^{-3}	4.688×10^{-3}	1.838×10^{-2}	6.393×10^{-2}	2.002×10^{-1}	5.720×10^{-1}	1.506	3.685	
+5	3.784×10^{-5}	2.353×10^{-4}	1.220×10^{-3}	5.406×10^{-3}	2.092×10^{-2}	7.198×10^{-2}	2.233×10^{-1}	6.323×10^{-1}	1.652	4.015	
+6	4.584×10^{-5}	2.796×10^{-4}	1.425×10^{-3}	6.225×10^{-3}	2.380×10^{-2}	8.097×10^{-2}	2.488×10^{-1}	6.985×10^{-1}	1.811	4.372	
+7	5.542×10^{-5}	3.316×10^{-4}	1.662×10^{-3}	7.159×10^{-3}	2.703×10^{-2}	9.098×10^{-2}	2.769×10^{-1}	7.709×10^{-1}	1.984	4.757	
+8	6.685×10^{-5}	3.925×10^{-4}	1.936×10^{-3}	8.228×10^{-3}	3.067×10^{-2}	1.021×10^{-1}	3.079×10^{-1}	8.502×10^{-1}	2.172	5.173	
+9	8.049×10^{-5}	4.638×10^{-4}	2.252×10^{-3}	9.432×10^{-3}	3.476×10^{-2}	1.145×10^{-1}	3.421×10^{-1}	9.370×10^{-1}	2.376	5.623	

BIBLIOGRAPHY

Publication No. 79 of the International Meteorological Organization, 1961, Definitions and Specifications of water vapour in the atmosphere.

851602

D d

APPENDIX 7

The Planck function

TABLE A.7.1

$$C_1 = (3{\cdot}7403 \pm 0{\cdot}0002) \times 10^{-5} \text{ erg cm}^2 \text{ s}^{-1}$$
$$C_2 = 1{\cdot}438\ 68 \pm 0{\cdot}000\ 01 \text{ cm deg.}$$

The normalizing factors for use with the table are:

$$\pi B = 5{\cdot}6698 \times 10^{-5}\theta^4 \text{ erg cm}^{-2} \text{ s}^{-1}$$
$$\pi(B_\lambda)_{\max} = 1{\cdot}287 \times 10^{-4}\theta^5 \text{ erg cm}^{-2} \text{ s}^{-1} \text{ cm}^{-1}$$
$$\pi(B_\nu)_{\max} = 5{\cdot}956 \times 10^{-16}\theta^3 \text{ erg cm}^{-2} \text{ s}^{-1} \text{ (c.p.s.)}^{-1}$$
$$\pi(\lambda B_\lambda)_{\max} = 4{\cdot}1736 \times 10^{-5}\theta^4 \text{ erg cm}^{-2} \text{ s}^{-1}$$

λT	$C_2/\lambda T = x$	$\int_0^\lambda B_\lambda\, d\lambda/B$	$B_\lambda/(B_\lambda)_{\max}$	$\lambda B_\lambda/(\lambda B_\lambda)_{\max}$	$B_\nu/(B_\nu)_{\max}$
cm deg	large x	$\dfrac{x^3 e^{-x}}{6{\cdot}4939}$	$\dfrac{x^5 e^{-x}}{21{\cdot}201}$	$\dfrac{x^4 e^{-x}}{4{\cdot}780}$	$\dfrac{x^3 e^{-x}}{1{\cdot}4214}$
0·01	143·85	0·0⁵⁶ 16	0·0⁵³ 98	0·0⁵⁴ 30	0·0⁵⁶ 70
0·02	71·925	0·0²⁶ 35	0·0²³ 52	0·0²⁴ 33	0·0²⁵ 15
0·03	47·950	0·0¹⁶ 27	0·0¹³ 18	0·0¹⁴ 17	0·0¹⁵ 12
0·04	35·962	0·0¹¹ 19	0·0⁹ 683	0·0¹⁰ 84	0·0¹¹ 79
0·05	28·770	0·0⁸ 130	0·0⁶ 297	0·0⁷ 458	0·0⁸ 536
0·06	23·975	0·0⁷ 932	0·0⁴ 145	0·0⁵ 267	0·0⁶ 375
0·07	20·550	0·0⁵ 186	0·0³ 206	0·0⁴ 444	0·0⁵ 726
0·08	17·981	0·0⁴ 165	0·001 38	0·0³ 339	0·0⁴ 635
0·09	15·983	0·0⁴ 872	0·005 63	0·001 56	0·0³ 329
0·10	14·385	0·0³ 321	0·016 44	0·005 07	0·001 18
0·12	11·988	0·002 14	0·072 64	0·026 87	0·007 54
0·14	10·275	0·007 80	0·186 27	0·080 41	0·026 31
0·16	8·9906	0·019 74	0·345 18	0·170 31	0·063 69
0·18	7·9917	0·039 38	0·520 26	0·288 77	0·121 58
0·20	7·1925	0·066 79	0·683 44	0·421 44	0·196 59
0·25	5·7540	0·161 45	0·946 12	0·729 36	0·426 22
0·30	4·7950	0·273 36	0·997 12	0·922 41	0·646 87
0·35	4·1100	0·383 06	0·922 71	0·995 74	0·814 74
0·40	3·5962	0·481 02	0·800 10	0·986 74	0·922 72
0·45	3·1967	0·564 44	0·671 36	0·931 61	0·979 96
0·50	2·8770	0·633 86	0·554 67	0·855 19	0·999 52
0·60	2·3975	0·737 89	0·373 78	0·691 55	0·969 93
0·70	2·0550	0·808 12	0·253 94	0·548 15	0·896 89
0·80	1·7981	0·856 31	0·175 98	0·434 08	0·811 77
0·90	1·5983	0·890 04	0·124 71	0·346 13	0·728 20

TABLE A.7.1 (*cont.*)

λT	$C_2/\lambda T = x$	$\int_0^\lambda B_\lambda \, d\lambda/B$	$B_\lambda/(B_\lambda)_{\max}$	$\lambda B_\lambda/(\lambda B_\lambda)_{\max}$	$B_\nu/(B_\nu)_{\max}$
1·00	1·4385	0·914 18	0·090 37	0·278 70	0·651 50
1·20	1·1988	0·945 06	0·050 40	0·186 52	0·523 30
1·40	1·0275	0·962 86	0·030 10	0·129 97	0·425 36
1·60	0·8991	0·973 77	0·019 02	0·093 79	0·350 87
1·80	0·7992	0·980 81	0·012 57	0·089 74	0·293 53
2·0	0·7192	0·985 56	0·008 62	0·053 18	0·248 59
3·0	0·4795	0·995 24	0·001 94	0·017 98	0·126 06
4·0	0·3596	0·997 90	0·0^3 655	0·008 08	0·075 60
5·0	0·2877	0·998 90	0·0^3 279	0·004 30	0·050 25
10	0·1438	0·9^3 848	0·0^4 188	0·0^3 579	0·013 54
20	0·0719	0·9^4 801	0·0^5 122	0·0^4 751	0·003 51
50	0·0288	0·9^5 878	0·0^7 318	0·0^5 491	0·0^3 574
100	0·0144	0·9^6 848	0·0^8 200	0·0^6 618	0·0^3 145
	small x	$1 - 0{\cdot}05134 x^3$	$0{\cdot}04717 x^4$	$0{\cdot}2092 x^3$	$0{\cdot}7035 x^2$

BIBLIOGRAPHY

The numerical data in this appendix and in Chapter 2 are from

ALLEN, C. W., 1955, *Astrophysical quantities*. University of London Press.

A detailed review of the fundamental physical constants and nearly 500 pages of numerical values of the Planck function are given by

PIVOVONSKY, M., and NAGEL, M. R., 1961, *Tables of blackbody radiation functions*. New York: Macmillan.

APPENDIX 8

The exponential integrals

$$E_n(x) = \int_1^\infty \frac{e^{-\omega x}}{\omega^n}\, d\omega,$$

$$E_1(0) = \infty,$$

$$E_n(0) = \frac{1}{n-1}.$$

Two recurrence formulae:

$$E_n'(x) = -E_{n-1}(x),$$

$$nE_{n+1}(x) = e^{-x} - xE_n(x).$$

The following expansion always converges, but is only useful for small values of the argument

$$E_1(x) = -\gamma - \log|x| + \sum_{n=1}^\infty (-1)^{n-1}\frac{x^n}{n\,.\,n!}$$

where $\gamma = 0{\cdot}577\ 215\ 6$ is Euler's constant.

From the recurrence relations

$$(n-1)!\,E_n(x) = (-x)^{n-1}E_1(x) + e^{-x}\sum_{s=0}^{n-2}(n-\varepsilon-2)!\,(-x)^s.$$

An asymptotic expansion for large x is

$$E_n(x) \sim \frac{e^{-x}}{x}\left(1 - \frac{n}{x} + \frac{n(n+1)}{x^2} - \ldots\right).$$

TABLE A.8.1

Numerical values of the exponential integrals

x	$E_1(x)$	$E_2(x)$	$E_3(x)$
0	∞	1·0000	0·5000
0·01	4·0379	0·9497	0·4903
·02	3·3547	·9131	·4810
·03	2·9591	·8817	·4720
·04	2·6813	·8535	·4633
0·05	2·4679	0·8278	0·4549
·06	2·2953	·8040	·4468
·07	2·1508	·7818	·4388
·08	2·0269	·7610	·4311
·09	1·9187	·7412	·4236
0·10	1·8229	0·7225	0·4163
·20	1·2227	·5742	·3519
·30	0·9057	·4691	·3000
·40	0·7024	·3894	·2573
0·50	0·5598	0·3267	0·2216
·60	·4544	·2762	·1916
·70	·3738	·2349	·1661
·80	·3106	·2009	·1443
·90	·2602	·1724	·1257
1·00	0·2194	0·1485	0·1097
·25	·1464	·1035	·0786
·50	·1000	·0731	·0367
·75	·0695	·0522	·0412
2·00	0·0489	0·0375	0·0301
·25	·0348	·0272	·0221
·50	·0249	·0198	·0163
·75	·0180	·0145	·0120
3·00	0·0130	0·0106	0·0089
·25	·0095	·0078	·0066
·50	·0070	·0058	·0049

BIBLIOGRAPHY

KOURGANOFF, V., 1952, *Basic methods in transfer problems*. Oxford University Press.

APPENDIX 9

The Ladenberg and Reiche function

$$L(u) = u \exp(-u)\{I_0(u) + I_1(u)\}$$

u	0	1	2	3	4	5	6	7	8	9
0·0	0·0000	0·0099	0·0198	0·0295	0·0392	0·0488	0·0583	0·0676	0·0769	0·0861
·1	·0952	·1042	·1132	·1220	·1308	·1395	·1482	·1567	·1652	·1735
·2	·1818	·1900	·1982	·2063	·2143	·2223	·2302	·2380	·2457	·2534
·3	·2610	·2685	·2760	·2834	·2908	·2981	·3053	·3125	·3196	·3267
·4	·3337	·3406	·3475	·3543	·3611	·3678	·3745	·3811	·3877	·3942
0·5	0·4007	0·4071	0·4135	0·4198	0·4261	0·4324	0·4386	0·4447	0·4508	0·4569
·6	·4629	·4689	·4748	·4807	·4865	·4923	·4981	·5038	·5095	·5152
·7	·5208	·5264	·5319	·5374	·5429	·5483	·5537	·5591	·5644	·5697
·8	·5749	·5801	·5853	·5905	·5956	·6007	·6058	·6108	·6158	·6208
·9	·6258	·6307	·6356	·6404	·6452	·6500	·6548	·6596	·6643	·6690
1·0	0·6737	0·6783	0·6829	0·6875	0·6921	0·6966	0·7012	0·7057	0·7101	0·7146
·1	·7190	·7234	·7278	·7322	·7365	·7408	·7451	·7494	·7536	·7578
·2	·7620	·7662	·7704	·7746	·7787	·7828	·7869	·7910	·7950	·7990
·3	·8030	·8070	·8110	·8150	·8189	·8228	·8267	·8306	·8345	·8384
·4	·8422	·8460	·8498	·8536	·8574	·8612	·8649	·8686	·8723	·8760
1·5	0·8797	0·8834	0·8870	0·8907	0·8943	0·8979	0·9015	0·9051	0·9086	0·9122
·6	·9157	·9193	·9228	·9263	·9298	·9332	·9367	·9402	·9436	·9470
·7	·9504	·9538	·9572	·9606	·9639	·9673	·9706	·9740	·9773	·9806
·8	·9839	·9872	·9904	·9937	·9969	1·0002	1·0034	1·0066	1·0098	1·0130
·9	·0162	·0194	·0226	·0257	·0289	·0320	·0351	·0383	·0414	·0445
2·0	1·0476	1·0506	1·0537	1·0568	1·0598	1·0629	1·0659	1·0689	1·0719	1·0750
·1	·0780	·0809	·0839	·0869	·0899	·0928	·0958	·0987	·1016	·1046
·2	·1075	·1104	·1133	·1162	·1191	·1220	·1248	·1277	·1305	·1334
·3	·1362	·1391	·1419	·1447	·1475	·1503	·1531	·1559	·1587	·1615
·4	·1642	·1670	·1698	·1725	·1753	·1780	·1807	·1835	·1862	·1889
2·5	1·1916	1·1943	1·1970	1·1997	1·2023	1·2050	1·2077	1·2103	1·2130	1·2156
·6	·2183	·2209	·2235	·2262	·2288	·2314	·2340	·2366	·2392	·2418
·7	·2444	·2470	·2495	·2521	·2547	·2572	·2598	·2623	·2649	·2674
·8	·2699	·2725	·2750	·2775	·2800	·2825	·2850	·2875	·2900	·2925
·9	·2949	·2974	·2999	·3024	·3048	·3073	·3097	·3122	·3146	·3171
3·0	1·3195	1·3219	1·3243	1·3268	1·3292	1·3316	1·3340	1·3364	1·3388	1·3412
·1	·3436	·3459	·3483	·3507	·3531	·3554	·3578	·3601	·3625	·3649
·2	·3672	·3695	·3719	·3742	·3765	·3789	·3812	·3835	·3858	·3881
·3	·3904	·3927	·3950	·3973	·3996	·4019	·4042	·4064	·4087	·4110
·4	·4132	·4155	·4178	·4200	·4223	·4245	·4268	·4290	·4312	·4335
3·5	1·4357	1·4379	1·4402	1·4424	1·4446	1·4468	1·4490	1·4512	1·4534	1·4556
·6	·4578	·4600	·4622	·4644	·4666	·4687	·4709	·4731	·4753	·4774
·7	·4796	·4817	·4839	·4861	·4882	·4903	·4925	·4946	·4968	·4989
·8	·5010	·5032	·5053	·5074	·5095	·5116	·5137	·5159	·5180	·5201
·9	·5222	·5243	·5264	·5285	·5305	·5326	·5347	·5368	·5389	·5409

TABLE A.9.1 (*cont.*)

u	0	1	2	3	4	5	6	7	8	9
4·0	1·5430	1·5451	1·5471	1·5492	1·5513	1·5533	1·5554	1·5574	1·5595	1·5615
·1	·5636	·5656	·5677	·5697	·5717	·5738	·5758	·5778	·5798	·5818
·2	·5839	·5859	·5879	·5899	·5919	·5939	·5959	·5979	·5999	·6019
·3	·6039	·6059	·6079	·6099	·6118	·6138	·6158	·6178	·6197	·6217
·4	·6237	·6256	·6276	·6296	·6315	·6335	·6354	·6374	·6393	·6413
4·5	1·6432	1·6452	1·6471	1·6490	1·6510	1·6529	1·6548	1·6567	1·6587	1·6606
·6	·6625	·6644	·6663	·6683	·6702	·6721	·6740	·6759	·6778	·6797
·7	·6816	·6835	·6853	·6872	·6891	·6910	·6929	·6948	·6967	·6986
·8	·7005	·7023	·7042	·7061	·7079	·7098	·7117	·7135	·7154	·7173
·9	·7191	·7210	·7228	·7247	·7265	·7284	·7302	·7321	·7339	·7357
5	1·7376	1·7558	1·7739	1·7918	1·8095	1·8270	1·8444	1·8616	1·8786	1·8955
6	·9123	·9288	·9453	·9616	·9778	·9938	2·0097	2·0255	2·0412	2·0568
7	2·0722	2·0875	2·1027	2·1178	2·1328	2·1477	·1625	·1771	·1917	·2062
8	·2206	·2349	·2491	·2632	·2772	·2912	·3050	·3188	·3325	·3461
9	·3597	·3731	·3865	·3998	·4130	·4262	·4393	·4523	·4653	·4781
10	2·4910	2·5037	2·5164	2·5290	2·5416	2·5541	2·5665	2·5789	2·5912	2·6035
11	·6157	·6278	·6399	·6519	·6639	·6758	·6877	·6995	·7113	·7230
12	·7347	·7463	·7579	·7694	·7809	·7923	·8037	·8150	·8263	·8375
13	·8487	·8599	·8710	·8821	·8931	·9041	·9150	·9259	·9368	·9476
14	·9584	·9691	·9798	·9905	3·0011	3·0117	3·0223	3·0328	3·0433	3·0537
15	3·0641	3·0745	3·0848	3·0951	3·1054	3·1156	3·1258	3·1360	3·1461	3·1562
16	·1663	·1763	·1863	·1963	·2063	·2162	·2261	·2359	·2457	·2555
17	·2653	·2750	·2847	·2944	·3041	·3137	·3233	·3328	·3424	·3519
18	·3614	·3708	·3803	·3897	·3990	·4084	·4177	·4270	·4363	·4456
19	·4548	·4640	·4732	·4823	·4914	·5006	·5096	·5187	·5277	·5367
20	3·5457	3·6344	3·7210	3·8055	3·8883	3·9693	4·0487	4·1266	4·2030	4·2781
30	4·3519	4·4244	4·4958	4·5660	4·6352	4·7034	4·7706	4·8369	4·9022	4·9667
40	5·0304	5·0933	5·1554	5·2168	5·2775	5·3374	5·3968	5·4554	5·5135	5·5709
50	5·6277									

BIBLIOGRAPHY

KAPLAN, L. D., and EGGERS, D. F., 1956, 'Intensity and line width of the 15-micron band, determined by a curve-of-growth method', *J. Chem. Phys.* **25,** p. 876.

APPENDIX 10

The error function and the exponential function

TABLE A.10.1

The error function

$$\operatorname{erf} x = \frac{2}{\sqrt{\pi}} \int_0^x e^{-t^2} dt.$$

For small x
$$\operatorname{erf} x = \frac{2}{\sqrt{\pi}} \left\{ x - \frac{x^3}{1!\,3!} + \frac{x^5}{2!\,5!} - \frac{x^7}{3!\,7!} \cdots \right\}.$$

For large x
$$1 - \operatorname{erf} x \sim \frac{e^{-x^2}}{x\sqrt{\pi}} \left\{ 1 - \frac{1}{2x^2} + \frac{1.3}{(2x^2)^2} - \frac{1.3.5}{(2x^2)^3} + \cdots \right\}.$$

x		0	1	2	3	4	5	6	7	8	9
0·0	0·0	000	113	226	338	451	564	676	789	901	*013
·1	0·1	125	236	348	459	569	680	790	900	*009	*118
·2	0·2	227	335	443	550	657	763	869	974	*079	*183
·3	0·3	286	389	491	593	694	794	893	992	*090	*187
·4	0·4	284	380	475	569	662	755	847	937	*027	*117
0·5	0·5	205	292	379	465	549	633	716	798	879	959
·6	0·6	039	117	194	270	346	420	494	566	638	708
·7	0·6	778	847	914	981	*047	*112	*175	*238	*300	*361
·8	0·7	421	480	538	595	651	707	761	814	867	918
·9	0·7	969	*019	*068	*116	*163	*209	*254	*299	*342	*385
1·0	0·8	427	468	508	548	586	624	661	698	733	768
·1	0·8	802	835	868	900	931	961	991	*020	*048	*076
·2	0·9	103	130	155	181	205	229	252	275	297	319
·3	0·9	340	361	381	400	419	438	456	473	490	507
·4	0·95	23	39	54	69	83	97	*11	*24	*37	*49
1·5	0·96	61	73	84	95	*06	*16	*26	*36	*45	*55
·6	0·97	63	72	80	88	96	*04	*11	*18	*25	*32
·7	0·98	38	44	50	56	61	67	72	77	82	86
·8	0·98	91	95	99	*03	*07	*11	*15	*18	*22	*25
·9	0·99	28	31	34	37	39	42	44	47	49	51
2·0	0·995	32	52	72	91	*09	*26	*42	*58	*73	*88
·1	0·997	02	15	28	41	53	64	75	85	95	*05
·2	0·998	14	22	31	39	46	54	61	67	74	80
·3	0·998	86	91	97	*02	*06	*11	*15	*20	*24	*28
·4	0·999	31	35	38	41	44	47	50	52	55	57
2·5	0·999	59	61	63	65	67	69	71	72	74	75
·6	0·999	76	78	79	80	81	82	83	84	85	86
·7	0·999	87	87	88	89	89	90	91	91	92	92
·8	0·9999	25	29	33	37	41	44	48	51	54	56
2·9	0·9999	59	61	64	66	68	70	72	73	75	77

TABLE A.10.2

The exponential function (e^{-x})

x	e^{-x}	x	e^{-x}	x	e^{-x}	x	e^{-x}
0·00	1·0000	0·45	0·6376	0·90	0·4066	1·35	0·2592
·01	0·9900	·46	·6313	·91	·4025	·36	·2567
·02	·9802	·47	·6250	·92	·3985	·37	·2541
·03	·9704	·48	·6188	·93	·3946	·38	·2516
·04	·9608	·49	·6126	·94	·3906	·39	·2491
0·05	0·9512	0·50	0·6065	0·95	0·3867	1·40	0·2466
·06	·9418	·51	·6005	·96	·3829	·41	·2441
·07	·9324	·52	·5945	·97	·3791	·42	·2417
·08	·9231	·53	·5886	·98	·3753	·43	·2393
·09	·9139	·54	·5827	·99	·3716	·44	·2369
0·10	0·9048	0·55	0·5769	1·00	0·3679	1·45	0·2346
·11	·8958	·56	·5712	·01	·3642	·46	·2322
·12	·8869	·57	·5655	·02	·3606	·47	·2299
·13	·8781	·58	·5599	·03	·3570	·48	·2276
·14	·8694	·59	·5543	·04	·3535	·49	·2254
0·15	0·8607	0·60	0·5488	1·05	0·3499	1·50	0·2231
·16	·8521	·61	·5434	·06	·3465	·51	·2209
·17	·8437	·62	·5379	·07	·3430	·52	·2187
·18	·8353	·63	·5326	·08	·3396	·53	·2165
·19	·8270	·64	·5273	·09	·3362	·54	·2144
0·20	0·8187	0·65	0·5220	1·10	0·3329	1·55	0·2122
·21	·8106	·66	·5169	·11	·3296	·56	·2101
·22	·8025	·67	·5117	·12	·3263	·57	·2080
·23	·7945	·68	·5066	·13	·3230	·58	·2060
·24	·7866	·69	·5016	·14	·3198	·59	·2039
0·25	0·7788	0·70	0·4966	1·15	0·3166	1·60	0·2019
·26	·7711	·71	·4916	·16	·3135	·61	·1999
·27	·7634	·72	·4868	·17	·3104	·62	·1979
·28	·7558	·73	·4819	·18	·3073	·63	·1959
·29	·7483	·74	·4771	·19	·3042	·64	·1940
0·30	0·7408	0·75	0·4724	1·20	0·3012	1·65	0·1920
·31	·7334	·76	·4677	·21	·2982	·66	·1901
·32	·7261	·77	·4630	·22	·2952	·67	·1882
·33	·7189	·78	·4584	·23	·2923	·68	·1864
·34	·7118	·79	·4538	·24	·2894	·69	·1845
0·35	0·7047	0·80	0·4493	1·25	0·2865	1·70	0·1827
·36	·6977	·81	·4449	·26	·2837	·71	·1809
·37	·6907	·82	·4404	·27	·2808	·72	·1791
·38	·6839	·83	·4360	·28	·2780	·73	·1773
·39	·6771	·84	·4317	·29	·2753	·74	·1755
0·40	0·6703	0·85	0·4274	1·30	0·2725	1·75	0·1738
·41	·6637	·86	·4232	·31	·2689	·76	·1720
·42	·6570	·87	·4190	·32	·2671	·77	·1703
·43	·6505	·88	·4148	·33	·2645	·78	·1686
·44	·6440	·89	·4107	·34	·2618	·79	·1670

TABLE A.10.2 (*cont.*)

x	e^{-x}	x	e^{-x}	x	e^{-x}	x	e^{-x}
1·80	0·1653	2·25	0·1054	2·70	0·06721	3·15	0·04285
·81	·1637	·26	·1044	·71	·06654	·16	·04243
·82	·1620	·27	·1033	·72	·06587	·17	·04200
·83	·1604	·28	·1023	·73	·06522	·18	·04159
·84	·1588	·29	·1013	·74	·06457	·19	·04117
1·85	0·1572	2·30	0·1003	2·75	0·06393	3·20	0·04076
·86	·1557	·31	·09926	·76	·06329	·21	·04036
·87	·1541	·32	·09827	·77	·06266	·22	·03996
·88	·1526	·33	·09730	·78	·06204	·23	·03956
·89	·1511	·34	·09633	·79	·06142	·24	·03916
1·90	0·1496	2·35	0·09537	2·80	0·06081	3·25	0·03877
·91	·1481	·36	·09442	·81	·06020	·26	·03839
·92	·1466	·37	·09348	·82	·05961	·27	·03801
·93	·1451	·38	·09255	·83	·05901	·28	·03763
·94	·1437	·39	·09163	·84	·05843	·29	·03725
1·95	0·1423	2·40	0·09072	2·85	0·05784	3·30	0·03688
·96	·1409	·41	·08982	·86	·05727	·31	·03652
·97	·1395	·42	·08892	·87	·05670	·32	·03615
·98	·1381	·43	·08804	·88	·05613	·33	·03579
·99	·1367	·44	·08716	·89	·05558	·34	·03544
2·00	0·1353	2·45	0·08629	2·90	0·05502	3·35	0·03508
·01	·1340	·46	·08543	·91	·05448	·36	·03474
·02	·1327	·47	·08458	·92	·05393	·37	·03439
·03	·1313	·48	·08374	·93	·05340	·38	·03405
·04	·1300	·49	·08291	·94	·05287	·39	·03371
2·05	1·1287	2·50	0·08208	2·95	0·05234	3·40	0·03337
·06	·1275	·51	·08127	·96	·05182	·41	·03304
·07	·1262	·52	·08046	·97	·05130	·42	·03271
·08	·1249	·53	·07966	·98	·05079	·43	·03239
·09	·1237	·54	·07887	·99	·05029	·44	·03206
2·10	0·1225	2·55	0·07808	3·00	0·04979	3·45	0·03175
·11	·1212	·56	·07730	·01	·04929	·46	·03143
·12	·1200	·57	·07654	·02	·04880	·47	·03112
·13	·1188	·58	·07577	·03	·04832	·48	·03081
·14	·1177	·59	·07502	·04	·04783	·49	·03050
2·15	0·1165	2·60	0·07427	3·05	0·04736	3·50	0·03020
·16	·1153	·61	·07353	·06	·04689	·51	·02990
·17	·1142	·62	·07280	·07	·04642	·52	·02960
·18	·1130	·63	·07208	·08	·04596	·53	·02930
·19	·1119	·64	·07136	·09	·04550	·54	·02901
2·20	0·1108	2·65	0·07065	3·10	0·04505	3·55	0·02872
·21	·1097	·66	·06995	·11	·04460	·56	·02844
·22	·1086	·67	·06925	·12	·04416	·57	·02816
·23	·1075	·68	·06856	·13	·04372	·58	·02788
·24	·1065	·69	·06788	·14	·04328	·59	·02760

TABLE A.10.2 (*cont.*)

x	e^{-x}	x	e^{-x}	x	e^{-x}	x	e^{-x}
3·60	0·02732	3·75	0·02352	3·90	0·02024	4·50	0·01111
·61	·02705	·76	·02328	·91	·02004	·60	·01005
·62	·02678	·77	·02305	·92	·01984	·70	·009095
·63	·02652	·78	·02282	·93	·01964	·80	·008230
·64	·02625	·79	·02260	·94	·01945	·90	·007447
3·65	0·02599	3·80	0·02237	3·95	0·01925	5·00	0·006738
·66	·02573	·81	·02215	·96	·01906	·10	·006097
·67	·02548	·82	·02193	·97	·01887	·20	·005517
·68	·02522	·83	·02171	·98	·01869	·30	·004992
·69	·02497	·84	·02149	·99	·01850	·40	·004517
3·70	0·02472	3·85	0·02128	4·00	0·01832	5·50	0·004087
·71	·02448	·86	·02107	·10	·01657	·60	·003698
·72	·02423	·87	·02086	·20	·01500	·70	·003346
·73	·02399	·88	·02065	·30	·01357	·80	·003028
·74	·02375	·89	·02045	·40	·01228	·90	·002739

BIBLIOGRAPHY

D'ANS, J., and LAX, E., 1949, *Taschenbuch für Chemiker und Physiker*, 2nd ed. Berlin: Springer (Table A.10.2); and

JAHNKE, E., and EMDE, F., 1945, *Tables of functions*. New York: Dover (Table A.10.1).

APPENDIX 11

The Elsasser function $E(y,u)$

$$E(y,u) = \int_{-\frac{1}{2}}^{+\frac{1}{2}} \exp\left[-2\pi y u \frac{\sinh 2\pi y}{\cosh 2\pi y - \cos 2\pi x}\right] dx$$

TABLE A.11.1

$\log_{10} u$	$\log_{10} y$												
	0	-0.2	-0.4	-0.6	-0.8	-1.0	-1.2	-1.4	-1.6	-1.8	-2.0	-2.2	-2.4
-1.5	0.819803	0.882179	0.923986	0.951422	0.969151	0.980480	0.987671	0.992216	0.995088	0.996901	0.998046	0.998768	0.999222
-1.25	.702345	.800180	.868912	.915393	.946005	.965756	.978348	.986328	.991369	.994557	.996564	.997849	.998648
-1.0	.533489	.672748	.779021	.854938	.906613	.940540	.962342	.976204	.984976	.990522	.994017	.996225	.997617
-0.75	.327154	.494204	.641798	.757900	.841751	.898540	.935548	.959223	.974242	.983751	.989737	.993526	.995912
-0.5	.137119	.285620	.455299	.613677	.740791	.831771	.892574	.931887	.956942	.972824	.982839	.989174	.993164
-0.25	.029209	.107790	.248226	.425665	.597505	.733211	.828071	.890574	.930716	.956223	.972364	.982558	.988996
0	0.001867	0.019086	0.085420	0.228257	0.421162	0.602472	0.739704	0.833225	0.894115	0.933025	0.957700	0.973300	0.983149
$+0.25$		0.000883	0.013262	0.080820	0.243257	0.449996	0.629960	0.760129	0.846978	0.903021	0.938706	0.961130	0.975573
0.5			0.000529	.014694	.103825	.292971	.502665	.671077	.788405	.865454	.914854	.946211	.966040
0.75				0.000876	.026680	.152201	.361635	.563239	.714958	.817674	.884390	.927064	.953903
1.0					0.002818	.053621	.218853	.436054	.622565	.755982	.844510	.901565	.937750
1.25						.009639	.099128	.296179	.509169	.676843	.792512	.868168	.916591
1.5						0.000532	.027379	.162295	.377192	.577164	.724935	.824284	.888580
1.75							0.003209	.061940	.238134	.456400	.638821	.766765	.851541
2.0								0.012711	0.115335	0.320145	0.530375	0.692162	0.802726
2.25								0.000866	.035641	.184674	.402508	.597320	.738889
2.5									.005068	.076825	.264119	.481019	.656585
2.75									0.000186	.018276	.136381	.347295	.553160
3.0										0.001647	.046995	.210050	.429006
3.25											.008074	.094610	.291546
3.5											0.000411	.025806	.159558
3.75												0.002952	.060695
4.0													0.012373

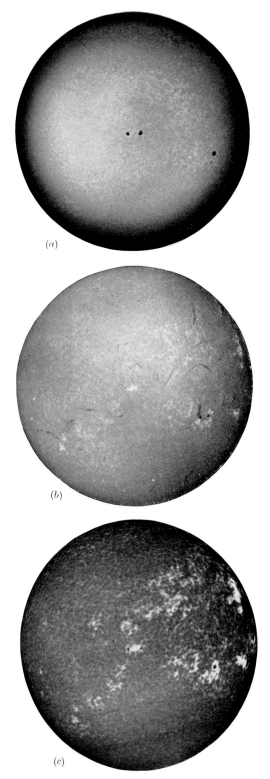

(a)

(b)

(c)

FIG. A.11. Simultaneous pictures of the sun in white and monochromatic light. (a) White light, (b) calcium K, (c) hydrogen Hα.

The intervals in Table A.11.1 are close enough to permit interpolation to four-figure accuracy. The table may be extended by means of the asymptotic forms

$$y \to \infty \quad \text{equation (4.65)},$$
$$u \to \infty \quad \text{equation (4.67)},$$
$$u \to 0 \quad \text{equation (4.13)},$$
$$y \to 0 \quad \text{equation (4.83)}.$$

BIBLIOGRAPHY

Table A.11.1 has been extracted from an unpublished report (1952).

Other tabulations have been given by

KAPLAN, L. D., 1953, 'Regions of validity of various absorption-coefficient approximations', *J. Met.* **10,** p. 100; and

WARK, D. Q., and WOLK, M., 1960, 'An extension of a table of absorption for Elsasser bands', *Mon. Weath. Rev.* **88,** p. 249.

APPENDIX 12

Rayleigh scattering coefficients

THE following computations are based on (7.39) and the refractive indices given in Appendix 4, for 0° C and $p = 1013\cdot25$ mb. The optical depth τ^* is based upon a total of $2\cdot149 \times 10^{25}$ molecules cm^{-2} in a vertical column of the atmosphere. Entries are close enough to permit accurate interpolation on the basis of the fourth-power law of wavelength.

TABLE A.12.1

Rayleigh scattering coefficients for dry air at s.t.p.

λ	s_n	s_m	s_v	τ^*
(μ)	(cm^2)	(cm^2 g^{-1})	(cm^{-1})	
0·2	$3\cdot551 \times 10^{-25}$	$7\cdot382 \times 10^{-3}$	$9\cdot542 \times 10^{-6}$	7·630
0·25	1·258	2·616	3·382	2·704
0·3	$5\cdot676 \times 10^{-26}$	1·180	1·525	1·220
0·35	2·951	$6\cdot135 \times 10^{-4}$	$7\cdot929 \times 10^{-7}$	$6\cdot341 \times 10^{-1}$
0·4	1·689	3·512	4·540	3·630
0·45	1·038	2·158	2·789	2·231
0·5	$6\cdot735 \times 10^{-27}$	1·400	1·810	1·447
0·55	4·563	$9\cdot486 \times 10^{-5}$	1·226	$9\cdot805 \times 10^{-2}$
0·6	3·202	6·657	$8\cdot604 \times 10^{-8}$	6·880
0·65	2·313	4·810	6·217	4·971
0·7	1·713	3·562	4·605	3·682
0·75	1·296	2·695	3·484	2·786
0·8	$9\cdot989 \times 10^{-28}$	2·077	2·684	2·147
0·9	6·212	1·292	1·670	1·335
1·0	4·065	$8\cdot452 \times 10^{-6}$	1·092	$8\cdot736 \times 10^{-3}$

BIBLIOGRAPHY

PENNDORF, R., 1957, 'Tables of the refractive index for standard air and the Rayleigh scattering coefficient for the spectral region between 0·2 μ and 20·0 μ and their application to atmospheric optics', *J. Opt. Soc. Am.* **47,** p. 176.

APPENDIX 13

The optical properties of liquid water and ice

Fig. A.2 shows n and k_v ($= 4\pi n'/\lambda$) for pure *liquid water*, throughout the relevant section of the electromagnetic spectrum. There is consistency between different experimenters as regards n. Aside from the spectral region between $27\,\mu$ and $6200\,\mu$, where results have been interpolated, the refractive indices are probably reliable.

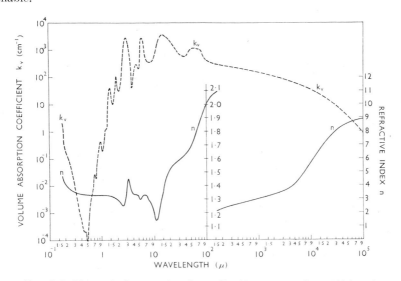

FIG. A.2. The optical properties of pure liquid water, at about 20° C and 1 atmosphere pressure.

This claim cannot be made with equal confidence for k_v. Difficulties in the technique of measurement, and the wide range of absorptivity from transparency in the visible spectrum to near-metallic opacity in the infra-red spectrum, combine to give uncertainties which may sometimes amount to ± 50 per cent.

In the microwave spectrum both n and k_v are temperature-dependent. The situation at other wavelengths has not been satisfactorily investigated.

The optical data on *ice* are more complex, and not well established excepting in the visible and microwave regions of the spectrum. Ice is both birefringent and dichroic, although in the visible spectrum the various coefficients do not differ sufficiently to be of importance. The refractivity and absorptivity for one H_2O molecule is similar in both solid and liquid phases for this spectral region. Dorsey (1940) records no refractive indices in the infra-red spectrum, but gives reflectivities which are less than or equal to 5 per cent for all wavelengths between $1\,\mu$ and $16\,\mu$. The increase of refractive index observed for liquid water at wavelengths greater than $100\,\mu$ appears not to take place for ice if we judge by microwave results which give $n = 1\cdot78$ independently of wavelength and temperature.

The general level of molecular absorption coefficients for ice and liquid water appear to be similar throughout the infra-red spectrum, but bands are shifted, and are markedly dichroic, so that absorptivities at fixed wavelengths may differ considerably between the solid and liquid phase. The analogy does not exist at all wavelengths, however, for in the centimetre band k_v for ice is about 10^{-3} times that for water, and its temperature coefficient is even larger than for water.

BIBLIOGRAPHY

Data on both ice and liquid water published before 1940 are listed by

DORSEY, N. E., 1940, *Properties of ordinary water-substance in all its phases*: *water-vapour, water and all the ices*. New York: Reinhold.

The refractivity of liquid water from $0\cdot2\ \mu$ to $27\ \mu$ is taken from the data listed in this book. The absorption coefficient of liquid water from $0\cdot2\ \mu$ to $4\ \mu$ follows Aschkinass, as listed by Dorsey. The sketchy information on ice, in all but the microwave region, is also taken from the same source.

In the microwave region of the spectrum, a review of more recent data on both water and ice has been made by

GUNN, K. L. S., and EAST, T. W. R., 1954, 'The microwave properties of precipitation particles', *Quart. J. R. Met. Soc.* **80**, p. 522.

Absorption coefficients of liquid water in the important spectral interval $4\ \mu$ to $90\ \mu$ are reviewed and discussed by

McDONALD, J. E., 1960, 'Absorption of atmospheric radiation by water films and water clouds', *J. Met.* **17**, p. 232.

APPENDIX 14

The physical state of the sun

A.14.1. The quiet sun

ASTROPHYSICS can now give a satisfactory account of most features of the solar emission likely to influence the earth's atmosphere; it will be the object of this Appendix to outline briefly the current ideas in this field.

The sun is a typical main-sequence, dwarf star of apparent visual magnitude −26·86 and spectral type G 1 v. Its probable chemical composition by mass is 64 per cent hydrogen, 32 per cent helium, and about 4 per cent heavier elements —including almost every known, stable element. The total energy emitted, which it derives from thermonuclear reactions between hydrogen atoms in the interior, is equal to that radiated by a black body with a temperature $\theta_e = 5785°$ K.

The visible disk of the sun, or *photosphere*, has a diameter of $1·3920 \times 10^{11}$ cm, and a mean angular diameter of $31'·988$ at the mean solar distance ($R_\odot = 1·4960 \times 10^{13}$ cm). The ellipticity of the earth's orbit causes the angular diameter to vary $\pm 1·7$ per cent from winter to summer. The photosphere is not evenly illuminated (see Fig. A.3), but shows *limb darkening*, or decrease of intensity towards the periphery. It is covered with small, rapidly changing, cellular structures with a predominant diameter of about 1000 km, known as *granules* (Fig. A.4). They have a lifetime of about 4 minutes, and vertical velocities near 1 km s^{-1} have been measured.

There is no discontinuity between the photosphere and the overlying solar 'atmosphere'. This gaseous envelope emits about as much light as the full moon, and can only be observed during a total eclipse or if extremely careful instrumental precautions are taken to cut off light from the photosphere. The *chromosphere*, which occupies the lowest 10 000 to 20 000 km, has at its base a relatively low temperature ($\sim 4500°$ K). A typical exposure taken in hydrogen $H\alpha$† emission shows it to consist of a dense mass of leaning towers or *spicules*, which may be associated with the underlying surface granulation.

The faint light of the *corona* has been traced almost all the way from the sun to the zodiacal light; there is little doubt that they are the same phenomenon. Three components can be distinguished: the K-component, caused by electron scattering, is polarized; the F-component is unmodified photospheric radiation scattered from dust particles, and is the inner component of the zodiacal light; the E-component consists of unusual emission lines from which we infer a coronal temperature of 10^6 °K (see Fig. A.5).

A.14.2. The solar spectrum

The solar spectrum consists of a continuous emission with a superposed line structure which, relative to the continuum, can appear either in emission or in absorption. The visible and infra-red spectrum of the photosphere shows absorption lines only (the *Fraunhofer* spectrum), the strongest of which are caused by neutral H, Na, Mg, Fe, Ca, Si, and singly ionized Ca and Mg. At $0·185\,\mu$ the

† $H\alpha$ is the first member of the Balmer series with a wavelength of $0·6563\,\mu$.

photospheric spectrum changes and most of the lines at shorter wavelengths are exhibited in emission. Light from the chromosphere and corona has emission lines in all parts of the spectrum.

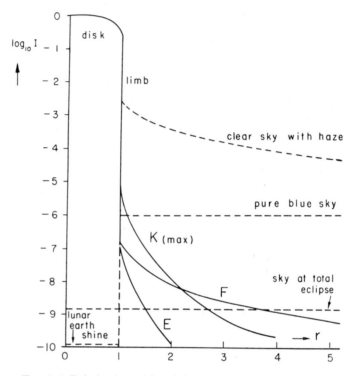

FIG. A.5. Relative intensities of the components of coronal light.

K = continuous light due to electron scattering; F = inner zodiacal light; E = combined light of emission lines; r = distance from sun in units of the solar radius. After van de Hulst (1953).

Most of the solar energy is carried by the continuum. This owes its existence to unquantized electronic transitions; both bound-free (ionizing) and free-free transitions taking part. The most important single contributor is hydrogen, both in its neutral state and in the form of negative ions. Neutral hydrogen has ionization continua starting at the limits of the Lyman, Balmer, Paschen, Brackett, and Rydberg series; the Balmer discontinuity at $0.3646\,\mu$ is particularly prominent in observational data (see, for example, the limb-darkening curves in Fig. A.9). Metallic atoms contribute strongly to the continuum in the ultra-violet spectrum, and are responsible for the maximum mass absorption coefficient at $0.15\,\mu$, shown in Fig. A.6. Through the visible and infra-red spectrum, however, the continuum emission is by negative hydrogen ions. Between $0.4\,\mu$ and $1.7\,\mu$, the main contribution is from the bound-free, ionizing transition; at longer wavelengths the absorption coefficient varies as (wavelength)^2, and is caused by a free-free transition.

The effect of the smoothed-out absorption of Fraunhofer lines is indicated by the curve marked 'line absorption' in Fig. A.6. At wavelengths less than $0.3\,\mu$ more than half of the continuum radiation is reabsorbed in lines. Under these circumstances empirical separation of lines and continuum in the observational data is no longer possible. At infra-red wavelengths line absorption can be neglected for practical purposes.

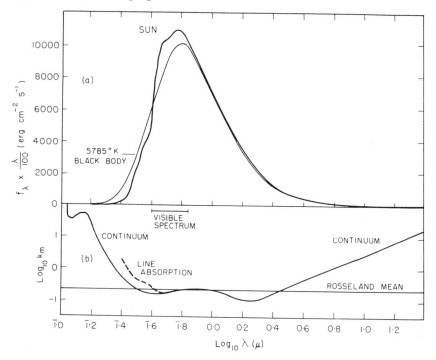

Fig. A.6. Solar energy distribution and photospheric absorption coefficients.

(a) Energy on a linear scale for the sun and for a black body at 5785° K. (b) Mass absorption coefficient at the surface of the photosphere for a temperature of 5785° K. λ = wavelength; f_λ = solar irradiation at mean solar distance. After Allen (1958).

The variation of absorption coefficient with wavelength implies that the terrestrial observer will 'see in' to different solar levels at different wavelengths. Material responsible for forming spectral lines gives an opacity additional to that of the continuum, and radiation from line centres therefore originates nearer to the observer than that from the continuum at neighbouring wavelengths, and may emit more or less strongly depending upon the temperature difference between the two levels. Thus, when radiation originates from a region with a positive temperature gradient, the lines stand out in emission; if it originates from a region with negative temperature gradient, then absorption lines are observed.†

† This statement is oversimplified and only applies to a system in thermodynamic equilibrium. See § 2.2 for discussion.

These considerations suggest that the emission lines below $0.185\,\mu$ are distinguished from Fraunhofer lines only by originating from a region where temperature increases with height. This can happen because of the rapid increase of absorption coefficient with decreasing wavelength in the far ultra-violet spectrum, so that below $0.185\,\mu$ we 'see in' only to the chromosphere and corona rather than to the photosphere. It is consistent with this picture that when careful observations separate the weak chromospheric light at wavelengths above $0.185\,\mu$, the spectrum resembles the Fraunhofer absorption spectrum in reverse. The

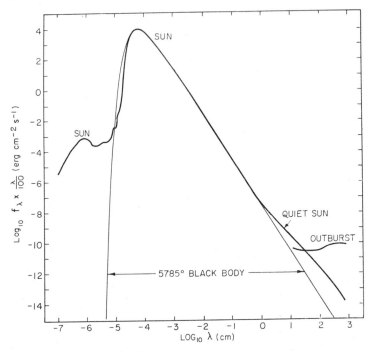

FIG. A.7. Comparison between the energy distribution of the sun and
a 5785° K black body. After Allen (1958).

scattered spectrum of the F-corona is the same as that of the photosphere, while the K-corona shows no lines at all. The E-corona exhibits an unusual spectrum of very highly ionized atoms; the strongest line, for example, is the $0.5303\,\mu$ line of iron, thirteen times ionized.

A.14.3. The intensity of solar radiation

The flux of radiative energy through unit area perpendicular to the sun's rays is the irradiation f. Its value at mean solar distance (the *solar constant*) is 1.40×10^6 erg cm^{-2} s^{-1} ($\equiv 2.0$ cal cm^{-2} min^{-1}). Methods of measuring this quantity and its spectral distribution (defined by f_λ or f_ν, where $f = \int_0^\infty f_\lambda\,d\lambda = \int_0^\infty f_\nu\,d\nu$, λ and ν being wavelength and frequency respectively) will be discussed in Part 2. In Fig. A.6 (a) the spectral energy distribution has been depicted in a way which gives area proportional to energy $(f_\lambda \lambda/100 \times d\log\lambda = (2\cdot303/100)f_\lambda\, d\lambda)$, and is compared with that of a black body with the same total energy. The two curves are sufficiently close for a black-body approximation to be useful for some purposes. In the far ultra-violet spectrum $(\lambda < 0\cdot1\,\mu)$ or in the radio-wave region $(\lambda > 1\text{ cm})$, however, the solar emission differs greatly from that of a 5785° K black body (see Fig. A.7) and can be very variable under disturbed conditions. The data used in Fig. A.7 are given in numerical form in Table A.14.1. Such

TABLE A.14.1

Irradiation from the quiet sun

After Allen (1958)

X-ray spectrum		Ultra-violet, visible, and infra-red spectrum				Radio spectrum	
λ	f_λ	λ	f_λ	λ	f_λ	λ	f_λ
(Å)	$(10^{-5}\text{ erg cm}^{-2}$ $\text{Å}^{-1}\text{ s}^{-1})$	(μ)	$(\text{erg cm}^{-2}$ $\text{Å}^{-1}\text{ s}^{-1})$	(μ)	$(\text{erg cm}^{-2}$ $\text{Å}^{-1}\text{ s}^{-1})$	(cm)	$(\text{erg cm}^{-2}$ $\text{cm}^{-1}\text{ s}^{-1})$
10	4	0·20	0·5	0·55	195	0·6	$2\cdot8\times10^{-5}$
20	23	0·22	3·0	0·60	183	1·5	$1\cdot1\times10^{-6}$
50	110	0·24	6	0·65	164	3	$9\cdot7\times10^{-8}$
70	120	0·26	12	0·70	146	6	$1\cdot1\times10^{-8}$
100	70	0·28	24	0·75	128	15	$7\cdot7\times10^{-10}$
200	11	0·30	55	0·80	113	30	$1\cdot1\times10^{-10}$
		0·32	75			60	$1\cdot8\times10^{-11}$
300	10	0·34	97	0·90	89		
500	8			1·0	72	150	$9\cdot1\times10^{-13}$
700	10	0·35	105	1·1	59·5	300	$6\cdot0\times10^{-14}$
800	20	0·36	109	1·2	49	600	$4\cdot2\times10^{-15}$
900	60	0·37	112				
		0·38	115	1·4	32·5		
1000	40	0·39	123	1·6	22·3		
1100	90	0·40	153	1·8	15·2		
1200	280	0·41	176	2·0	10·8		
1400	540	0·42	186	2·5	4·97		
1600	1 600						
1800	10 000	0·44	203	3	2·63		
2000	50 000	0·46	215	4	0·93		
		0·48	214	5	0·41		
		0·50	206	6	0·21		
				7	0·12		
				10	0·023		

data contain an arbitrary element unless the method of averaging is defined. In Fig. A.8 is shown a comparison between these figures and independent measurements with higher spectral resolution. At very high resolution the spectrum would appear to be yet again very different. The differences between the two curves in Fig. A.8 indicate the magnitude of the uncertainties involved in these important astronomical data.

The foregoing data refer to the radiation from the entire solar disk, but since the limb darkening varies with wavelength they do not correspond to any particular point on the disk. The intensity at any point can, however, be inferred from the limb-darkening curves in Fig. A.9.

The constancy of the solar intensity is a matter of debate. Most stars show measurable variations in intensity, the sun being one of the most steady sources in the sky. No variations have been established in the continuum radiation which accounts for most of the energy, although changes on the time-scale of a century or more cannot be ruled out on the grounds of direct evidence. Individual lines show intensity variations, however, e.g. calcium K and hydrogen $H\alpha$ in the

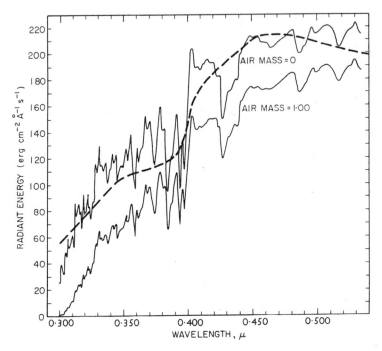

Fig. A.8. Solar spectral energy distribution from $0\cdot3\,\mu$ to $0\cdot55\,\mu$.

The broken line corresponds to the figures in Table A.14.1, while the full lines give measurements made at Sunspot, New Mexico (altitude 9200 ft), under clear conditions in June, by Stair and Johnston (1956).

visible spectrum (see §A.14.4), and there are very large variations in the far ultra-violet where emission lines originate in the chromosphere and corona. It is also possible that small variations occur in the blue region of the spectrum, where Fraunhofer absorption is intense.

The properties of the nearly constant continuum emission can be summarized in terms of a model of the sun. In Chapter 2 we express the emission from a star or planet in thermodynamic equilibrium in terms of the temperature as a function of optical path. From these expressions and the observed intensity distribution across the solar disk a relationship between τ_λ and θ can be obtained. From observations at different wavelengths τ_λ can be determined as a function of λ for constant θ (i.e. for one level in the sun). It is usual to form the Rosseland mean from these data, and to quote the temperature as a function of this parameter.

Some results are shown in Fig. A.6 (b) (interpreted in terms of an absorption coefficient) and Table A.14.2. The relation between τ_λ and λ for wavelengths longer than 0.5μ corresponds very closely with that predicted for negative hydrogen ions. In Table A.14.2, $\bar{\tau}_R$ is related to the pressure, density, and distance

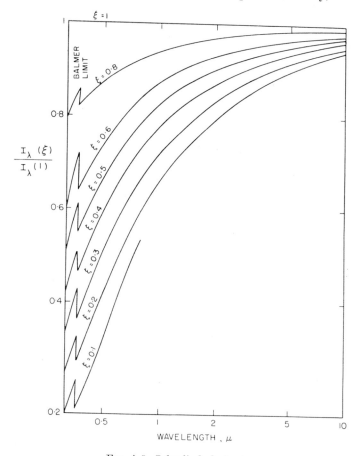

Fig. A.9. Solar limb darkening.

ξ is the cosine of the zenith angle at the sun's surface. $\xi = 1$ corresponds to the centre of the disk and $\xi = 0$ to the limb. After Allen (1955).

below the level at which $\bar{\tau}_R = 0.004$, corresponding to the visible limb of the photosphere. From these data we can recompute the solar intensity as a function of wavelength and angle of emission, with the advantage that gaps in the original data can thus be filled and faulty data detected.

Table A.14.3 gives a speculative model of the undisturbed chromosphere and corona. From the discussion of line formation in § A.14.2 we can identify the level of origin of emission lines as either in the chromosphere or corona.

TABLE A.14.2

A model of the undisturbed photosphere

After Allen (1955)

$\bar{\tau}_R$	Temperature	$\log_{10}$ (*gas pressure, dyne cm^{-2}*)	Depth	$\log_{10}$ (*density, g cm^{-3}*)
	(°K)		(km)	
0·004	4300	3·8	0	$\bar{8}$·4
0·01	4450	4·01	−40	$\bar{8}$·6
0·02	4580	4·18	−80	$\bar{8}$·72
0·05	4800	4·39	−125	$\bar{8}$·91
0·1	5010	4·55	−160	$\bar{7}$·05
0·2	5280	4·73	−200	$\bar{7}$·21
0·5	5800	4·91	−260	$\bar{7}$·35
1	6300	5·04	−300	$\bar{7}$·44
2	7000	5·16	−330	$\bar{7}$·52
5	8300	5·25	−380	$\bar{7}$·6
10	9700	5·5	−430	$\bar{7}$·7

TABLE A.14.3

Model of the undisturbed chromosphere and corona

After van de Hulst (1953)

Solar layer	Height	Temperature	State of hydrogen
	(km)	(°K)	
Outer corona . .	1 400 000	1 500 000	Ionized
Medium corona . .	700 000	1 500 000	,,
Inner corona . .	21 000	1 000 000	,,
Upper chromosphere .	14 000	300 000	,,
	7 000	25 000	,,
Lower chromosphere .	4 000	6 100	Neutral
	1 000	5 300	,,
	500	4 900	,,
Limb . . .	0	4 500	,,

A.14.4. The disturbed sun

The most striking visual disturbances on the photosphere are the *sunspots*, patches varying in diameter from a few thousand to a hundred thousand kilometres, with an emission temperature about 1500° K lower than that of the undisturbed surface. A particularly large group, which could be seen by the unaided eye, is shown in Fig. A.3, and details of the central *umbra* and the structured *penumbra* can be seen in Fig. A.4. The fraction of the photosphere covered by spots is never more than 0·2 per cent and their average persistence is about 6 days.

The Wolf sunspot number (R) is an index of solar activity based solely on sunspots,
$$R = a(10g+s),$$
where

a = a reduction constant of order unity, depending upon the telescope employed,

g = number of groups of spots,

s = number of individual spots.

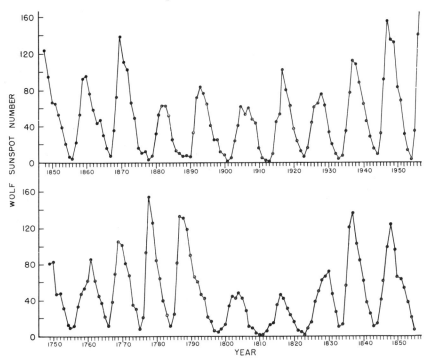

FIG. A.10. The sunspot cycle.

After Menzel (1959).

In Fig. A.10 the change of sunspot number over more than 200 years is plotted, showing the large variations in activity with a period close to 11·2 years. At times of minimum activity it is common to observe no spots at all for many days, while at maximum activity they can always be detected in considerable numbers. The number of sunspots is only one characteristic feature which changes in this rhythmic manner. Just after the minimum, spots first appear near latitude 27° in both hemispheres. As the cycle proceeds they drift inwards and disappear close to latitude 8°. They are rarely observed at latitudes higher than 30° or lower than 5°.

Other types of disturbance are associated with sunspots. When a spot is near the limb it can be seen to be surrounded by a large network of enhanced photospheric emission, patches of which are called *faculae*. These have a longer lifetime than the associated spot group, appearing before and disappearing later.

When the sun is viewed in monochromatic light from a single chemical element, other photospheric disturbances can be seen. Fig. A.11 illustrates typical features in hydrogen $H\alpha$ and calcium K light. The spots are visible on the monochromatic images, but the surrounding bright areas, known as *flocculi*, are now the most prominent features, and they occur also in high latitudes where spots are unknown. Occasionally a hydrogen flocculus near a spot will brighten up, in extreme cases to the extent that the brightening is visible to the eye in integrated 'white' light. These brightenings are known as *solar flares*, and their importance lies in the great increases of hydrogen Lyman α and other ultra-violet radiations, which influence the ionization of the upper atmosphere.

Prominences are photospheric eruptions, extending into the chromosphere, which are often to be seen on the limb of the sun if precautions are taken to cut out the intense photospheric light. Many different forms occur, but a typical prominence might be 30 000 km high, 200 000 km long, with a temperature of 5000° K. Since this temperature is less than that of the photosphere, prominences can be seen in photographs taken in $H\alpha$ light as *dark filaments* on a bright background. More prominences occur at sunspot maximum that at sunspot minimum but they are not closely related to sunspots.

Coronal disturbances are more directly associated with the sunspot cycle. In visible light the corona appears more jagged at sunspot maximum than at sunspot minimum. Solar radio noise, emitted by the corona, shows a marked variation with the sunspot cycle also following short-period changes in sunspot number. The intensity of noise from occasional *outbursts* suggests local condensations with temperatures of 10^7 °K, which also give rise to marked increases in the far ultra-violet and X-ray emission of the sun.

BIBLIOGRAPHY

The data quoted in this appendix are mainly taken from

ALLEN, C. W., 1955, *Astrophysical quantities*. London: Athlone Press.

Other material may be found in the following books and reviews, the last of which is a semi-popular exposition:

VAN DE HULST, H. C., 1953, 'The chromosphere and corona', in *The Sun*, ed. G. P. Kuiper. Chicago Univ. Press, p. 207;
MINNAERT, M., 1953, 'The photosphere' in *The Sun*, ed. G. P. Kuiper, Chicago Univ. Press, p. 88;
ALLEN, C. W., 1958, 'Solar radiation', *Quart. J. R. Met. Soc.* **84**, p. 307;
SHAW, J. H., 1962, 'The radiation environment of interplanetary space', *App. Optics*, **1**, p. 87; FRIEDMAN (1960, § 1.3); and
MENZEL, D. H., 1959, *Our sun*. Harvard Univ. Press.

Reference was also made in the text to

STAIR, R., and JOHNSTON, R. G., 1956, 'Preliminary spectroactinometric measurements of the solar constant', *J. Res. Nat. Bur. Stan.* **67**, p. 205.

AUTHOR INDEX

SUBJECT INDEX

Absorption area, 25.
  atmospheric, 4.
  coefficient, 27, 37.
    Rosseland mean, 58, 349 f.
    Planck mean, 58, 348 f.
    effective, 348 ff.
    Chandrasekhar mean, 58.
    for non-homogeneous paths, 233.
    variation with temperature, 247.
Aerosol, 15.
Air, data for, 385.
  refractive index of, 389.
  scattering coefficients of, 414.
Airy's theory of the rainbow, 311.
Albedo, 1.
Ammonia gas, 10.
  line profile, 112.
  temperature dependence, 114.
Amount of matter, 24.
  effective, 238 ff.
Anderson's theory of line broadening, 43.
Angular momentum, vectorial addition of,
  85 ff.
  vibrational, 79.
Anharmonicities, 81 ff.
Argon, 10.
  depolarization factor, 294.
Ascent curve, 249 ff.
Asymmetric top, 73 ff.
Atomic number, 83.

Babinet's principle, 297 ff.
Band, combination, 81 ff.
  difference, 81 ff.
  head, 93.
  intensity, 41, 82.
  models, 122 ff.
    comparison of regular and random,
      154.
    empirical, 163.
    gaps in, 161 ff.
    restrictions on, 163 ff.
    superposition of, 158.
    three parameter, 159 ff.
    with continuum, 164 f.
    with modulated contour, 165 f.
  overtone, 81 ff.
  P-branch, 90.
  parallel, 90.
  parameters, 133 ff.
  perpendicular, 90.
  Q-branch, 90.
  R-branch, 90.

  structure, random, 124 ff.
    regular, 125 ff.
  systems, 68 ff.
Band, Σ–Σ, 90.
  upper state, 81 ff.
  vibration-rotation, 68 ff.
Beer's law, 142.
Black body radiation, 29.
Boltzmann's law, 32.
Bouguet's law, 23.
Boundary conditions, 44, 53, 54.
Boundary layer, atmospheric, 369.
  radiative-diffusive, 350 ff.

Carbon dioxide, absence of ν_1, 81.
  absorption by, 3 f.
  as a linear molecule, 79.
  bands, 207 ff.
    absorption area, 205.
    Coriolis interactions in, 201.
    electronic, 196 f.
    empirical constants for, 206.
    fundamentals, 143, 199.
    intensities, 201.
    l-type doubling in, 200.
    multiplication property of, 123 f.
    under high resolution, 202 f.
    under low resolution, 204.
    visible and photographic infra-red,
      209.
  computation of $\tilde{a}$, 256.
  concentration, 10, 14.
  Curtis–Godson approximation for $15\,\mu$
    bands, 243.
  depolarization factor for, 294.
  doublet model for, 147.
  effect of weak lines, 171.
  ϵ^* for, 250.
  Fermi resonance in, 82.
  flux and flux divergence computations,
    245.
  in the lower stratosphere, 341.
  isotopes, 199.
  line-wing shape, 117.
  natural lifetime of vibrational states,
    41 ff.
  population of upper states, 81.
  solar absorption by, 248.
  solar heating by, 330.
  symmetry, 70, 199.
  terrestrial radiation, 330.
  vibrational relaxation times, 41 ff.

PRINTED IN GREAT BRITAIN AT THE UNIVERSITY PRESS, OXFORD
BY VIVIAN RIDLER, PRINTER TO THE UNIVERSITY